Nanofillers

This book explores the modifying effects of various nanofillers on mechanical and physical properties of polymer nanocomposites. Looking at the four basic aspects of processing, characterization, properties, and applications, it analyzes how their features can allow for innovative multifunction within industry.

Covering design, production, and manufacture, this book focuses on meeting end-use requirements and the fabrication of materials. The importance of mindful design and the use of an appropriate synthesis method is the primary lens through which theory and practice are discussed. This volume looks at the various synthesis methods available for organic nanofillers and what characterizes them. Properties including mechanical, thermal, electrical, and tribological are thoroughly examined, along with the various computational techniques used to determine them. With important sustainable properties, nanofillers are essential to meeting the increasing demand for biodegradable and environmentally friendly materials. This book details the role nanofillers have to play in sustainability, alongside economic factors such as efficient manufacturing processes.

This book will appeal to both academic and industrial engineers involved with nanofillers in a variety of industries, including automotive, aerospace, and biomedical engineering.

Nanofillers
Fabrication, Characterization and Applications of Organic Nanofillers

Edited by
Vijay Chaudhary, Bhasha Sharma, Shashank Shekhar, and Partha Pratim Das

CRC Press
Taylor & Francis Group
Boca Raton London New York

CRC Press is an imprint of the
Taylor & Francis Group, an **informa** business

First edition published 2024
by CRC Press
6000 Broken Sound Parkway NW, Suite 300, Boca Raton, FL 33487-2742

and by CRC Press
4 Park Square, Milton Park, Abingdon, Oxon, OX14 4RN

CRC Press is an imprint of Taylor & Francis Group, LLC

ISBN: 9781032245843 (hbk)
ISBN: 9781032245850 (pbk)
ISBN: 9781003279372 (ebk)

DOI: 10.1201/9781003279372

Typeset in Times
by codeMantra

Contents

Preface...vii

Editors..ix

Contributors..xi

Chapter 1   Organic Nanofillers and Their Classification Employed in Polymers 1

Prakash Chander Thapliyal

Chapter 2   Dimension-Based Nanofillers: Synthesis and Characterization......... 19

Qazi Adfar, Mohammad Aslam, and Shrikant S. Maktedar

Chapter 3   Mechanical, Thermal, Electrical, Optical, and Magnetic
Properties of Organic Nanofillers 71

*Tousif Reza, Rokib Uddin, Mohammad Fahim Tazwar,
Md Enamul Hoque, and Yashdi Saif Autul*

Chapter 4   Standard Reinforcement Methods, Mechanism, Compatibility,
and Surface Modification of Nanofillers in Polymers...................... 115

Kriti Sharma, G. L. Devnani, Sanya Verma, and Mohit Nigam

Chapter 5   Nanocellulose-Based Green Fillers for Elastomers—Influence
of Geometry and Chemical Modification on Properties of
Nanocomposites 145

*Milanta Tom, Sabu Thomas, Bastien Seantier, Yves Grohens,
P. K. Mohamed, S. Ramakrishnan, and Job Kuriakose*

Chapter 6   Recent Advances in Organic Nanofillers-Derived Polymer
Hydrogels.. 173

*M. Iqbal Ishra, B. S. S. Nishadani,
M. G. R. Tamasha Jayawickrama, and
M. M. M. G. P. G. Manthilaka*

Chapter 7   Environmental Impact in Terms of Nanotoxicity and
Limitations of Employing Organic Nanofillers in Polymers 199

*Habibul Islam, Md Enamul Hoque, Shek Md Atiqure Rahman,
and Faris M. Al-Oqla*

Chapter 8   Natural Nanofillers Family, Their Properties, and Applications
                    in Polymers...233

                    Aswathy Ajayan and Maida Mary Jude

Chapter 9   Formulation of Polymer Nano-Composite NPK Fertilizer
                    [Cellulose-*Graft*-Poly (Acrylamide)/Nano-Hydroxyapatite/
                    Soluble Fertilizer] and Evaluation of Its Nutrient Release...............257

                    Kiplangat Rop, George N. Karuku, and Damaris Mbui

Chapter 10 Natural Nanofibers Classification, Their Properties, and
                    Applications..281

                    P. A. Nizam and Sabu Thomas

Chapter 11 Biomedical Applications of Organic Nanofillers in Polymers.........297

                    Basma M. Eid

Index...323

Preface

The present edited book volume named *Nanofillers: Fabrication, Characterization and Applications of Organic Nanofillers* will provide a systematic knowledge of various aspects of nanofillers and their influence on the performance of polymer composites. This edited book will cover various methods of nanofiller synthesis and development of nanocomposites along with their characterization techniques to study various properties (mechanical, tribological, thermal, electrical, etc.). It also provides an insight on a wide spectrum of potentially advanced applications ranging from automotive and aerospace to biomedical and packaging, etc.

This book will also address the advantages of nanofillers to promote applications of organic nanofillers and nanocomposites. This book will also address the limitations of using nanofiller to promote sustainability and the eco-friendly environment, the economic aspects of using nanofillers and their future perspectives.

The contents of this book will be beneficial for students attending mechanical engineering, civil engineering, material science, chemistry, physics, and researchers both working in industry and academia. In short, this book will focus on the synthesis, characterization, and application of characterization and applications of organic nanofillers.

This book consists of 11 chapters whose details are as follows:

Chapter 1 discusses in detail about the organic nanofillers and their classification employed in polymers. In this chapter, the use of organic nanofillers in making polymer nanocomposites dealing with the major present developments has been discussed.

Chapter 2 is dedicated to the fabrication and characterization of nanofillers. Apart from that, the potential applications have also been discussed.

Chapter 3 discusses the mechanical, thermal, electrical, optical and magnetic properties of organic nanofillers. Apart from those, the various physical and chemical properties have also been discussed in the present chapter.

Chapter 4 covers the standard reinforcement methods, mechanism, compatibility and surface modification of nanofillers in polymers. This chapter is finally concluded with the benefits, drawbacks and potential applications of nanofillers/nanofiller-reinforced polymer composite.

Chapter 5 describes the nanocellulose-based green fillers for elastomers - influence of geometry and chemical modification on properties of nanocomposites. In addition, this chapter looks inward on how the reinforcement of nanocellulose can yield a material with improved performance characteristics

Chapter 6 is focused on the recent advances in organic nanofillers derived polymer hydrogels. This chapter gives an idea that organic nanofiller-derived polymer hydrogels are applied in modification of hydrogel materials, logical systems to release substance, matrix for health monitoring, antibacterial coatings, wound dressing, drug delivery, tissue engineering, and various other applications.

Chapter 7 explains about the environmental impact in terms of nanotoxicity and limitations of employing organic nanofillers in polymers. This chapter tries to give an overview of how nanoparticles can enter the environment and create adverse effects on different components of the environment.

Chapter 8 discusses natural nanofillers family, their properties, and applications in polymers. Various natural nanofiller families, their key properties, and various uses are discussed in this chapter.

Chapter 9 discusses about the formulation of polymer nanocomposite NPK fertilizer [cellulose-graft-poly (acrylamide)/nano-hydroxyapatite/soluble fertilizer] and evaluating its nutrient release. In this chapter, data revealed reduced chances of leaching losses with concomitant pollution to the groundwater and toxic effects to the plant roots as well as synchronized nutrient release to cater for the requirement by crops.

Chapter 10 focuses on the classification of natural nanofibers, their properties, and applications. This chapter focuses on several types of natural fibers, their properties, and their use in diverse industries

Chapter 11 discusses the biomedical applications of organic nanofillers in polymers. This chapter highlights the role of organic nanofiller namely cellulose nanoparticles (CN), chitosan nanoparticles (CsNPs) and lignin nanoparticles (LNPs) in developing the polymer matrix composite for biomedical applications.

All the three editors are thankful to almighty God. Apart from this, Dr. Vijay Chaudhary, Dr. Bhasha Sharma, and Mr. Partha Pratim Das are also thankful to their family members for the support extended during the editing of this book. Dr. Shashank Shekhar is also thankful to her parents and wife for the encouragement and support extended during the entire duration of editing this book.

We editors are also thankful to all our contributors who submitted their chapters in the present volume of this book.

We are also thankful to entire team of CRC Press (Taylor & Francis Group) for publishing this book in fastest possible time and in the most efficient manner.

Dr. Vijay Chaudhary

Dr. Bhasha Sharma

Dr. Shashank Shekhar

Mr. Partha Pratim Das

Editors

Dr. Vijay Chaudhary is currently working as an Assistant Professor in the Department of Mechanical Engineering, Amity School of Engineering and Technology (A.S.E.T.), Amity University Uttar Pradesh, Noida (India). He obtained his B. Tech. in 2011 from Department of Mechanical Engineering, Uttar Pradesh Technical University, Lucknow, India and then obtained his M. Tech (Hons) in 2013 from Department of Mechanical Engineering, Madan Mohan Malviya Engineering College, Gorakhpur, India. He obtained his Ph.D. in 2019 from Department of Mechanical Engineering, Netaji Subhas University of Technology, University of Delhi, India. His research areas of interest include processing and characterization of Polymer composites, tribological analysis of biofiber-based polymer composites, water absorption of biofiber-based polymer composites, and surface modification techniques related to polymer composite materials. Dr. Chaudhary has over 8 years of teaching and research experience. He has published more than 60 research papers in peer-reviewed international journals as well as in reputed international and national conferences. He has published 16 book chapters with reputed publishers. More than 25 students have completed their Summer Internships, B.Tech. Projects and M.Tech. Dissertations under his guidance. Currently, he is working in the field of biocomposites, nanocomposites, and smart materials.

Dr. Bhasha Sharma is currently working as an Assistant Professor in Department of Chemistry, Shivaji College, University of Delhi, India. She received her BSc (2011) in Polymer Sciences from the University of Delhi. Dr. Sharma completed her Ph.D. in Chemistry in 2019 under the guidance of Prof Purnima Jain from the University of Delhi. She has more than 7 years of teaching experience. She has published more than 40 research publications in reputed international journals. Her recently edited book titled *Graphene-Based Biopolymer Nanocomposites* has been published in Springer Nature. Her authored book *3D Printing Technology for Sustainable Polymers* and edited books *Biodegradability of Conventional Plastics: Opportunities, Challenges, and Misconceptions*, and *Sustainable Packaging: Gaps, Challenges, and Opportunities* has been accepted in Wiley, Elsevier and Taylor Francis, respectively. Her research interests revolve around sustainable polymers for packaging applications, environmentally benign approaches for biodegradation of plastic wastes, fabrication of bionanocomposites and finding strategies to ameliorate the electrochemical activity of biopolymers.

Dr. Shashank Shekhar is currently working as an Assistant Professor at Netaji Subhas University of Technology and is also associated with the Quantum Research Centre of Excellence as an Associate Director in the Department of Renewable Energy. He completed his PhD in Chemistry at the University of Delhi. Dr. Shekhar has been working on biopolymers and Schiff base metal complexes for the last 5 years and has published more than 20 articles in reputed international journals. He has 6 years of research and teaching experience. Presently, he is working on several projects including the circular economy approach to plastic waste, synthesis of nanomaterials and nanocomposites for energy harnessing, biodegradation of plastic wastes, electrochemical analysis of resultant biodegradable nanocomposites for employment in super-capacitor applications, and polymer technology for packaging applications.

Mr. Partha Pratim Das is pursuing Master of Technology (Materials Science and Metallurgical Engineering) in the Department of Materials Science and Metallurgical Engineering at Indian Institute of Technology Hyderabad (IITH), India. Currently, he is working in the field of active food packaging to extend the shelf-life of fresh produce with Cellulose and Composites Research Group at IIT Hyderabad. He completed his B.Tech in Mechanical Engineering with first-class distinction in the year 2021 from Amity University Uttar Pradesh, Noida, India. During his B.Tech, he worked on various projects with the Indian Institute of Technology Guwahati and Indian Oil Corporation Limited, Guwahati, Assam. He has presented several research papers at national and international conferences and published a good number of research papers in SCI journals and book chapters to his credit with reputed publishers including Springer, CRC press (T&F), and Elsevier. He also served as a reviewer in Materials Today: Proceedings and Applied Composite Materials, Springer. In the year 2020, he received the Innovative researcher of the year award. He is also a Certified Executive of Lean Management and Data Practitioner (Minitab and MS-Excel) from the Institute for Industrial Performance and Engagement (IIPE), Faridabad, India. He is a community associate member at American Chemical Society (ACS). His area of research includes natural fibre-based composites, processing and characterization of polymer matrix composites, nanofiller-based composites for various applications and biodegradable food packaging.

Contributors

Qazi Adfar
Vishwa Bharti Degree College
Rainawari, India

Aswathy Ajayan
Post Graduate and Research
    Department of Chemistry
Sree Sankara College
Kalady, India

Faris M. Al-Oqla
Department of Mechanical Engineering
Hashemite University
Zarqa, Jordan

Mohammad Aslam
Materials Chemistry & Engineering
    Research Laboratory, Department of
    Chemistry
National Institute of Technology
Srinagar, India

Yashdi Saif Autul
Department of Mechanical and
    Materials Engineering
Worcester Polytechnic Institute (WPI)
Worcester, Massachusetts

G. L. Devnani
Department of Chemical Engineering
Harcourt Butler Technical University
Kanpur, India

Basma M. Eid
Institute of Textile Research and
    Technology
National Research Centre
Giza, Egypt

Yves Grohens
Université Bretagne Sud
Lorient, France

Md Enamul Hoque
Department of Biomedical Engineering
Military Institute of Science and
    Technology (MIST)
Dhaka, Bangladesh

M. Iqbal Ishra
Postgraduate Institute of Science
University of Peradeniya
Peradeniya, Sri Lanka
and
MRSC
Institute of Materials Engineering and
    Technopreneurships
Kandy, Sri Lanka

Habibul Islam
Department of Biomedical Engineering
Military Institute of Science and
    Technology (MIST)
Dhaka, Bangladesh

M. G. R. Tamasha Jayawickrama
Postgraduate Institute of Science
University of Peradeniya
Peradeniya, Sri Lanka
and
MRSC
Institute of Materials Engineering and
    Technopreneurships
Kandy, Sri Lanka

Maida Mary Jude
Post Graduate and Research
Department of Chemistry
Sree Sankara College
Kalady, India

George N. Karuku
Department of Land Resources and
    Agricultural Technology
University of Nairobi
Nairobi, Kenya

Job Kuriakose
Global R&D Centre, Asia
Apollo Tyres Ltd.
Chennai, India

Shrikant S. Maktedar
Biofuels Research Laboratory
Department of Chemistry
National Institute of Technology
Srinagar, India

M. M. M. G. P. G. Manthilaka
Postgraduate Institute of Science
University of Peradeniya
Peradeniya, Sri Lanka
and
MRSC
Institute of Materials Engineering and
    Technopreneurships
Kandy, Sri Lanka

Damaris Mbui
Department of Land Resources and
    Agricultural Technology
University of Nairobi
Nairobi, Kenya

P. K. Mohamed
Global R&D Centre, Asia
Apollo Tyres Ltd.
Chennai, India

Mohit Nigam
Department of Chemical Engineering
RBSETC, Bichpuri
Agra, India

B. S. S. Nishadani
Postgraduate Institute of Science
University of Peradeniya
Peradeniya, Sri Lanka
and
MRSC
Institute of Materials Engineering and
    Technopreneurships
Kandy, Sri Lanka

P. A. Nizam
School of Chemical Science
Mahatma Gandhi University
Kottayam, India

S. Ramakrishnan
Global R&D Centre, Asia
Apollo Tyres Ltd.
Chennai, Tamil Nadu, India

Shek Md Atiqure Rahman
Sustainable and Renewable Energy
Engineering
University of Sharjah
Sharjah, United Arab Emirates

Tousif Reza
Department of Mechanical Engineering
Military Institute of Science and
    Technology (MIST)
Dhaka, Bangladesh

Kiplangat Rop
Department of Chemistry
University of Nairobi
Nairobi, Kenya

Bastien Seantier
Université Bretagne Sud
Lorient, France

Kriti Sharma
Department of Chemical Engineering
Harcourt Butler Technical University
Kanpur, India

Mohammad Fahim Tazwar
Department of Mechanical Engineering
Military Institute of Science and
    Technology (MIST)
Dhaka, Bangladesh

Prakash Chander Thapliyal
Advanced Structural Composites and
    Durability Group
CSIR-Central Building Research
    Institute
Roorkee, India

Sabu Thomas
School of Energy Materials
Mahatma Gandhi University
Kottayam, India

Milanta Tom
School of Energy Materials
Mahatma Gandhi University
Kottayam, India

Rokib Uddin
Department of Mechanical Engineering
Military Institute of Science and
    Technology (MIST)
Dhaka, Bangladesh

Sanya Verma
Biochemical Engineering and
    BioTechnology
Indian Institute of Technology
New Delhi, India

1 Organic Nanofillers and Their Classification Employed in Polymers

Prakash Chander Thapliyal
CSIR-Central Building Research Institute

CONTENTS

1.1 Introduction 1
1.2 Types of Nanofillers 2
1.3 Organic Nanofillers 3
    1.3.1 Nanocellulose 3
    1.3.2 Nano Lignins 7
    1.3.3 Nano Chitosan 7
    1.3.4 Miscellaneous 9
1.4 Limitations of Organic Nanofillers 11
1.5 Future Scope of Organic Nanofillers 11
Acknowledgments 12
References 12

1.1 INTRODUCTION

A lot of literature is available on research works where studies have been made on the impact of the inclusion of nanofillers into polymers. Nanofillers are materials of diverse chemical characteristics containing at least one element between 1 and 100 nm. Characteristically, such polymer nanocomposites display a great deal superior mechanical, thermal and other properties than conventional polymer composites. A distinctive crossing point formed involving polymer and nanofiller demonstrates enhanced performances of such nanofiller having composites. Due to the large explicit surface area of nanofillers, figure of filler-polymer matrix contact sites increases in finely dispersed polymer nanocomposites, playing a significant role in controlling the properties (See Table 1.1). Adhesive properties similar to the coefficient of wetting, interfacial free energy, etc. are mainly impacted by the nanofillers (Guchait et al., 2022). The insertion of nanofiller in polymeric coatings can significantly advance their corrosion resistance, thermal stability, flame retardancy, transparency and resistance to organic solvents and decrease the inclination for the coating to blister and/or delaminate (Kabeb et al., 2019; Wang et al., 2006).

DOI: 10.1201/9781003279372-1

TABLE 1.1
Pros and Cons of Use of Nanofillers

Pros	Cons
• Reinforcement	
• Chemical resistance	
• Mechanical properties	• Dispersion problems
(tensile strength, toughness, and stiffness)	• Processability
• Dimensional stability	• Sedimentation
• Thermal expansion	• Optical clarity
• Gas barrier	• Coloration
• Flame retardant	
• Ablation resistance	

Innovative trends and perspectives in nanotechnology have facilitated conduit en route for employing nanofillers like dynamic agents in packaging manufacturing. This resolution applies toward making of biodegradable packaging composites depending on biopolymers as well as nanofillers. Such kind of nanocomposite material is incapable of partially substituting man-made packaging materials because of its fragile mechanical properties, excessive hydrophilicity and vulnerability to disintegration. On the other hand, they may be proven alternatives to plastic materials and be able to apply into areas wherever plastic revival is not efficiently viable (Majeed et al., 2013). At present, there is mounting concern in grouping of biopolymer with nanofiller-based materials. Nanofillers comprise diverse figures and dimensions; however, their individual particle dimension must be under 100 nm (George and Ishida, 2018). Nanofillers are commonly added to polymers en route for attaining well-designed properties not possessed in polymer itself. On the whole, such appropriate additives are flame retardants along with conductive nanofillers (Moczo and Pukanszky, 2008).

In this chapter, an overview of the research that has appeared in the literature is presented for the use of organic nanofillers in making polymer nanocomposites, dealing with the major present developments and progress in this field.

1.2   TYPES OF NANOFILLERS

Natural fibers are believed to be an important as well as strong substitute to synthetic silicates and carbon-based fibers; nevertheless, the incompatibility of natural fiber with polymer matrices along with the higher dampness absorption fraction of natural fiber confines their applications. To surmount face treatment of natural fiber and nanofiller addition have turn out to be the most important aspects toward improving performance of natural fiber reinforced polymer composites. The shape and characteristics of the nanofiller will also determine the process to be engaged for the making of the polymer nanocomposites.

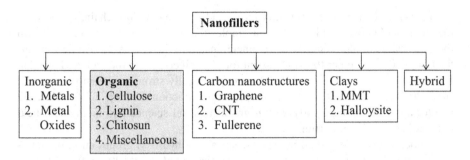

FIGURE 1.1 Classification of nanofillers.

There are mainly four types of nanofillers on the basis of chemistry: clays, inorganic, organic and carbon nanostructure (see Figure 1.1). Among these are organic nanofillers, which may include natural biopolymers, e.g. chitosan and cellulose (Youssef and El-Sayed, 2018; Sothornvit, 2019).

1.3 ORGANIC NANOFILLERS

Organic nanofillers have been used to improve the performance of the biopolymers (Othman, 2014). Natural biopolymer nanofibrils composed of molecules such as collagen, cellulose and chitin are repeatedly used as organic nanofillers owing to their biodegradability, biocompatibility, availability, durability and mechanical properties (Ling et al., 2018).

An extensive diversity of organic and inorganic nanofillers has been joined with diverse polymers to achieve polymer nanocomposites planned with precise properties for every type of function. Organic nanofillers include many polymer nanofibers and natural fibers, e.g. cellulose, flax and wood (Miao and Hamad, 2019; Zhu et al., 2019).

1.3.1 Nanocellulose

Material researchers in academia and industry are looking for the ways to replace synthetic nanofillers with materials from natural and biodegradable sources. There are many sources of getting a variety of organic nanofillers. Advantageous cellulose properties comprise biodegradability, toughness, biocompatibility, renewability, water insolubility, low-cost, colorless, non-toxicity, lightweight, high mechanical strength and durability (George and Sabapathi, 2015), and such striking properties create it an appropriate option to fossil fuel-based polymers.

Cellulose is the most abundant renewable polymer resource and has been growing in use owing to its fine thermal as well as chemical strength, higher hydrophilicity and exceptional biocompatibility, making it a potential material for diverse applications such as biomedical applications, nano-drug release systems and food packaging. Cellulose nanofibers (CNF) have admirable mechanical strength, good optical properties, thermal and dimensional stability and their renewable nature with the occurrence of hydroxyl functionality by the side of the cellulose chain, resulting in the creation of network of intra- and intermolecular bonds (Mariano et al., 2014) with the establishment of van der Waals connections amid the chains. Its Young's modulus is a great deal higher than that meant for glass fibers, i.e. in the region of 70 GPa. It is analogous to Kevlar (60–125 GPa) and potentially stronger in comparison with steel (200–220 GPa) and in general depends on extraction methods. CNCs illustrate greater applications especially within green material science and biomedical field (Lizundia et al., 2016; Fortunati et al., 2017).

Nanocellulose is identified by reactive surface of hydroxyl groups and can be tailored in the direction of diverse surface properties (Khalil et al., 2012a; Phanthong et al., 2018). Three types of cellulose are used as a nanofiller in biopolymeric films, i.e. nanofibrillated cellulose, nanocrystalline cellulose and bacterial nanocellulose, which differ in their morphology, quantity of crystallinity, grain dimension, etc. Bacterial nanocellulose (BNC) is primarily obtained from the commencing cultures of gram-negative bacteria Gluconacetobacter xylinus and has a higher molecular weight and higher crystallinity as compared to plant-based cellulose (Niamsap et al., 2019).

Auad et al. (2008) used CNCs to boost the mechanical strength of polyurethane-based SMP and achieved a 53% raise in the tensile strength. Furthermore, nanocomposite showed shape recoveries greater than 95%, signifying strengthening CNCs has no effect on the shape memory.

Earlier, Bitinis et al. (2013) integrated CNCs with a mixture of poly (lactic acid) and natural rubber and achieved a fine distribution of the nanofillers into the mix in course of casting and extrusion processes. Similarly, Mariano et al. (2017) have done non-covalent amendment of CNCs in the midst of two diverse poly(Lactide) (PLLA) dependent surfactants to get better nanofiller–matrix compatibility. Rueda et al. (2013) evaluated the mechanical as well as thermal behaviors of polyurethane-based SMP reinforced by way of various CNC fillings (1.5–30 weight %) and found an increase in material stiffness with temperature along with enhanced strength. Lately, non-toxic electroactive hydrogel was fabricated through reinforcing of polyvinyl alcohol by CNCs having superior displacement in electric field (Jayaramudu et al., 2018).

Wu et al. (2014) made-up water-triggered SMP nanocomposites having paramount shape recall of 98 in addition to 99% for shape fixity and recovery by way of 23 weight % CNCs accordingly; Young's modulus improved by 4400% while elongation at break improved by 400%.

Using nanocellulose, rod-like core–shell nanocellulose/polyaniline nanocomposites have been prepared via in situ polymerization. In comparison with clean polyaniline, these nanocomposites illustrate better film-forming characteristics along with exceptional electrochromic properties, i.e. quicker reaction instance (1.5 s for bleaching along with 1.0 s for coloring), superior optical contrast (62.9%), higher coloration effectiveness (206.2 cm^2/C) and amazing switching steadiness of over 500 cycles (Zhang et al., 2017).

Kaboorani et al. (2017) prepared a UV-curable acrylic coating system by adding cellulose nanocrystal (CNC) and showed CNC is an ideal reinforcing nanofiller that can positively influence the mechanical properties of coating systems.

Shankar and Rhim (2016) developed a method for preparing nanocellulose (NC) starting from microcrystalline cellulose (MCC) and tested the properties of agar-based composite films (1, 3, 5 and 10 weight % based on agar). NC can be used as a reinforcing agent in support of developing films for biodegradable food packaging materials.

Addition of CNF resulted in 44% enhancement in compressive stiffness of foams compared with virgin foam. Further, it was discovered that 0.1% cellulose nanofiller loading can result in maximum enhancement in compressive stiffness of the foams. Using nanofillers to improve foam properties encompasses the potential to decrease polyol content as well as shift in direction of greener foams (Ghasemi et al., 2021).

Xu et al. (2013) prepared CNCs and used them as organic nanofillers by dispersing them into polycarbonate and studied their control on mechanical, optical and thermal properties of resulting composite films. Due to excellent thinning out of nanofillers in polymeric matrix, powerful hydrogen bonds resulted amid carbonyl groups of the polymer chain along with the hydroxyl functionality of CNCs, leading to a reinforcement outcome both on mechanical and thermal properties. Furthermore, nanofillers in polymer nanocomposites efficiently stalled main thermal degradation routes of polycarbonate concerning chain scission by the side of carbonate linkage along with the rearrangement of carbonate groups. When compared with neat polycarbonate, nanocomposite films with 3% by weight CNCs have enhancement of 27.3% in Young's modulus, 30.6% in tensile strength and 3.3% in maximum decomposition temperature although films stay transparent.

Cellulose-based organic nanofillers were brought in with polymers to fabricate transparent composite film. CNCs are introduced into polycarbonates by Jagadeesh et al. (2021) in organic solvent by means of solvent switchover. Hydroxyl groups in CNC were hydrogen bonded with carbonyl group of polycarbonates (Xu et al., 2013).

Prihatiningtyas et al. (2019) showed that the incorporation of CNCs modified the sponge-like membrane to self-assembled structure allowing for an excellent separation of water from NaCl. Experiments also demonstrated that the incorporation of 3% CNCs into cellulose triacetate membrane effectively enhanced water flux more than three times.

Nanocellulose hydrogels are comparatively fresh group of engineerable, renewable, biodegradable and non-toxic nanofillers. These gels are exceedingly hydrated (99.9%–95% water) and outline a three-dimensional permeable network having fine mechanical properties and finding applications in cell culture, drug delivery, tissue engineering, diagnostics and separation (Curvello et al., 2019).

CNCs were openly incorporated within aqueous piperazine solution meant for fabricating thin film nanocomposite nanofiltration membranes, which exhibit rougher and more hydrophilic surfaces compared with membranes devoid of CNCs, ensuing in drastic boost of water permeation flux (106.9 Lm^{-2}/h) while keeping rational salt elimination to sodium sulfate (98.3%) and magnesium sulfate (96.1%) (Huang et al., 2019).

Organic filler like CNCs were used as nanofiller within biodegradable polymer matrix since very long. Rescignano et al. (2014) modulated thermal, physical in addition to biocompatibility properties of poly(vinyl alcohol) (PVA) by combining CNCs with nanoparticles of poly(D, L-lactide-co-glycolide) (PLGA) and loading them by way of bovine serum albumin fluorescein isothiocyanate conjugate (FITC-BSA). These systems join high mechanical strength and biocompatibility of polymeric PLGA nanoparticles with PVA/CNC systems respectively.

Comparable data concerning the consequence of reinforcement part of cellulosic nanostructures in polymer matrix were obtained by means of erstwhile cellulosic nanostructures extracted commencing on or after diverse natural resources in addition to united with diverse ecological matrices. CNC and cellulose nanofibers obtained from sunflower stalks were used in gluten-associated nanocomposites for growing mechanical and barrier properties (Fortunati et al., 2016), while for flax and phormium fibers (Fortunati et al., 2013) and barley derivatives (Fortunati et al., 2015) into PVA, or Posidonia oceanica in grouping with ZnO nanoparticles, Luzi et al., (2017a) was also carried out. Luzi et al. (2016) studied poly(lactic acid) with poly(butylene succinate) blends as well as bionanocomposite films loaded with 1–3 weight % CNCs obtained from Carmagnola carded hemp fibers.

Luzi et al. (2016) observed during their studies a heartening synergic outcome of poly(butylene succinate) (PBS) together with cellulose nanocrystals (CNC & s-CNC) in stipulations of barrier properties; CNC in addition to s-CNC resulted in a decrease of oxygen and water vapor permeability due to the presence of PBS, which enhances the crystallinity quantity.

CNC was added to copolyesters having butylene succinate (BS) and triethylene succinate (TES) (Fortunati et al., 2017) and results showed how copolymerization and nanofiller addition together produced slim extruded films with tunable properties by changing quantity of nanofiller (s-CNC) meeting the needs for a broad variety of applications.

Cellulose nanofibrils (CNF), CNCs and cellulose filaments (CelFil) were used as nanofillers at 0.1–0.8 weight % loading levels (Ghasemi et al., 2021). It was found that the addition of 0.1 weight % CNF resulted in a 44% enhancement in the compressive modulus of foams.

After incorporating 5 weight % CNCs, the mechanical strength of carboxymethyl cellulose nanocomposite film was amplified linearly and CNC from barley straw, wheat straw and rice straw results in improvement of tensile strength by 42.6%, 25.2% and 45.7% respectively (Oun and Rhim, 2016).

CNCs were extracted from kiwi Actinidia deliciosa and used at the same time as nanoreinforcement in poly(vinyl alcohol) blended through chitosan systems. Also, the presence of carvacrol resulted in antimicrobial activity (Luzi et al., 2017a).

Perumal et al. (2018) studied the thermal effect of rice stalk CNC-filled CS/PVA biocomposites and showed that CNC is a valuable substance to augment thermal steadiness by strong interaction flanked through CNC functional groups and CS/PVA polymer chain blends.

Sugar palm nanocrystalline cellulose reinforced sugar palm starch composites were developed (Ilyas et al., 2018). The nanofiller concentration was varied from 0 to 1 weight % and mounting nanofiller concentration led to decline in mass loss under 300^0C, proving the thermal stability improvement of nanofiller addition. Further eggshell nanoparticles filled hemp/epoxy composites showed higher thermal stability (Bhoopathi and Ramesh, 2020).

Kenaf cellulose nano-fibrillated fiber (CNF) incorporation in PLA improved the thermal stability when compared to neat polylactic acid composite owing to enhanced distribution with PLA along with creation of strong attachment (Rizal et al., 2021).

1.3.2 Nano Lignins

Lignin is the second most abundant and widely accepted aromatic polymer on the globe and is a three-dimensional, cross-linked molecule with varieties of substituted phenols. Manufacturing organic molecules such as hydrocarbons, polyols, alcohols, acids, ketones and phenols can be obtained from lignin. Lignins can be thought of as another lignocellulosic resource to be integrated within different polymers (Thakur et al., 2014). In current existence, diverse lignin nanoparticles (LNPs) from various resources were produced (Beisl et al., 2017a; Tian et al., 2017). However, lignin's tendency to self-aggregate harmfully influences distribution in formulations (Beisl et al., 2017b).

Antioxidant action was obtained for grocery wrapping purposes once bio-based nanoparticles have been used as active nanofillers in various biopolymers (Yang et al., 2018). It was earlier found that there was a connection flanked by antibacterial and antioxidant property of lignin given that the free-radical scavenging action of lignin was set up to be firmly associated with phenolic structures on oxygen-containing reactive free radicals (Dizhbite et al., 2004).

Several researchers studied lignin nanoparticles (LNPs) once they are integrated in diverse polymers for packaging (Armentano et al., 2018). Two dissimilar PLA mixes, one having 3 weight % LNPs and other having 3 weight % LNPs along with 1 weight % CNCs tested adjacent to bacterial plant pathogen Pseudomonas syringae pv. tomato (Yang et al., 2016a, 2016b) while action of pure PVA, CH or mixed PVA/CH films and nanocomposites having LNPs (Yang et al., 2016c) against gram-positive and gram-negative bacteria plant pathogens revealing lesser CFU inside first 3 hours. Yang et al. were too deliberate on the effects of having LNPs on mechanical as well as thermal behavior of wheat gluten (WG) with glycidyl methacrylate-grafted polylactic acid (Yang et al., 2015a, 2015b). Results also established a common decline in light transmittance during visible light spectrum following LNP incorporation from 89% for neat WG and headed for 56% for WG containing 3 weight % LNPs while in UV range, i.e. beneath 400 nm, transmittance less than 2% reached for LNP containing PVA in addition to WG samples.

Gupta et al. (2010) developed and assessed poly(dl-lactide-co-glycolide) nanoparticles for sparfloxacin ophthalmic release to get better precorneal habitation moment as well as visual infiltration. The resulting nanosuspension had a mean particle size of 180–190 nm appropriate for ophthalmic use with zeta potential of –22 mV. In vitro discharge from this nanosuspension showed an extensive discharge contour of sparfloxacin.

1.3.3 Nano Chitosan

Like cellulose, chitin is also a long-chain polysaccharide consisting of numerous N-acetylglucosamine molecules, which are glucose derivatives. Chitin is mainly obtained from animal sources like octopi, squid, insects, crabs, prawns, lobsters and also from fungi. Just similar to cellulose, chitin can form nanowhiskers and nanofibrils, having several applications in bio-nanocomposite films.

After cellulose, chitin is the next most profuse polysaccharide occurring in nature. Chitin nanocrystals or chitin whiskers (CHWs) can also be isolated under definite conditions such as acidolysis from amorphous domains. CHWs collectively with other organic nanofillers such as cellulose and starch illustrate lots of advantages above conventional inorganic nanofillers like effortless accessibility, non-toxicity, low density, biodegradability and uncomplicated alteration. Hence, CHWs have been extensively used as substitutes on behalf of inorganic nanofillers in reinforcing polymer nanocomposites (Zeng et al., 2012).

The chitin and chitosan nanoparticles are accepted organic nanofillers for the reason that of their fine film-forming aptitude, non-toxic nature, active surface area and biocompatible nature. The combustion of oil palm side products that is to say EFB, fibers along with nutshells, gives organic nanoparticles. These nanoparticles have typical span amid 93.9 and 192.20 nm, while width involving 18.17 and 43.45 nm (Khalil et al., 2012a).

Inorganic nanomaterials are known to harden nylon with decrease in its toughness as well as ductility, and they are not eco-friendly. Hao et al. (2020) studied the smooth modification of nylon's mechanical strength from rigid to hard by means of CNCs in addition to chitosan nanowhiskers (CSWs) as biorenewable nanofillers. It was found that by way of CSWs, *in situ*-incorporated nanofiller with 0.4 weight % loading strengthened nylon and led to 1.9 fold raise in its Young's modulus (2.6 GPa) plus 1.7 times boost in ultimate tensile strength (106 MPa), whereas solution blended nanofiller by means of a 0.3 weight % loading toughened polymer with a 2.1 fold augment.

Chitosan has inherent antibacterial activity, and hence, chitosan nanofillers were expected to improve antibacterial properties. To obtain nanofiller from chitosan, it was first emulsified by surfactants and then cross-linked by means of sodium tripolyphosphate (TPP) to outline antibacterial nanofillers (Pan et al., 2019).

Poonguzhali et al. (2017) prepared Chitosan/Poly (vinyl pyrrolidone)/nanocellulose (CPN) nanocomposites by solution casting method and have shown composite with 3% CPN had antibacterial property and can be used as wound healing material.

Chitosan and chitin nanoparticles have received interest as nanofillers owing to striking surface area, non-toxicity, biocompatibility in addition to film-forming capacity (Hassan et al., 2016; Olivera et al., 2016; Ma et al., 2018; Azarifar et al., 2019; Jamroz et al., 2019; Vahedikia et al., 2019). Further, their stability can be enhanced by scheming the environment such as by modifying the constitution by way of chemical agents. Presently, meager solubility of chitosan nanoparticles is a big problem in encapsulation of hydrophobic drugs (Naskar et al., 2019).

The chitosan thymol nanofillers packed CS-quinoa composites were weathered on behalf of antibacterial performance against one gram-negative along with two gram-positive bacteria (Medina et al., 2019). Researcher (Jannatyha et al., 2020) compared antimicrobial properties of nanochitosan (NCH)/CMC as well as nanocellulose (NCL)/CMC nanocomposites. Good antimicrobial properties were observed owing to accumulation of nanofillers within NCH/CMC nanocomposites.

1.3.4 MISCELLANEOUS

Rafieian et al. (2014) studied thermal, mechanical and morphological properties of nanocomposite films prepared commencing WG matrix with cellulose nanofibrils. A coconut crust nanofine particles packed jute mat/epoxy composite was formed for diverse structural applications, and it was observed that 15 weight % nanofiller addition resulted in the highest TS of 38.7 MPa, 15.17% superior than without nanofiller (Jagadeesh et al., 2021).

Polytetrafluoroethylene nanoparticles in Ni matrix were introduced by electrodeposition, and resulting composite films have shown good water repellency and solid lubrication owing to low surface free energy in addition to frication coefficient (Wang et al., 2004). Similarly Ni-P-TiO_2-PTFE nanocomposite coatings by way of extensive array of surface energy components were prepared by Liu and Zhao (2011) using an electroless plating.

The inclusion of MMT/GO hybrid nanofillers into PAN nanofibers showed a clear enhancement of up to 30°C for starting decomposition temperature plus 1.32 times bigger tensile strength than pure PAN demonstrating that hybrid nanofiller is capable contender in getting better thermal and mechanical polymeric properties (Wang et al., 2014).

Yuan et al. (2016) developed nanocomposite multifunctional coatings with self-cleaning, anticorrosion and antiwear properties on top of aluminum substrate by spraying core–shell structured polyaniline/functionalized carbon nanotubes (PANI/fCNTs) into ethylene tetrafluoroethylene (ETFE) matrix en route for creating lotus leaf like structures. This coating demonstrates steady non-wetting performance over broad temperature range of less than 400°C and exceptional self-cleaning capability to avert contagion by sludge, concentrated H_2SO_4 and ethylene glycol.

Elkhouly et al. (2020) reported use of organic nanoparticles derived from date seeds as nanofiller for polyethylene terephthalate (PET) in making of polymer nanocomposites, which showed better chemical stability. Additionally, main properties like hardness, compressive strength and wear resistance were enhanced and optimized with a DSN reinforcement of 0.75 weight %.

Yang et al. (2019) synthesized three-dimensional COF (3D-COF) nanofiller for preparing nanocomposite membrane on the road to carbon dioxide separation and showed that 3D-COF was competent in sinking aging process of membrane through capitalizing on amine functional groups in addition to vast surface area to freeze and rigidify polyimide polymer chains and thus avoiding collapse of free volume.

Recently, researchers have developed polyethersulfone (PES) membrane by means of covalent organic frameworks (COFs) nanofiller having average dimension of 30 nm causing a drop in surface roughness and an improvement in hydrophilicity improving their flux and fouling resistance significantly. Water flux improved 2.6 times by adding 0.5 weight % COF, and this membrane can be used for dye separation along with judicious salt separation with appropriate antifouling properties (Vatanpour and Paziresh, 2022).

Sen and Kumar (2010) prepared coir-fiber-based fire retardant nanofiller for use in epoxy resin. Initially, coir fiber brominated using saturated bromine water and afterward reacted by stannous chloride solution. On air drying, it was powdered to nanodimension and mixed fine with epoxy resin for synthesis of composites. Smoke density as well as limiting oxygen index of coir–epoxy nanocomposites has improved considerably (Saba et al., 2014).

Recently, antioxidant 2D covalent organic framework dopamine (COFDOPA) nanofiller was developed and incorporated into plasticized poly(lactic acid) matrix by means of 15 weight % of acetyl tributyl citrate to build up non-migratory sustainable wrapping. Bionanocomposites loaded with low amounts (0.5 weight % and 1 weight %) resulted in transparency and showed fine interfacial adhesion, better crystallinity and thermo-mechanical performance (Arroyo et al., 2020; Prihatiningtyas et al., 2019).

Saba et al. (2016) found that tensile strength as well as toughness of kenaf toughened epoxy composites enhanced by adding oil palm nanoparticles, great surface area behaving as stage in support of connecting amid matrix with fiber. Oil palm nanoparticles also enhanced interface physical adhesion along with appropriate stress transfer under loading.

Date palm nanofillers have found in packaging and dental applications in recent times (Ghori et al., 2018). Adel et al (2018) obtained oxidized nanocellulose from fibers of date palm sheath and used it as a packaging additive in chitosan films. Salih et al. (2018) formed maxillary denture bases from poly(methyl methacrylate) (PMMA) composites reinforced with nanoscale pressed date powder. Arifin et al. (2018) make use of rice husk agricultural waste and prepared organic nanofillers (325 mesh < 45 μm) for the development of polymer nanocomposites.

Wang et al. (2020) synthesized aminophend/formaldehyde polymeric nanospheres (APFNSs) by means of sol-gel technique from 3-aminophenol and formaldehyde and used them as nanofillers through incorporation into polyamide (PA) sheets to construct thin film nanocomposite (TFN) membranes. The water flux of TFN-As membrane modified as a result of APFNSs nanofillers is 71.3 L/m^{2}h twice of pure TFC membrane, and salt rejection keeps over 96%. This effort demonstrated that aminophend/formaldehyde polymeric nanospheres (APFNSs) can be promising as nanofiller for budding novel TFN membranes with superior desalination performance.

A novel category of poly(ether sulfone) (PES) supported MMM containing metal–organic framework (MOF) nanofillers of HKUST-1 were blended among poly(methyl methacrylate-co-methacrylic acid) (PMMA-co-MAA) copolymer to give HKUST-1@mPES MMM for use in ultrafiltration and competent water treatment applications. The ensuing HKUST-1@mPES MMM exhibited pure water permeability (PWP) nearly three times higher than that of mPES membrane san HKUST-1 nanofillers (Lin et al., 2019).

Liao et al. (2020) used asymmetric organic nanofiller resorcinol formaldehyde nanobowls (RFB), into the polyamide membrane resulting in superior nanofiltration performance of the prepared membrane. Water fluxes of TFN-B$_{0.12}$ membrane were 119.74 ± 4.28 for sodium sulfate solution and 141.46 ± 5.92 L/(m^{2}h) for magnesium chloride solution, one and half times higher than that of thin film composite (TFC) membrane.

The use of aramid nanofibers (ANFs) as multifunctional nanofillers improved polyethylene oxide (PEO)-LiTFSI electrolytes through the hydrogen bond interactions. Thus, ANF tailored electrolytes demonstrated better room temperature conductivity of 8.8×10^{-5} S/cm and displayed better mechanical strength, electrochemical firmness, thermostability along with interfacial resistance against lithium dendrites (Liu et al., 2020).

Even et al. (2011) revealed peptide nanotubes (PNTs) as nanofillers for epoxy with an increase of 70% in shear strength in addition to 450% in peel strength in comparison with unmodified epoxy. In fact, these effects go beyond reinforcement effect of several well-known inorganic nanofillers.

Stimuli-response nanocomposites respond to electric currents, light, temperature, magnetic field, pH changes, etc. (Ponnamma et al., 2019; Zarnegar et al., 2019; Li et al., 2020;). Chen et al. (2020) reported production of a shape-memory nanocomposite with a rapid light response and self-healing performance from poly(ε-caprolactone)/thermoplastic polyurethane and used polydopamine nanospheres as active nanofiller providing the system a memory response shape effect to light with 100% ratio. Moreover, these nanocomposites showed superb self-healing effects in light irradiation for only 150s and exhibited exceptional mechanical properties (tensile strength of 1.6 MPa), which makes them perfect for biomedical uses.

1.4 LIMITATIONS OF ORGANIC NANOFILLERS

A widespread test intended for every organic nanofiller to achieve the best possible performance of a nanocomposite is the homogeneous distribution of nanofillers in the polymer matrix (Melly et al., 2020).

On the other hand, CNC-based materials have some restrictions and drawbacks, such as moisture uptake/swelling, agglomeration of nanofillers, trouble in redispersing agglomerated particles and inappropriateness amid hydrophobic polymers (See Table 1.1).

Earlier researchers indicated that in the bulk polymer nanocomposites, huge disarray was omnipresent and recommended incorporation of high aspect ratio nanofillers unproductive owing to the following: (i) troubles with agglomeration and dispersion and (ii) the prospective for customized molecular self-assembly concerning a superior extent of 'cooperation' amid the nanofiller and matrix (Schaefer and Justice, 2007).

The addition of organic nanofillers resulted in drop of water vapor permeability of alginate-based nanocomposites (Abdollahi et al., 2013). Owing to the analogous polysaccharide structures, there were fine interfacial connections, and as a result, the tensile strength along with the elongation of nanocomposite films improved. Lastly, these outcomes suggested that organic CNP had the capability to create entirely renewable and natural nanocomposites compared with inorganic nanoclay fillers in the case of carbohydrate biopolymers.

1.5 FUTURE SCOPE OF ORGANIC NANOFILLERS

Single-component organic nanofillers have revealed a few restrictions in their performance, which can be overcome through hybrid nanofillers with two or more dissimilar components.

Organic nanofillers have enhanced prospects for lessening constraints in the use of nanocomposite films in food packaging.

Polymer nanocomposites demonstrate stupendous performance and bulk-processing capabilities compared with usual composites because they are recyclable, lightweight and efficient even at small nanofiller loading. In particular, organic nanofillers with a high aspect ratio and low density are appealing for the reason that their consistent dispersal in a polymer medium provides an exceptional matrix/filler interface for improving thermal and/or mechanical properties of polymers, which in turn leads to lighter, cheaper, more environmentally friendly and more sustainable polymer nanocomposites. Nanocellulose, CNCs, rod-like shaped, can be found in the natural structure of cellulose fibers, which have a high aspect ratio and are almost defect-free, renewable and biocompatible. Adding together, CNCs are stiffer than Kevlar, and their mechanical properties are comparable to other reinforcements, even those with a low density. This is why there is still growing interest in sustainable cellulose materials.

Hence, there is a need for a more methodical and rigorous approach for nanocomposite researchers guided by a development in multistage theoretical modeling. Innovative experimental techniques need to be developed and applied to better probe and understand fine structures and conformations at the polymer–nanofiller interface. Also, there is a need to study possible environmental health and safety implications.

ACKNOWLEDGMENTS

Author wishes to express gratitude to the Director of the CSIR-Central Building Research Institute, Roorkee, for his constant encouragement.

REFERENCES

Abdollahi, M., Alboofetileh, M., Rezaei, M., & Behrooz, R. (2013). Comparing physico-mechanical and thermal properties of alginate nanocomposite films reinforced with organic and/or inorganic nanofillers. *Food Hydrocolloids*, 32(2), 416–424.

Adel, A., El-Shafei, A., Ibrahim, A., & Al-Shemy, M. (2018). Extraction of oxidized nanocellulose from date palm (Phoenix dactylifera L.) sheath fibers: Influence of CI and CII polymorphs on the properties of chitosan/bionanocomposite films. *Ind. Crops Prod.*, 124, 155–165.

Arifin, B., Aprilia, S., Alam, P. N., Mulana, F., Amin, A., Anaska, D. M., & Putri, D. E. (2018). Characterization nanofillers from agriculture waste for polymer nanocomposites reinforcement. *MATEC Web of Conferences*, 156, 05020.

Armentano, I., Puglia, D., Luzi, F., Arciola, C. R., Morena, F., Martino, S., & Torre, L. (2018). Nanocomposites based on biodegradable polymers. *Materials*, 11(5), 795. doi:10.3390/ma11050795

Arroyo, P. G., Arrieta, M. P., Garcia, D. G., Rodriguez, R. C., Fombuena, V., Mancheno, M. J., & Segura, J. L. (2020). Plasticized poly(lactic acid) reinforced with antioxidant covalent organic frameworks (COFs) as novel nanofillers designed for non-migrating active packaging applications. *Polymer*, 196, 122466.

Auad, M. L., Contos, V. S., Nutt, S., Aranguren, M. I., & Marcovich, N. E. (2008). Characterization of nanocellulose reinforced shape memory polyurethanes. *Polym Int.*, 57, 651–659.

Azarifar, M., Ghanbarzadeh, B., Khiabani, M. S., Basti, A. A., Abdulkhani, A., Noshirvani, N., & Hosseini, M. (2019). The optimization of gelatin-CMC based active films containing chitin nanofiber and Trachyspermum ammi essential oil by response surface methodology. *Carbohydr. Polym.*, 208, 457–468.

Beisl, S., Friedl, A., & Miltner, A. (2017a). Lignin from micro- to nanosize: Applications. *Int. J. Mol. Sci.*, 18, 2367.

Beisl, S., Miltner, A., & Friedl, A. (2017b). Lignin from micro- to nanosize: Production Methods. *Int. J. Mol. Sci.*, 18, 1244.

Bhoopathi, R., & Ramesh, M. (2020). Influence of eggshell nanoparticles and effect of alkalization on characterization of industrial hemp fibre reinforced epoxy composites. *J. Polym. Environ.*, 28, 2178–2190.

Bitinis, N., Verdejo, R., Bras, J., Fortunati, E., Kenny, J. M., Torre, L., & López-Manchado, M. A. (2013). Poly(lactic acid)/natural rubber/cellulose nanocrystal bionanocomposites. Part I. Processing and morphology. *Carbohyd Polym.*, 96(2), 611–620.

Chen, Y., Zhao, X., Luo, C., Yang, M. B., & Yin, B. (2020). A facile fabrication of shape memory polymer nanocomposites with fast light-response and self-healing performance. *Compos. Part A: Appl. Sci. Manuf.*, 135, 105931.

Curvello, R., Raghuwanshi, V. S., & Garnier, G. (2019). Engineering nanocellulose hydrogels for biomedical applications. *Adv. Coll. Interface Sci.*, 267, 47–61.

Dizhbite, T., Telysheva, G., Jurkjane, V., & Viesturs, U. (2004). Characterization of the radical scavenging activity of lignins-Natural antioxidants. *Bioresour. Technol.*, 95, 309–317.

Elkhouly, H. I., Rushdi, M. A., & Magied, R. K. A. (2020). Eco-friendly date-seed nanofillers for polyethylene terephthalate composite reinforcement. *Mater. Res. Express*, 7, 025101 https://doi.org/10.1088/2053-1591/ab6daa

Even, N., Abramovich, L. A., Buzhansky, L., Dodiuk, H., & Gazit, E. (2011). Improvement of the mechanical properties of epoxy by peptide nanotube fillers. *Small*, 7(8), 1007–1011.

Fortunati, E., Benincasa, P., Balestra, G. M., Luzi, F., Mazzaglia, A., Del Buono, D., Puglia, D., & Torre, L. (2015). Revalorization of barley straw and husk as precursors for cellulose nanocrystals extraction and their effect on PVA_CH nanocomposites. *Ind. Crop. Prod.*, 92, 201–217.

Fortunati, E., Gigli, M., Luzi, F., Dominici, F., Lotti, N., Gazzano, M., Cano, A., Chiralt, A., Munari, A., Kenny, J. M., et al. (2017). Processing and characterization of nanocomposite based on Poly(butylene/triethylene succinate) copolymers and cellulose nanocrystals. *Carbohydr. Polym.*, 165, 51–60.

Fortunati, E., Luzi, F., Jiménez, A., Gopakumar, D.A., Puglia, D., Thomas, S., Kenny, J. M., Chiralt, A., & Torre, L. (2016). Revalorization of sunflower stalks as novel sources of cellulose nanofibrils and nanocrystals and their effect on wheat gluten bionanocomposite properties. *Carbohydr. Polym.*, 149, 357–368.

Fortunati, E., Puglia, D., Luzi, F., Santulli, C., Kenny, J. M., & Torre, L. (2013). Binary PVA bio-nanocomposites containing cellulose nanocrystals extracted from different natural sources: Part I. *Carbohydr. Polym.*, 97, 825–836.

George, J., & Ishida, H. (2018). A review on the very high nanofiller-content nanocomposites: Their preparation methods and properties with high aspect ratio fillers. *Prog. Polym. Sci.*, 86, 1–39.

George, J., & Sabapathi, S. N. (2015). Cellulose nanocrystals: Synthesis, functional properties, and applications. *Nanotechnol Sci Appl*, 8, 45–54.

Ghasemi, S., Amini, E. N., Tajvidi, M., Kiziltas, A., Mielewski, D. F., & Gardner, D. J. (2021). Flexible polyurethane foams reinforced with organic and inorganic nanofillers. *J. Appl. Polym. Sci.*, 138, e49983.

Ghori, W., Saba, N., Jawaid, M., & Asim, M. (2018). A review on date palm (phoenix dactylifera) fibers and its polymer composites. *IOP Conf. Series: Materials Science and Engineering*, 368012009.

Guchait, A., Saxena, A., Chattopadhyay, S., & Mondal, T. (2022). Influence of nanofillers on adhesion properties of polymeric composites. *ACS Omega*, 7, 3844–3859.

Gupta, H., Aqil, M., Khar, R. K., Ali, A., Bhatnagar, A., & Mittal, G. (2010). Sparfloxacin-loaded PLGA nanoparticles for sustained ocular drug delivery. *Nanomed Nanotechnol Biol Med.*, 6, 324–333.

Hao, L. T., Eom, Y., Tran, T. H., Koo, J. M., Jegal, J., Hwang, S. Y., Oh, D. X., & Park, J. (2020). Rediscovery of nylon upgraded by interactive biorenewable nano-fillers. *Nanoscale*, 12, 2393–2405.

Hassan, E. A., Hassan, M. L., Abou-zeid, R. E., & El-Wakil, N. A. (2016). Novel nanofibrillated cellulose/chitosan nanoparticles nanocomposites films and their use for paper coating. *Ind. Crops Prod.*, 93, 219–226.

Huang, S., Wu, M. B., Zhu, C. Y., Ma, M. Q., Yang, J., Wu, J., & Xu, Z. K. (2019). Polyamide nanofiltration membranes incorporated with cellulose nanocrystals for enhanced water flux and chlorine resistance. *ACS Sustainable Chem. Eng.*, 7(14), 12315–12322.

Ilyas, R. A., Sapuan, S. M., Ishak, M. R., & Zainudin, E. S. (2018). Development and characterization of sugar palm nanocrystalline cellulose reinforced sugar palm starch bionanocomposites. *Carbohydr. Polym.*, 202, 186–202.

Jagadeesh, P., Puttegowda, M., Rangappa S. M., & Siengchin, S. (2021). Influence of nanofillers on biodegradable composites: A comprehensive review. *Polym. Compos*, 42(11), 5691.

Jamroz, E., Kulawik, P., & Kopel, P. (2019). The effect of nanofillers on the functional properties of biopolymer based films: A review. *Polymers*, 11(4), 675. doi:10.3390/polym11040675

Jannatyha, N., Shojaee-Aliabadi, S., Moslehishad, M., & Moradi, E. (2020). Comparing mechanical, barrier and antimicrobial properties of nanocellulose/CMC and nanochitosan/CMC composite films. *Int. J. Biol. Macromol.*, 164, 2323–2328.

Jayaramudu, T., Ko, H. U., Kim, H. C., Kim, J. W., Muthoka, R. M., & Kim J. (2018). Electroactive hydrogels made with polyvinyl alcohol/cellulose nanocrystals. *Materials*, 11(9), 1615.

Kabeb, S. M., Hassan, A., Mohamad, Z., Sharer, Z., Mokhtar, M., & Ahmad, F. (2019). Exploring the effects of nanofillers of epoxy nanocomposite coating for sustainable corrosion protection. *Chem. Eng. Trans.*, 72, 121–126.

Kaboorani, A., Auclair, N., Riedl, B., & Landry, V. (2017). Mechanical properties of UV-cured cellulose nanocrystal (CNC) nanocomposite coating for wood furniture. *Prog. Org. Coat.*, 104, 91–96.

Khalil, H. A., Mahayuni, A. R., Rudi, D., Almulali, M. Z., & Abdullah, C. K. (2012). Ash from palm biomass. *BioResources*, 7(4), 5771–5780.

Khalil, H. P. S. A., Bhat, A. H. I., & Yusra, A. F. (2012). Green composites from sustainable cellulose nanofibrils: A review. *Carbohydr. Polym.*, 87, 963–979.

Li, M., Shi, Y., Gao, H., & Chen, Z. (2020). Bio-inspired nanospiky metal particles enable thin, flexible, and thermo-responsive polymer nanocomposites for thermal regulation. *Adv. Funct. Mater.*, 30, 1910328.

Liao, Z., Fang, X., Li, Q., Xie, J., Ni, L., Wang, D., Sun, X., Wang, L., & Li, J. (2020). Resorcinol-formaldehyde nanobowls modified thin film nanocomposite membrane with enhanced nanofiltration performance. *J. Membr. Sci.*, 594, 117468.

Lin, Y., Wu, H. C., Yasui, T., Yoshioka, T., & Matsuyama, H. (2019). Development of an HKUST-1 nanofiller templated poly(ether sulfone) mixed matrix membrane for a highly efficient ultrafiltration process. *ACS Appl. Mater. Interfaces*, 11(20), 18782–18796.

Ling, S., Chen, W., Fan, Y., Zheng, K., Jin, K., Yu, H., Buehler, M. J., & Kaplan, D. L. (2018). Biopolymer nanofibrils: Structure, modeling, preparation, and applications. *Prog. Polym. Sci.*, 85, 1–56.

Liu, C., & Zhao, Q. (2011). Influence of surface-energy components of Ni-P-TiO2-PTFE nanocomposite coatings on bacterial adhesion. *Langmuir*, 27(15), 9512–9519.

Liu, L., Lyu, J., Mo, J., Yan, H., Xu, L., Peng, P., Li, J., Jiang, B., Chu, L., & Li, M. (2020). Comprehensively-upgraded polymer electrolytes by multifunctional aramid nanofibers for stable all-solid-state Li-ion batteries. *Nano Energy*, 69, 104398.

Lizundia, E., Fortunati, E., Dominici, F., Vilas, J. L., León, L. M., Armentano, I., Torre, L., & Kenny, J. M. (2016). PLLA-grafted cellulose nanocrystals: Role of the CNC content and grafting on the PLA bionanocomposite film properties. *Carbohydr. Polym.*, 142, 105–113.

Luzi, F., Fortunati, E., Giovanale, G., Mazzaglia, A., Torre, L., & Balestra, G. M. (2017a). Cellulose nanocrystals from Actinidia deliciosa pruning residues combined with carvacrol in PVA_CH films with antioxidant/antimicrobial properties for packaging applications. *Int. J. Biol. Macromol.*, 104, 43–55.

Luzi, F., Fortunati, E., Jiménez, A., Puglia, D., Chiralt, A., & Torre, L. (2017b). PLA nanocomposites reinforced with cellulose nanocrystals from posidonia oceanica and ZnO nanoparticles for packaging application. *J. Renew. Mater.*, 5, 103–115.

Luzi, F., Fortunati, E., Jiménez, A., Puglia, D., Pezzolla, D., Gigliotti, G., Kenny, J. M., Chiralt, A., & Torre, L. (2016). Production and characterization of PLA_PBS biodegradable blends reinforced with cellulose nanocrystals extracted from hemp fibres. *Ind. Crops Prod.*, 93, 276–289.

Ma, Q., Liang, T., Cao, L., & Wang, L. (2018). Intelligent poly (vinyl alcohol)-chitosan nanoparticles-mulberry extracts films capable of monitoring pH variations. *Int. J. Biol. Macromol.*, 108, 576–584.

Majeed, K., Jawaid, M., Hassan, A., Bakar, A. A., Khalil, H. P. S. A., Salema, A. A., & Inuwa, I. (2013). Potential materials for food packaging from nanoclay/natural fibres filled hybrid composites. *Mater. Des.*, 46, 391–410.

Mariano, M., El Kissi, N., & Dufresne, A. (2014). Cellulose nanocrystals and related nanocomposites: Review of some properties and challenges. *J. Polym. Sci. Part B Polym. Phys.*, 52, 791–806.

Mariano, M., Pilate, F., de Oliveira, F. B., Khelifa, F., Dubois, P., Raquez, J. M., & Dufresne, A. (2017). Preparation of cellulose nanocrystal-reinforced poly(lactic acid) nanocomposites through noncovalent modification with PLLA-based surfactants. *ACS Omega*, 2(6), 2678–2688.

Medina, E., Caro, N., Abugoch, L., Gamboa, A., Díaz-Dosque, M., & Tapia, C. (2019). Chitosan thymol nanoparticles improve the antimicrobial effect and the water vapour barrier of chitosan-quinoa protein films. *J. Food Eng.*, 240, 191–198.

Melly, S. K., Liu, L., Liu, Y., & Leng, J. (2020). Active composites based on shape memory polymers: Overview, fabrication methods, applications, and future prospects. *J. Mater. Sci.*, 55, 10975–11051.

Miao, C., & Hamad, W. Y. (2019). Critical insights into the reinforcement potential of cellulose nanocrystals in polymer nanocomposites. *Curr. Opin. Solid State Mater. Sci.*, 23, 100761.

Moczo, J., & Pukanszky, B. (2008). Polymer micro and nanocomposites: Structure, interactions, properties. *J. Ind. Eng. Chem.*, 14(5), 535–563.

Naskar, S., Sharma, S., & Kuotsu, K. (2019). Chitosan-based nanoparticles: An overview of biomedical applications and its preparation. *J. Drug Delivery Sci. Technol.*, 49, 66–81.

Niamsap, T., Lam, N. T., & Sukyai, P. (2019). Production of hydroxyapatite-bacterial nanocellulose scaffold with assist of cellulose nanocrystals. *Carbohydr. Polym.*, 205, 159–166.

Olivera, S., Muralidhara, H. B., Venkatesh, K., Guna, V. K., Gopalakrishna, K., & Kumar, Y. (2016). Potential applications of cellulose and chitosan nanoparticles/composites in wastewater treatment: A review. *Carbohydr. Polym.*, 153, 600–618.

Othman, S. H. (2014). Bio-nanocomposite materials for food packaging applications: Types of biopolymer and nano-sized filler. *Agric. Agric. Sci. Procedia*, 2, 296–303.

Oun, A. A., & Rhim, J. W. (2016). Isolation of cellulose nanocrystals from grain straws and their use for the preparation of carboxymethyl cellulose-based nanocomposite films. *Carbohydr. Polym.*, 150, 187–200.

Pan, C., Qian, J., Fan, J., Guo, H., Gou, L., Yang, H., & Liang, C. (2019). Preparation nanoparticle by ionic cross-linked emulsified chitosan and its antibacterial activity. *Colloids
Surf. A Physicochem. Eng. Aspects*, 568, 362–370.

Perumal, A. B., Sellamuthu, P. S., Nambiar, R. B., & Sadiku, E. R. (2018). Development of
polyvinyl alcohol/chitosan bio-nanocomposite films reinforced with cellulose nanocrystals isolated from rice straw. *Appl. Surf. Sci.*, 449, 591–602.

Phanthong, P., Reubroycharoen, P., Hao, X., Xu, G., Abudula, A., & Guan, G. (2018).
Nanocellulose: Extraction and application. *Carbon Resour. Convers.*, 1, 32–43.

Ponnamma, D., Parangusan, H., Tanvir, A., & AlMa'adeed, M. A. A. (2019). Smart and robust
electrospun fabrics of piezoelectric polymer nanocomposite for self-powering electronic
textiles. *Mater. Des.*, 184, 108176.

Poonguzhali, R., Basha, S. K., & Kumari, V. S. (2017). Synthesis and characterization of chitosan-PVP-nanocellulose composites for in-vitro wound dressing application. *Int. J. Biol.
Macromol.*, 105, 111–120.

Prihatiningtyas, I., Volodin, A., & Bruggen, B. V. (2019). Cellulose nanocrystals as organic
nanofillers for cellulose triacetate membranes used for desalination by pervaporation.
Ind. Eng. Chem. Res., 58(31), 14340–14349.

Rafieian, F., Shahedi, M., Keramat, J., & Simonsen, J. (2014). Thermomechanical and morphological properties of nanocomposite films from wheat gluten matrix and cellulose
nanofibrils. *J. Food Sci.*, 79, N100–N107.

Rescignano, N., Fortunati, E., Montesano, S., Emiliani, C., Kenny, J. M., Martino, S., &
Armentano, I. (2014). PVA bio-nanocomposites: A new take-off using cellulose nanocrystals and PLGA nanoparticles. *Carbohydr. Polym.*, 99, 47–58.

Rizal, S., Sadasivuni, K. K., Atiqah, M. S. N., Olaiya, N. G., Paridah, M. T., Abdullah, C. K.,
Alfatah, T., Mistar, E. M., & Khalil, H. P. S. A. (2021). The role of cellulose nanofibrillated fibers produced withcombined supercritical carbon dioxide and high-pressure-
homogenization process as reinforcement material inbiodegradable polymer. *Polym.
Compos.*, 42, 1795–1808.

Rueda, L., Saralegui, A., d'Arlas, B. F., Zhou, Q., Berglund, L. A., Corcuera, M. A.,
Mondragon, I., & Eceiza, A. (2013). Cellulose nanocrystals/polyurethane nanocomposites. Study from the viewpoint of microphase separated structure. *Carbohyd. Polym.*,
92(1), 751–757.

Saba, N., Paridah, M. T., Abdan, K., & Ibrahim, N. A. (2016). Effect of oil palm nano filler on
mechanical and morphological properties of kenaf reinforced epoxy composites. *Constr.
Build. Mater.*, 123, 15–26.

Saba, N., Tahir, P. M., & Jawaid, M. (2014). A review on potentiality of nano filler/natural fiber
filled polymer hybrid composites. *Polymers*, 6, 2247–2273. doi:10.3390/polym6082247

Salih, S. I., Oleiwi, J. K., & Mohamed, A. S. (2018). Investigation of mechanical properties
of PMMA composite reinforced with different types of natural powders. *ARPN J. Eng.
Appl. Sci.*, 13, 2–21.

Schaefer, D. W., & Justice, R. S. (2007). How nano are nanocomposites? *Macromolecules*,
40 (24), 8501–8517.

Sen, A. K., & Kumar, S. (2010). Coir-fiber-based fire retardant nano filler for epoxy composites. *J. Therm. Anal. Calorim.*, 101, 265–271.

Shankar, S., & Rhim, J. W. (2016). Preparation of nanocellulose from micro-crystalline cellulose: The effect on the performance and properties of agar-based composite films.
Carbohydr. Polym., 135, 18–26.

Sothornvit, R. (2019). Nanostructured materials for food packaging systems: New functional
properties. *Curr. Opin. Food Sci.*, 2, 82–87.

Thakur, V. K., Thakur, M. K., Raghavan, P., & Kessler, M.R. (2014). Progress in green polymer
composites from lignin for multifunctional applications: A review. *ACS Sustain. Chem.
Eng.*, 2, 1072–1092.

Tian, D., Hu, J., Bao, J., Chandra, R. P., Saddler, J. N., & Lu, C. (2017). Lignin valorization: Lignin nanoparticles as high-value bio-additive for multifunctional nanocomposites. *Biotechnol. Biofuels*, 10, 192.

Vahedikia, N., Garavand, F., Tajeddin, B., Cacciotti, I., Jafari, S. M., Omidi, T., & Zahedi, Z. (2019). Biodegradable zein film composites reinforced with chitosan nanoparticles and cinnamon essential oil: Physical, mechanical, structural and antimicrobial attributes. *Colloids Surf. B Biointerfaces*, 177, 25–32.

Vatanpour, V., & Paziresh, S. (2022). A melamine-based covalent organic framework nanomaterial as a nanofiller in polyether sulfone mixed matrix membranes to improve separation and antifouling performance. *J. Appl. Polym. Sci.*, 139(1), e51428.

Wang, F., Arai, S., & Endo, M. (2004). Electrochemical preparation and characterization of nickel/ultra-dispersed PTFE composite films from aqueous solution. *Mater. Trans.*, 45 (4), 1311–1316.

Wang, Q., Li, G., Zhang, J., Huang, F., Lu, K., & Wei, Q. (2014). PAN nanofibers reinforced with mmt/go hybrid nanofillers. *J. Nanomater.*, 2014, Article ID 298021, 10 pages. http://dx.doi.org/10.1155/2014/298021

Wang, Y., Zhang, H., Song, C., Gao, C., & Zhu, G. (2020). Effect of aminophend/formaldehyde resin polymeric nanospheres as nanofiller on polyamide thin film nanocomposite membranes for reverse osmosis application. *J. Membr. Sci.*, 614, 118496.

Wang, Z., Han, E., & Ke, W. (2006). Effect of nanoparticles on the improvement in fire-resistant and anti-ageing properties of flame-retardant coating. *Surf. Coat. Technol.*, 200(20), 5706–5716.

Wu, T., Frydrych, M., O'Kelly, K., & Chen, B. (2014). Poly(glycerol sebacate urethane)-cellulose nanocomposites with water-active shape-memory effects. *Biomacromolecules*, 15, 2663–2671.

Xu, W., Qin, Z., Yu, H., Liu, Y., Liu, N., Zhou, Z., & Chen, L. (2013). Cellulose nanocrystals as organic nanofillers for transparent polycarbonate films. *J. Nanopart. Res.*, 15(4), 1–8.

Yang, W., Dominici, F., Fortunati, E., Kenny, J. M., & Puglia, D. (2015a). Effect of lignin nanoparticles and masterbatch procedures on the final properties of glycidyl methacrylate-g-Poly(lactic acid) films before and after accelerated UV weathering. *Ind. Crops Prod.*, 77, 833–844.

Yang, W., Fortunati, E., Dominici, F., Giovanale, G., Mazzaglia, A., Balestra, G. M., Kenny, J. M., & Puglia, D. (2016a). Effect of cellulose and lignin on disintegration, antimicrobial and antioxidant properties of PLA active films. *Int. J. Biol. Macromol.*, 89, 360–368.

Yang, W., Fortunati, E., Dominici, F., Giovanale, G., Mazzaglia, A., Balestra, G. M., Kenny, J. M., & Puglia, D. (2016b). Synergic effect of cellulose and lignin nanostructures in PLA based systems for food antibacterial packaging. *Eur. Polym. J.*, 79, 1–12.

Yang, W., Fortunati, E., Gao, D., Balestra, G. M., Giovanale, G., He, X., Torre, L., Kenny, J. M., & Puglia, D. (2018). Valorization of acid isolated high yield lignin nanoparticles as innovative antioxidant/antimicrobial organic materials. *ACS Sustain. Chem. Eng.*, 6, 3502–3514.

Yang, W., Kenny, J. M., & Puglia, D. (2015b). Structure and properties of biodegradable wheat gluten bionanocomposites containing lignin nanoparticles. *Ind. Crops Prod.*, 74, 348–356.

Yang, W., Owczarek, J. S. S., Fortunati, E., Kozanecki, M., Mazzaglia, A., Balestra, G. M. M., Kenny, J. M. M., Torre, L., & Puglia, D. (2016c). Antioxidant and antibacterial lignin nanoparticles in polyvinyl alcohol/chitosan films for active packaging. *Ind. Crops Prod.*, 94, 800–811.

Yang, Y., Goh, K., Weerachanchai, P., & Bae, T. H. (2019). 3D covalent organic framework for morphologically induced high-performance membranes with strong resistance toward physical aging. *J. Membr. Sci.*, 574, 235–242.

Youssef, A. M., & El-Sayed, S. M. (2018). Bionanocomposites materials for food packaging applications: Concepts and future outlook. *Carbohydr. Polym.*, 193, 19–27.

Yuan, R., Wu, S., Yu, P. et al. (2016). Superamphiphobic and electroactive nanocomposite toward self-cleaning, antiwear, and anticorrosion coatings. *ACS Appl. Mater. Interfaces*, 8(19), 12481–12493.

Zarnegar, Z., Safari, J., & Zahraei, Z. (2019). Design, synthesis and antimicrobial evaluation of silver decorated magnetic polymeric nanocomposites. *Nano-Struct. Nano-Objects*, 19, 100368.

Zeng, J. B., He, Y. S., Li, S. L., & Wang, Y. Z. (2012). Chitin Whiskers: An overview. *Biomacromolecules*, 13(1), 1–11.

Zhang, S., Sun, G., He, Y., Fu, R., Gu, Y., & Chen, S. (2017). Preparation, characterization, and electrochromic properties of nanocellulose-based polyaniline nanocomposite films. *ACS Appl. Mater. Interfaces*, 9(19), 16426–16434.

Zhu, T. T., Zhou, C. H., Kabwe, F. B., Wu, Q. Q., Li, C. S., & Zhang, J. R. (2019). Exfoliation of montmorillonite and related properties of clay/polymer nanocomposites. *Appl. Clay Sci.*, 169, 48–66.

2 Dimension-Based Nanofillers
Synthesis and Characterization

Qazi Adfar
Vishwa Bharti Degree College

Mohammad Aslam and Shrikant S. Maktedar
National Institute of Technology

CONTENTS

2.1 Introduction 20
    2.1.1 NANO: The Small Size and Its Usefulness 21
2.2 Methods of Fabrication of Nanofillers 23
    2.2.1 Top–Down Methods 24
        2.2.1.1 Synthesis of Nanoelectromechanical Systems 24
        2.2.1.2 Lithographic Techniques 24
        2.2.1.3 Arc Discharge Method 26
        2.2.1.4 Laser Ablation Method 26
        2.2.1.5 Ball Milling Method 27
        2.2.1.6 Inert Gas Condensation 28
    2.2.2 Bottom–Up Methods 30
        2.2.2.1 Electrospinning 30
        2.2.2.2 Hydrothermal Synthesis 31
        2.2.2.3 Sol–Gel Method 33
        2.2.2.4 Electrochemical Method 34
        2.2.2.5 Molecular Beam Epitaxy 36
        2.2.2.6 Chemical Vapor Deposition Method 37
        2.2.2.7 Microwave Method 41
        2.2.2.8 Homogeneous Nucleation 41
2.3 Characterization Techniques for Chemical Analysis of Nanofillers 43
    2.3.1 X-Ray Diffraction 44
    2.3.2 Raman Spectroscopy 48
    2.3.3 XPS Technique: X-Ray Photoelectron Spectroscopy 53
    2.3.4 DLS Technique 56

DOI: 10.1201/9781003279372-2

19

2.4   Classification of Nanomaterials 60
      2.4.1   Zero Dimensional (0-D) Nanomaterials 60
      2.4.2   One-Dimensional Nanomaterials 61
      2.4.3   Two-Dimensional Nanomaterials 62
              2.4.3.1   Properties and Applications 62
      2.4.4   Three-Dimensional Nanomaterials 63
2.5   Functionally Graded Materials 65
      2.5.1   Introduction..65
      2.5.2   FGM Types 66
      2.5.3   FGM Applications 66
2.6   Conclusion 67
Bibliography 68

2.1   INTRODUCTION

In the course of the evolution of mankind, technologies have come and gone. Technologies are so important that their names are given to the ages they were known for, for example, the Stone Age, the Bronze Age, and the Iron Age, rather than the age of hunting, roaming for food, agriculture, urbanization, etc. One large difference between the present-day technology and the technology of the past lies in the time taken for the evolution of near-perfect or perfect technologies. The agrarian era, driven by the associated technologies of irrigation, tools, fertilizers, etc., took a few thousand years to evolve with the time period involved varying, as it depended on the geographical locations and the overall circumstances of particular regions.

The industrial age, which came just after the agrarian era, started around the 1800s and took about 150 years to improve upon. By the middle of the 19th century, it was in full swing in Britain and then rapidly spread to Europe and North America. This, in turn, was replaced by the information technology revolution, marked by unprecedented capabilities in gathering, storing, retrieving, and analyzing information, which finally shaped up into modern-day high-speed computing. We are still within that epoch, but the next revolution already appears to be on the horizon, and it has been named the nanorevolution (Figure 2.1).

Although technologies have come and gone, it is important to assess what they have given us. The industrial age of the 1900s gave us advanced agricultural practices like the use of chemical fertilizers, radio, TV, air conditioning cars, jets, planes, modern medicines, and fabrics. The information age has given us mobile phones, Internet, cable TV, e-mail, ATMS, administrative reforms, reduced distances, among other things, which has transformed our neighborhood completely. This age has removed the distinction between the king and the common man. The immense change in our lives has occurred due to science. Chemistry, on the whole, has been the driving force in the front, which has contributed to major changes in the 19th and 20th centuries. The large production of ammonia, sulfuric acid, dyes, polymers, plastics, petroleum products, etc., has changed the world. While this growth of chemistry was continuing, a major breakthrough occurred in 1947 in the field of physics, with the discovery of transistor. In the 1950s and beyond, the advent of semiconductor devices facilitated the development of consumer electronics. Everything that we

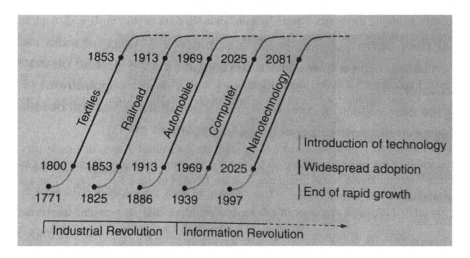

FIGURE 2.1 Technological revolutions in science. (Adapted from Nano-technology the science of small-Wiley.)

purchase today has an integrated circuit in it. Gadgets from toys to cars function with these circuits. This, along with computers, helped build an era of physics. Many predict that the next era would be of biology and materials. The next era seems to be all set to remove the barriers between humans and their surroundings. We can only wonder how the next era unfolds. With the evolution of nanoera in our laps, it is expected that more of the vast mysteries of nature are yet to be accomplished, which will open a new gateway for human beings to enter into a universe that they will have to rediscover with a new beginning.

Progress made in few decades has taken us deeper into the unending nature of matter as a whole, providing us with all kinds of new opportunities, possibilities, and abilities to achieve exciting and unimaginable technological advancements in multi-disciplinary sciences for the benefit of mankind and novel civilization that we humans will be carving for ourselves, with the help of nanoscience and nanotechnology.

Basic nanoscience research has been in existence for over 20 years. Since then, many products have made their way into the market place. Nanotechnology has the potential to impact all the commercial products of today and tomorrow. It has potential to change our lifestyles to a great extent, and it is already being influenced by it.

The current nanorevolution is attributed not only to the recognition of its potential value but also to the availability of the tools and methods that allow scientists to explore the world at the nanoscale and transform their discoveries and inventions into various applications that practically benefits our everyday life.

2.1.1 NANO: THE SMALL SIZE AND ITS USEFULNESS

The word "nano" is a Greek word meaning "dwarf." The term nanoscale is used to refer to objects with dimensions of the order of 1–100 nanometers (nm). A nanometer is one-billionth of a meter, 0.000000001, or 10^{-9} m. Hence, in the international system of units, the prefix "nano-" means one-billionth or 10^{-9} m. It is difficult to imagine just how small "nano" is; so, here are some examples:

- A sheet of paper is about 100,000 nm thick.
- There are 25,400,000 nm in 1 inch.
- A strand of human hair is approximately 80,000–100,000 nm wide.

Some more illustrations are given in Figures 2.2 and 2.3.

A nanometer is used to measure things that are very small. Atoms and molecules, the smallest pieces of everything around us, are measured in nanometers. For example, a water molecule is less than 1 nm, and our DNA is in the 2.5-nm range. The size of nanomaterials or nanofillers is similar to that of most biological molecules and structures. Nanoparticles exist in the natural world and are also created as a result

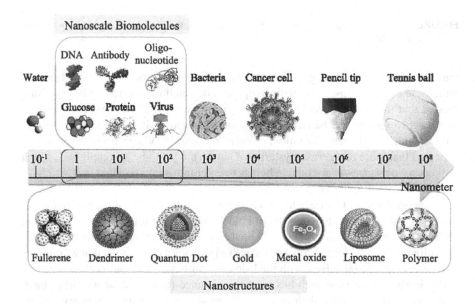

FIGURE 2.2   Nanoscale biomolecules and nanostructures. (Adapted from ResearchGate.)

FIGURE 2.3   Ten hydrogen atoms placed side by side to make 1 nm. (Adapted from Wiley.)

of human activities. Because of their submicroscopic size, they have unique material characteristics. The nanoscale effects on properties are as follows:

- They show better catalytic efficiency through a higher surface area-to-volume ratio.
- Their electrical conductivity is high.
- They have high magnetic coercivity and paramagnetic behavior.
- They show a spectral shift of the optical region.
- They show high permeability through biological barriers and are biocompatible.
- They exhibit high hardness and toughness of metals and alloys.

For their novel and unique properties, which do not exist in their bulk forms, nanoparticles are also engineered and designed for various applications. They are used in many commercial products and processes such as sunscreens, cosmetics, sports goods, stain-resistant clothing, tires, and electronics. Nanomaterials are also of paramount importance for biologists and medical students. Biological structures are within the nanosize range, which researchers are now able to manipulate and control. Therefore, nanomaterials can be used both in in vivo and in vitro biomedical research and applications. They are used in medicine for purposes of diagnosis, imaging, and drug delivery.

Nowadays, the role of nanomaterials or nanofillers has become so significant in technological advancements that various modern techniques have been developed to fabricate them, some of which are based on earlier concepts, while some are designed on modern lines.

2.2   METHODS OF FABRICATION OF NANOFILLERS

Nanofabrication involves the manufacture of nanofillers or nanostructures that are products with zero, one, two, or three dimensions in the nanometer range, which most commonly find applications in fields of industry, agriculture, food industry, and biomedical technology.

Two different paths are pursued for nanofabrication:

One is the top–down approach of miniaturizing, or in engineering terms, it is simply a scaling down process. This process is essentially subtractive. A material is removed from bulk blocks by grinding, ball milling, crushing, etc., or even carving a block of wood or sculpturing are examples of a top–down method. In the top–down method, there is a certain amount of waste materials.

The second approach is called the bottom–up method in which the molecular bulk is synthesized atom by atom, and it is similar to building a house brick by brick. Nanofabrication of nanofillers is carried out by either of these two methods. In general, the bottom–up approach is meant to be the synthesis of nanoparticles by means of chemical reactions among atoms/ions/molecules, whereas the top–down method involves the mechanical methods of crushing/breaking of bulk into several parts to form nanoparticles.

2.2.1 Top–Down Methods

2.2.1.1 Synthesis of Nanoelectromechanical Systems

The main advantage of the top–down approach in nanofabrications is the fabrication of integrated circuits, where the parts are both patterned and built in place, so that no assembly step is needed, e.g., nanoelectromechanical systems (NEMSs). In these systems, small mechanical components such as levers, springs, and fluid channels along with electronic circuits are embedded into a small chip. The starting materials in these fabrications are relatively large structures such as silicon crystals.

The increased sensitivity achieved by NEMSs leads to smaller and more efficient sensors to detect stresses, vibrations, forces at the atomic level, and chemical signals. Atomic force microscope (AFM) tips and other detection at nanoscale heavily depend on the NEMS (Figure 2.4).

2.2.1.2 Lithographic Techniques

The most common top–down approach involves lithographic patterning techniques, which use short-wavelength optical sources, such as extreme ultraviolet and X-rays. These are being developed to allow lithographic printing techniques to reach

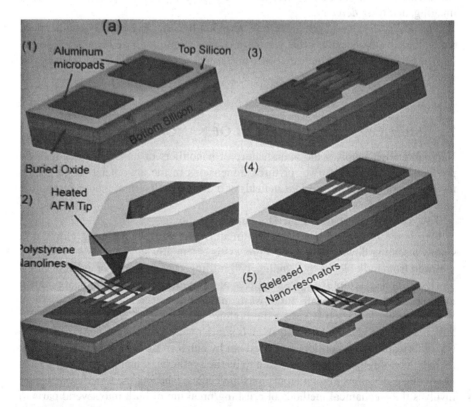

FIGURE 2.4 Tip-based nanofabrication of NEMS devices. (Adapted from springer link.) NEMS, nanoelectromechanical system.

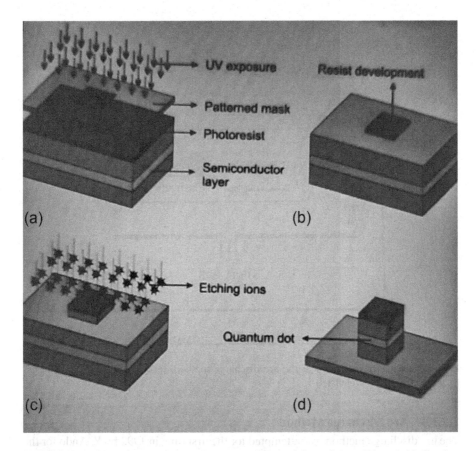

FIGURE 2.5 Photolithography process steps. (Adapted from ScienceDirect.) (a) Coating of photoresistant material on the substrate, mask placed over the upper layer and exposed to UV radiations. (b) Resist development and stripping. (c) Ion etching process. (d) Final QD structure after process.

dimensions from 10 to 100 nm. Scanning beam techniques such as electron beam lithography provides patterns down to about 20 nm (Figure 2.5).

Lithography has few more types depending on the kind of source used for the top–down process, viz., focused ion beam and neutral atom beam lithography, nano-imprint lithography, AFM nanolithography, and electron beam lithography.

Electron beam lithography has the advantage that it can write custom patterns with sub-10-nm resolution and low throughput; this, at the same time, limits its usage to photomask fabrication, low-volume production of semiconductor devices, and research development.

Overall, lithography is considered a hybrid approach because the etching process is a top–down approach, whereas the growth of nanolayers involves the bottom–up approach.

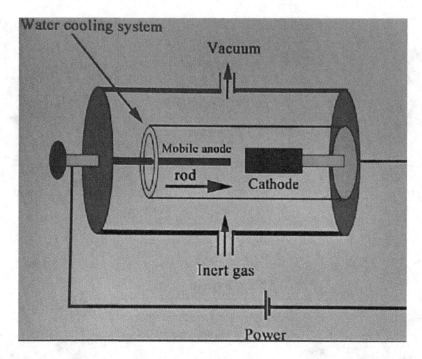

FIGURE 2.6 Simplest setup of the arc discharge method, technique for CNT synthesis. (Adapted from ResearchGate.)

2.2.1.3 Arc Discharge Method

The arc discharge method was attempted for the first time in 1992 by Y. Ando for the synthesis of CNTs. In this direct current, arc voltage is applied across two graphite electrodes immersed in an inert gas such as He. When pure graphite rods are used, fullerenes are deposited as a soot inside the chamber, and multiwalled carbon nanotubes (MWCNTs) are deposited on the cathode. When a graphite anode containing a metal catalyst (Fe or Co) is used with a pure graphite cathode, single-walled carbon nanotubes (SWCNTs) are generated in the form of soot (Figure 2.6).

2.2.1.4 Laser Ablation Method

The laser ablation method is a method for fabricating various kinds of nanoparticles, including semiconductor quantum dots, carbon nanotubes, nanowires, and core–shell nanoparticles. In this method, nanoparticles are generated by nucleation and growth of laser-vaporized species in a background gas, or in a vacuum chamber. The most common background gas in this process is oxygen. Laser ablation is commonly utilized to synthesize various carbon nanomaterials, such as CNTs and fullerences. Kroto, Curl, and Smalley synthesized fullerenes (C) by vaporizing graphite via pulsed laser irradiation. Sun et al. were one of the first to apply laser ablation to a carbon source to produce CDs. These nanoparticles only became CDs after surface passivation with poly(ethylene glycol) and poly(propionylethylene imine-co-ethylene imine).

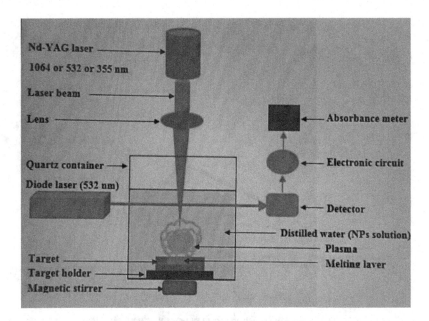

FIGURE 2.7 Synthesis of nanoparticles by laser ablation method. (Adapted from ResearchGate.)

The laser ablation technique provides an alternative to the conventional metal reduction method to form metal-based NPs. NPs produced by this method require no chemical/stabilizing agents during synthesis (Figure 2.7).

2.2.1.5 Ball Milling Method

Ball milling is a mechanical technique that is broadly used to grind powders into fine particles. In ball milling, reactants are broken by mechanical forces; hence, the term "mechanochemistry" has been introduced in this regard. Usually, ceramics, fling pebbles, and stainless steel are used as tiny rigid balls in a concealed container, which generate localized high pressure in the ball mill device. It is simply a size reduction technique that uses media in a rotating cylinder chamber to mill materials to a fine nanopowder, e.g., high-energy ball milling is used to prepare nickel–ferrite ($NiFe_2O_4$) nanopowders. High-energy ball milling (HEBM) of the mixture of alpha-NiO and alpha Fe_2O_3, followed by annealing at 1000°C, was carried out to synthesize nickel–ferrite nanoparticles.

The main advantages of this method are that it can be applied on the large scale also. A large number of elemental and metal oxide nanocrystals like Co, Cr, Al, Fe, Ag-Fe, and Fe can be prepared including a variety of intermetallic compounds Ni and Al. By using this method, ultra-five poly(methyl methacrylate) PMMA can also be synthesized.

The formation of nanostructures requires a ball velocity of >5 m/s and a frequency of >1000 cycles/min. However, these powders have a low surface dispersed size distribution, and they are highly reactive to oxygen hydrogen and nitrogen.

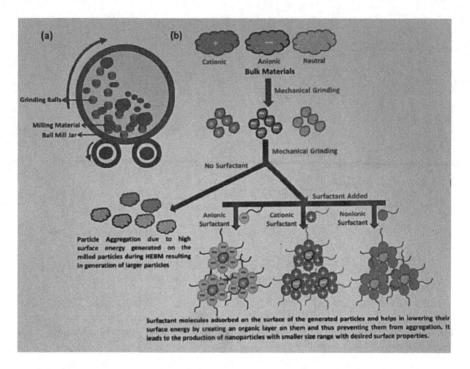

FIGURE 2.8 (a) HEBM system and (b) schematic representation of the NP synthesis using the HEBM method with and without surfactants. (Adapted from ResearchGate.) HEBM, high-energy ball milling.

The ball milling technique is used to synthesize NPs from a diverse range of materials, such as metals, metal oxides, ceramics, and polymers, as shown in Figure 2.8:

2.2.1.6 Inert Gas Condensation

Inert gas condensation involves the vaporization of metal or inorganic materials from a vaporizing source in the presence of an inert gas atmosphere, followed by rapid condensation of the vaporized atoms of the metal or inorganic material on a cold surface to form nanoparticles (Figure 2.9).

This system comprises an ultrahigh-vacuum (UHV) chamber, which acts as a container, having the metal or inorganic substance in a solid or liquid form. The container has one of the vaporization techniques deployed in them. When started, the contents in the container are converted into a gaseous state. Inert gas introduced into the UHV chamber. An outlet tube takes the vapor out of the chamber, which is kept cool by using liquid nitrogen. This allows the collection and rapid condensation of the particles. When the collection procedure is completed, nanoparticles and nanocomposites drop down into the collecting funnel and then into the chamber, where these nanoparticles and nanocomposites are treated with low- and high-pressure compaction units, respectively, to make sure the desired physical properties.

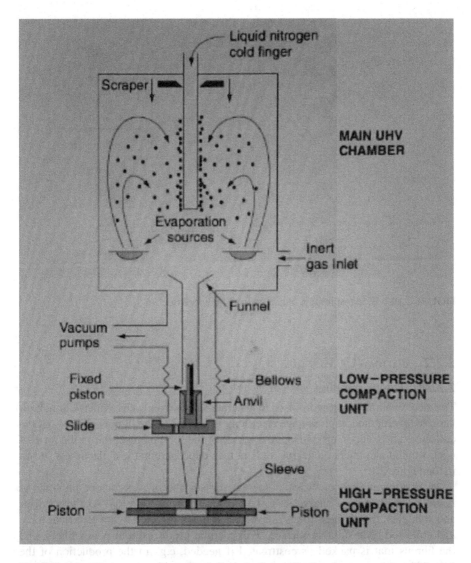

FIGURE 2.9 Inert gas condensation technique for synthesis of nanoparticles. (Adapted from nano graphi.com.)

The method can synthesize Co, Mn, Mo, Fe, and Zn like Fe-Cu and Fe-Ni. Different nanomaterials like intermetallic compounds, semiconductors, and ceramic composites are also prepared. The method has many advantages, and a major advantage is that the size and shape of nanoparticles can be controlled by different factors, such as the pressure of the chamber, the inert gas injected into the chamber, and the temperature.

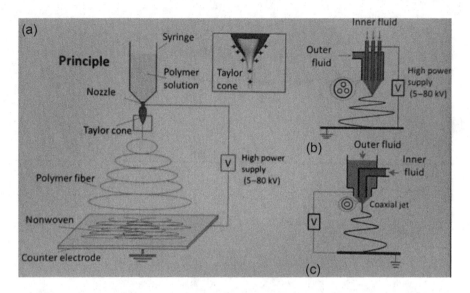

FIGURE 2.10   Electrospinning. (Adapted from ScienceDirect.)

2.2.2   BOTTOM–UP METHODS

2.2.2.1   Electrospinning

The electrospinning method is used to generate a nonwoven web of micro- or nanofibers. In this method, high-voltage electricity is applied to the liquid solution and the collector. The solution ejects from the nozzle of the solution reservoir in the form of a jet, which form fibers on drying and are then deposited on the collector, as shown in the Figure 2.10:

In electrospinning, electrical forces at the surface of a precursor polymer or metal solution cause a charge jet to be ejected from the solution. As the jet dries and solidifies, the charged fibers are accelerated and directed by an electrical force and collected on a suitable surface. This method produces amorphous fibers, and the fibrous mat is packed or posttreated if needed, e.g., in the production of the TiO_2/PVP composite, pyrolysis of PVP at 500°C yields porous TiO_2 nanofibers. Electrospinning is a simple and versatile method for preparation of long polymer fibers, e.g., polyacrylonitrile nanofibers and polyethylene oxides, which are embedded with nanoparticles or nanowires and organic–inorganic hybrid fibers, e.g., TiO_2 polyvinyl pyrrolidone (PVP) composite, used for a wide range of applications, including optics, microelectronics, and protective clothing. Nanofibers prepared using traditional single-needle electrospinning methods have found broad applications and have prospects in many fields, e.g., textile, filtration, energy, biomedicine, sensing, and protection. However, the solution flow rate during the single-needle electrospinning process generally varies from 0.1 to 0.3 mL/h, which results in very low nanofiber production. Therefore, to improve the yield of nanofibers and to take their production on large scale, researchers have developed multineedle

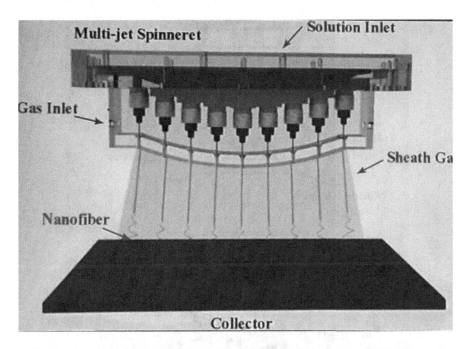

Multi-jet Spinneret

Solution Inlet

Gas Inlet

Sheath Ga

Nanofiber

Collector

FIGURE 2.11   Active generation of multiple jets for producing nanofibers with high quality and high throughput. (Adapted from ScienceDirect.)

electrospinning devices to obtain more jets and thus increase the yield of nanofibers, as shown in Figure 2.11.

2.2.2.2 Hydrothermal Synthesis

The term "hydrothermal" is purely of geological origin. It was first coined by British geologist Sir Roderick Murchison to describe the action of water under elevated temperature and pressure in bringing changes in the earth crust, leading to the formation of various rocks and minerals.

In the hydrothermal synthesis of nanomaterials, the hydrothermal method is based on the ability of water and aqueous solution to dilute the substances at high temperatures and pressures, which are practically insoluble under normal conditions. These include oxides, silicates, and sulfides. This method leads to the formation of single crystals, which depends on the solubility of minerals in hot water under high pressure. Crystal growth is performed in an apparatus consisting of a steel pressure vessel called an autoclave, in which a nutrient is supplied along with water/solvent. It is heated externally, and the pressure is generated in it internally (Figure 2.12).

The autoclave is washed with double-distilled water, and nutrient is placed in its lower part. It is filled with a specific amount of solvent and is heated in a furnace up to a desired temperature (not more than 300°C), and two temperature zones are created. The nutrient dissolves in the hotter zone, and the saturated aqueous solution in the lower part is transported to the upper part of the autoclave by connection motion of the solution. The cooler and denser solution in the upper part of the autoclave

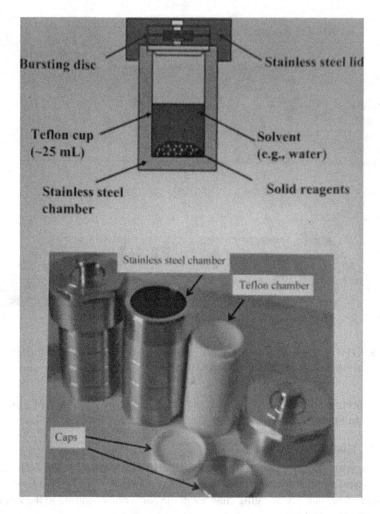

FIGURE 2.12 Teflon-lined stainless steel autoclave used in hydrothermal synthesis. (Adapted from ScienceDirect.)

descends, while the counter flow of the solution ascends. The solution in the auto-clave becomes supersaturated in the upper region, and the crystallization process takes place (Figure 2.13).

The main advantage of the hydrothermal method is that most of the materials can be dissolved in an appropriate solvent by modulating the temperature and pressure. This technique links all the important technologies like geotechnology, biotechnology, and nanotechnology. It has become one of the most important tools for advanced materials processing and find applications in electronics, optoelectronics, catalysis, ceramics, magnetic data storage, biomedicine, biophotonics, etc. The technique offers a unique method for coating of various compounds on metals, polymers, and ceramics as well as for the fabrication of powders and bulk ceramic bodies.

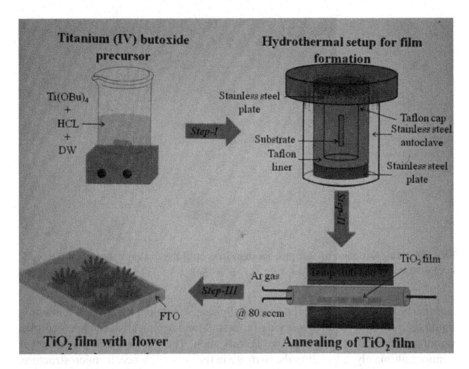

FIGURE 2.13 Experimental setup of TiO₂ thin film formation using hydrothermal method. (Adapted from ResearchGate.)

2.2.2.3 Sol–Gel Method

The interest in sol–gel processing can be traced back to the mid-1800 with the observation that the hydrolysis of tetraethyl orthosilicate under acidic conditions led to the formation of SiO_2 in the form of fibers and monoliths, and later, it was developed in 1950s for the production of radioactive powders of UO_2 and ThO_2 for nuclear fuels. In the 1990s, it became so important that about 35,000 papers were published worldwide through this process. Since then, an intensive and extensive research has been carried out in this regard. The area of technology covered by this method ranges from physics to biology.

The sol–gel process is a wet chemical technique also known as chemical solution deposition, which involves several steps that are as follows: hydrolysis, polycondensation, gelation, aging, drying, densification, and crystallization (as illustrated in Figure 2.14).

In this chemical procedure, a "sol" (i.e. a colloidal solution) is formed, which gradually evolves in the formation of a gel-like diphasic system containing both liquid and solid phases whose morphologies range from discrete particles to continuous polymer networks. In case the particle density is very low, a significant amount of liquid may need to be removed initially for the gel-like properties to be recognized. This can be accomplished in any number of ways. The simplest method is to allow time for sedimentation to occur and then pour off the remaining liquid. Centrifugation

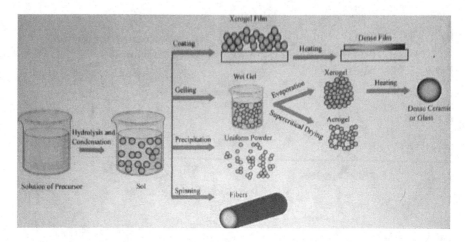

FIGURE 2.14 Different sol–gel process steps to control the final morphology of the product. (Adapted from ResearchGate.)

can also be used to accelerate the process of phase separation. Removal of remaining liquid (i.e., solvent) requires a drying process, which is typically accompanied by a significant amount of shrinkage and densification. The rate at which the solvent is removed ultimately determines the porosity in the gel and, hence, its microstructure. Afterward, thermal treatment, or firing process, is conducted, if necessary, in order to favor further polycondensation for enhancement of mechanical properties and structural stability via final sintering, densification, and grain growth.

The sol–gel method has the following advantages:

- It is a low-cost method and is energy-saving due to low annealing temperature.
- The obtained nanoparticles have good uniformity and high porosity. Even small quantities of dopants, such as organic dyes and rare earth elements, can be introduced into the sol to end up uniformly dispersed in the final product.
- Sol–gel-derived materials have diverse applications in optics, electronics, energy, space, biosensors, medicine (e.g., controlled drug release), reactive material, and separation technology (e.g., chromatography). They are used in ceramics, casting materials, or as a means of producing very thin films of metal oxides for various purposes, such as coating fibers and glasses to synthesize powders (e.g., microspheres and nanospheres).

2.2.2.4 Electrochemical Method

The electrochemical deposition technique has been widely used for the synthesis of metal nanoparticles. Electrochemical deposition occurs at the interface of an electrolyte solution containing the metal to be deposited and an electrically conductive metal substrate. It is a potentially low-cost technique and has high product yield.

It gives yield at low synthetic temperatures with high purity, contamination with by-products resulting from chemical reduction agents is avoided, and the products are easy to isolate from the precipitate. Besides all these advantages, the electrochemical method is also environmentally friendly. Therefore, electrochemical synthesis represents a highly efficient method for the fabrication of nanostructured energy materials and various nanostructures such as nanoparticles, nanorods, nanowires, nanotubes, nanosheets, dendritic nanostructures, and composite nanostructures.

The electrochemical techniques are quite interesting because they allow obtaining particles with high purity using fast and simple procedures. More importantly, by this method, the particle size and shape are controlled by modulating numerous parameters such as current density, solvent polarity, charge flow, and distance between electrodes and temperature. All these parameters are critical for controlling particle size and shape selectivity of the electrochemical synthesis process. For instance,

- In case of Pd, NPs size from about 1 up to 5 nm is tuned using temperature, charge flow, and current density-variable parameters. This approach has also been used to synthesize bimetallic NPs including Ni-Pd, Fe-Co, and Fe-Ni.
- Cobalt NPs with a tunable average particle size from 2 to 7 nm have been synthesized electrochemically in the presence of tetraalkyl ammonium salts.
- Ni- and Rh-structured catalysts were prepared by electrosynthesis of hydrotalcite-type compounds on FeCrAlY foam, followed by calcination. The film coating growth and the morphology of the final catalytic layer were controlled by the applied potential and deposition time. By tuning the deposition parameters, it was possible to prepare Ni- and Rh-based catalysts, which are very active in steam reforming and catalytic partial oxidation of methane. In this study, CdS NPs with an average size of 5 nm have also been synthesized by using the one-step electrochemical process.
- Noble metal nanoparticles have been intensely investigated due to their excellent optical, catalytic, and electric properties; for example, metallic silver NPs are technologically important because they show unique properties normally related to noble metals such as excellent conductivity, chemical stability, and nonlinear optical behavior; besides, they exhibit impressive antibacterial action.
- An electrochemical procedure, based on the dissolution of a metallic anode in an aprotic solvent, has been used to obtain silver nanoparticles ranging from 2 to 7 nm. By changing the current density and by using different kinds of counterelectrodes, different particle sizes of silver NPs were obtained.
- Different stabilizers have been used in electrochemical techniques, which include organic monomers as electrostatic stabilizers, and polymeric compounds as steric stabilizers; the latter has been shown to obtain more stable NPs, e.g., polyethylene glycol-stabilized silver has been found to be highly

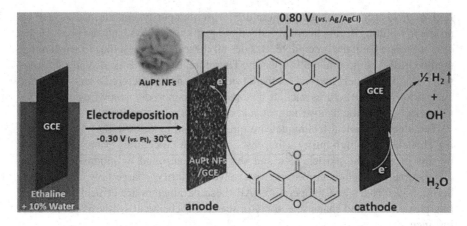

FIGURE 2.15 Schematic illustration of the preparation of AuPt NFs and the electrochemical synthesis of XO from XT. (Adapted from ResearchGate.)

stable and effectively controlling the size and shape of silver NPs throughout the electrochemical process.

- Flower-like AuPt alloy nanoparticles were successfully synthesized using a one-step electrochemical reduction method at a low potential of −0.30V (vs Pt) and a low temperature of 30°C. In this process, the deep eutectic solvent (DES) called a new-generation green solvent, acted as a solvent and shape-directing agent. Moreover, the electrode modified with the as-prepared nanomaterials were used as the anode for the electrochemical oxidation synthesis. The glassy carbon electrode modified with the Aupt nanoflowers was directly employed to the electro-oxidation of xanthene (XT) to xanthone (XO) under a constant low potential of 0.80V (vs Ag/AgCl) and room temperature, with high yield of XO. Moreover, the synthesis process was milder and more environment-friendly than conventional organic synthesis, as shown in Figure 2.15:

- Yet another interesting electrochemical synthesis was carried out by Ye et al. (2017). In this study, selenium NPs were prepared using a selenium powder-doped carbon paste electrode. Se NPs formed were spherical with a diameter of about 43, 60, and 85 nm and were well dispersed in the presence of sucrose, polyvinylpyrrolidone, and sodium dodecyl sulfonate, respectively. Surprisingly, when cetane trimethyl ammonium bromide was added to Se NPs modified with sodium dodecyl sulfonate, the spherical Se NPs arranged in a sea urchin-like form by electrostatic assembly, as shown in Figure 2.16.

2.2.2.5 Molecular Beam Epitaxy

Molecular beam epitaxy is a process in which a thin single crystal layer is deposited on a single crystal substrate using atomic or molecular beams generated in Knudsen cells contained in a UHV chamber. A Knudsen cell is an effusion evaporator source

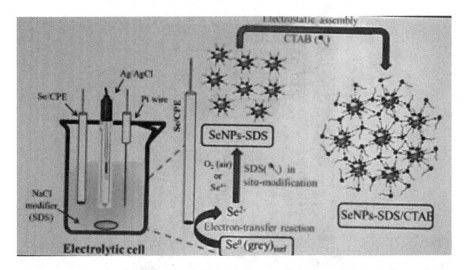

FIGURE 2.16 Electrochemical synthesis of Se NPs and the formation of sea urchin-like assembly due to electrostatic influences. (Adapted from ScienceDirect.com)

for relatively low-partial pressure elementary source (e.g., Ga. Al, Hg, and As) because in this, it is easy to control temperature. This setup is combined with a set of in situ tools such as Auger electron spectroscopy (AES) and/or the reflection high electron diffraction (RHEED) for characterization of deposited layers during growth. In a solid source MBE, ultrapure elements such as gallium and arsenic are heated in separate quasi-Knudsen effusion cells until they begin to slowly evaporate. The evaporated elements then condense on the wafer, where they may react with each other. In the case of gallium and arsenic, single crystal gallium arsenide is formed. The term beam simply means that evaporated atoms do not interact with each other or any other vacuum chamber gases until they reach the wafer due to a long mean force path of beam. The substance is rotated to ensure uniform growth over its surface. By operating mechanical shutters in front of the cells, it is possible to control which semiconductor on metal is deposited (Figure 2.17).

2.2.2.6 Chemical Vapor Deposition Method

Chemical vapor deposition (CVD) is a vacuum deposition method used to produce high-quality and high-performance solid materials of various structures. The process is often used in the semiconductor industry to produce thin films. In typical CVD, the wafer is exposed to one or more volatile processors, which react and/or decompose on the substrate surface to produce a desired deposit. Frequently volatile by-products are also produced, which are removed by gas flow through the reaction chamber.

Microfabrication processes widely use CVD to deposit materials in various forms, including nanocrystalline, polycrystalline, amorphous, and epitaxial films and coating. The method is also used to fabricate carbon nanofillers such as fibers, nanofibers, nanotubes CNTs, diamond, and graphene. Besides, numerous inorganic materials are synthesized, which have a wide range of applications.

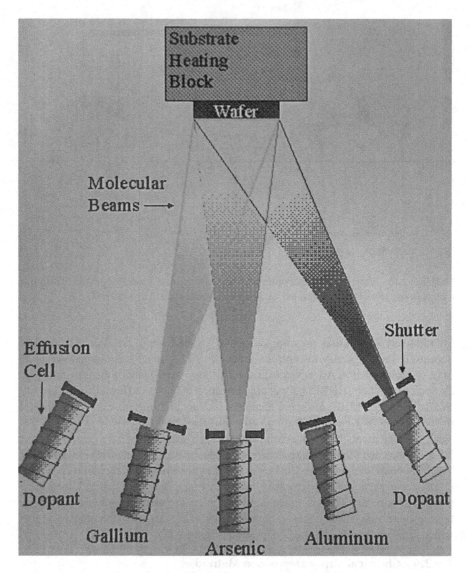

FIGURE 2.17 Molecular beam epitaxy showing growth with nearly monolayer precision. (Adapted from deep phase.com)

For example, in the preparation of CNTs, CVD involves the cleavage of a carbon atom containing gas continuously flowing through the catalyst nanoparticles to generate carbon atoms and subsequent synthesis of CNTs on the surface of the catalyst or the substrate. The reaction is carried out at a sufficiently high temperature in a tubular reactor.

Mechanism: When a hydrocarbon vapor is contacted with the heated metal nanoparticles, it is first decomposed to carbon and hydrogen. Hydrogen leaves with

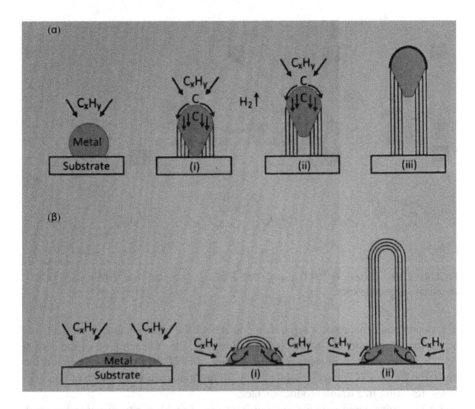

FIGURE 2.18 Synthesis of carbon nanotubes by catalytic CVD-illustrating two growth model diagrams for CNTs (alpha)-showing tip growth mechanism and (beta)-showing root growth mechanism. (Adapted from researchgate.net) CVD, chemical vapor deposition.

the passing carrier gas or reducing gas, and carbon dissolves in the metal catalyst. When the temperature reaches the carbon solubility limit of the metal, the decomposed carbon particles are precipitated and crystallize to CNTs. For growth of CNTs, currently, there are two widely accepted mechanisms, depending on the interaction between the catalyst and the substrate.

- When the catalyst interacts weakly with the substrate, it entirely promotes the growth of metal nanoparticles. When these are entirely covered with excess carbon, growth stops. This is called "tip growth," as shown in Figure 2.18.
- When the catalyst interacts strongly with the substrate, carbon precipitates from the top of the metal without pushing up the metal particles. This is called the "basic growth model," which is also called the "root growth model," as shown in Figure 2.18.

During the synthesis of CNTs, many parameters affect the final morphology and properties of CNTs such as reactor temperature, system pressure, flow rate of carrier

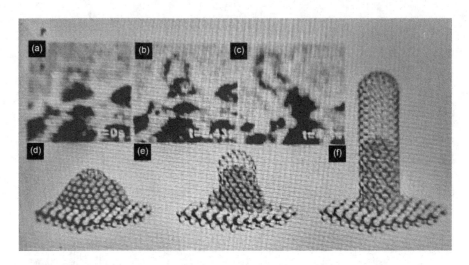

FIGURE 2.19 Growth of CNTs using CVD method. (Adapted from Intech open.) CVD, chemical vapor deposition.

gas, deposition time, reactor type, the geometry of the reactor, catalyst support, and active metal components in the catalyst. Moreover, the concentration of the carbon source gas also has an impact on the properties of CNTs. If the concentration is too high, too many CNTs can be wrapped by amorphous carbon and carbon nanoparticles, resulting in a rough product surface.

CVD is considered a viable method in terms of yield, purity, structural control, and growth parameters.

Some popular transition metal nanoparticles that are usually used as catalysts in CVD synthesis of CNTs are Fe, Pd, Pt, Au, Mn, Ti, W, Co, Mg, Al, In, Na, K, Cs, and Ni (Figure 2.19).

Typical growth temperatures are 550°C–750°C for MWCNTs and 850°C–1000°C for SWCNTs. Synthetic SWCNTs usually require nanosized particles as catalysts; however, MWCNTs can be produced without catalyst. At present, the preparation process of MWCNTs is quite mature, and its industrial production has been realized by using the CVD method.

It is possible to grow CNTs in plasma without using a catalyst, with an exception in the growth of CNTs on SiGe islands on the Si substrate using methane and hydrogen.

SWCNTs can be synthesized from methane, benzene, phenylacetylene, fullerenes, and cyclohexane as a carbon source.

For example, in 2002, Maruyam et al. reported synthesis of high-purity SWCNTs from ethanol on a double-impregnated zeolite matrix at low temperatures. In this, CNTs produced have no amorphous carbon.

In 2015, Zdrejek et al. demonstrated that CVD can be used to synthesize high-quality SWCNTs and MWCNTs at a range of temperatures using propane as a carbon source. By adjusting the growth temperature, the volume of CNTs could be controlled.

Besides, these CNTs can be successfully synthesized by using the CVD method from camphor (Kumar et al 2020.) with kerosene, liquefied petroleum, gas, natural gas, waste plastics, green grass, and other daily raw materials as a carbon source.

2.2.2.7 Microwave Method

Microwave synthesis has been employed in the synthesis of nanoparticles as it combines the advantage of speed and homogeneous heating of the precursor materials. Microwave irradiation has penetration characteristics, which makes it possible to homogeneously heat up the reaction solution. The result is uniform nucleation and rapid crystal growth, which lead to the formation of crystallites that have narrow size distribution. Compared to other methods, the synthesis of microwave irradiation has the advantage of short time of reaction, which is ascribed to the combined forces created by both electric and magnetic components of the microwave, which generate friction and collision of molecules. Microwaves are selective in approach, i.e., they are absorbed by molecules having an electric dipole.

Microwave energy has the potential to selectively heat either the solvent or the precursor molecules for nanomaterial preparation, e.g., the template of microwave synthesis of silver nitrate and starch solution (Nanofibers-by Brijesh Kumar). Starch acts as a reductant as well as a capping material to protect the nanoparticles at the surfaces and prevents the particles from aggregation.

- Zuliani et al. made use of a monomode microwave reactor, in which the magnetite metallic nickel nanoparticles were synthesized by using simple nickel chloride as a precursor, ethylene glycol as a solvent, and ethanol as a reducing agent. They achieved metallic nickel in 5 minutes at 250°C under MW conditions, which showed high catalytic activity for the hydrogenolysis of benzyl phenyl ether.
- May-Masnon et al. utilized microwave energy and nanotechnology, which improved the printability of cotton prints. This was carried out by making use of the screen printing technique. They pre-treated each cotton sample separately by microwave power in the range of 300–700W for a period of 1–9 minutes, followed by printing using the paste dye Remazol and silver nanoparticles of various concentrations. This was again followed by fixing with MW energy and further subjected to streaming or thermal fixation. The results indicated that the prints obtained by using microwave and Ag NPs were found to have better color, strength, fastness, antibacterial behavior, and surface morphology than conventional techniques.

Reports on microwave-assisted synthesis have been well documented since last five decades, which includes reactions illustrated in Figure 2.20.

2.2.2.8 Homogeneous Nucleation

When the concentration of a solute in a solvent exceeds its equilibrium solubility or the temperature decreases below the phase transformation point, a new phase appears. At this point, the solution possesses a high Gibbs free energy, and the

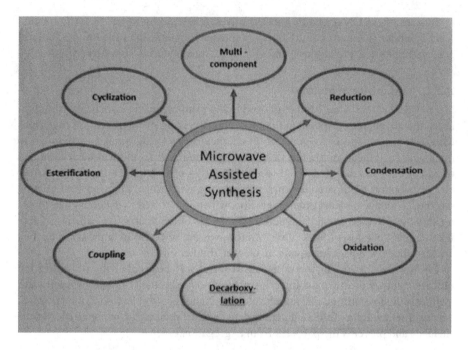

FIGURE 2.20 Various types of microwave-assisted synthesis. (Adapted from Intech Open.)

overall energy of the system would be reduced by segregating solute from the solution. This reduction of Gibbs free energy is the driving force for both nucleations and growth. Homogeneous nucleation is also responsible for the formation of particles from the vapor phase. In this process, vapor atoms or molecules are transformed into solid particles in a supersaturated vapor without the presence of foreign substances.

The formation of nuclei within its own melt with the help of foreign substances or substrate is known as heterogeneous nucleation. This probably occurs at structural inhomogeneties, e.g., container surfaces, impurities, grain boundaries, and dislocations, whereas homogeneous nucleation occurs when nuclei form uniformly throughout the parent phase.

Homogeneous nucleation is yet another method to synthesis various NPs, viz., Au, Ag, CdS, TiO_2, SiO_3, Zn, and Fe_2O_3 (Figure 2.21).

For example:

- Gold NPs are synthesized by gradually adding aqueous solution of sodium citrate to chloroauric acid at 80°C till the color of the solution turns red. This change in color indicates the formation of gold nanoparticles.
- Silver NPs are prepared by gradually adding aqueous solution of sodium citrate to aqueous $AgNO_3$ by stirring for 1 hour at 80°C. The clear solution turns gold yellow, indicating the formation of silver nanoparticles.

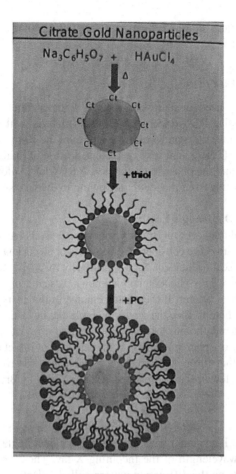

FIGURE 2.21 Schematic representation of homogeneous nucleation of Au nanoparticles on adding sodium citrate to chloroauric acid and subsequent heating. (Adapted from ResearchGate.)

2.3 CHARACTERIZATION TECHNIQUES FOR CHEMICAL ANALYSIS OF NANOFILLERS

Nanoscience and nanotechnology are considered the key technologies in the current century. Efforts are being made worldwide to create smart and intelligent nanomaterials that have innumerable applications in various products and systems, which will lead to an unprecedented level of performance in various sectors of different industries. The precise control of nanomaterials, viz., their dimensions, size, size distributions, dispersion at nanolevel, and their deposition on substrates that affects their properties and dynamic performances, needs highly sophisticated characterization techniques such as particle size analysis, electron microscopy (SEM/TEM/ HRTEM), AFM, X-ray diffraction (XRD), X-ray photoelectron spectroscopy, and dynamic light scattering (DLS) techniques. We will now discuss some of these instrumental techniques employed in field of nanotechnology that lie within the scope of this chapter.

2.3.1 X-Ray Diffraction

XRD is a technique used in materials science to determine the crystallographic structure of a material. XRD works by irradiating a material with incident X-rays and then measuring the intensities and scattering angles of the X-rays that leave the material. A primary use of XRD analysis is the identification of materials based on their diffraction patterns and the phase identification. XRD yields information on how the actual structure deviates from the ideal one, owing to internal stresses and defects.

- **Principle XRD (Bragg's Law):**
  The XRD method works on Bragg's principle. Crystals are a regular array of atoms, while X-rays are considered to be waves of electromagnetic radiations. Crystal atoms scatter incident X-rays, primarily through interaction with the atoms' electrons. This phenomenon is known as scattering, and the electron is known as the scatterer. If there is no change in the energy of the incident and scattered photons, then the scattering is elastic. Inelastic scattering occurs when the scattered photons lose energy. A regular array of scatterers produce a regular array of spherical waves. In the majority of directions, these waves cancel each other through destructive interference; however, they add constructively in a few specific directions, as determined by Bragg's law:

$$n\lambda = 2d \sin \theta$$

  where d is the interplanar spacing, θ is the incident angle, n is an integer, and λ is the wavelength of the incoming X-ray. These specific directions appear as spots on the diffraction pattern called reflections.

  X-rays are used to produce a diffraction pattern because their wavelength lambda ranges from 0.01 to 0.7 nm, and this range matches the lattice spacing in the crystal structures (Figure 2.22).

- **Generation of X-rays:**
  X-rays are generated in the source arm of the diffractometer, which comprises an electron gun, a target, and a slit. The electron gun consists of a source material that is surrounded by electrically controlled coils, as shown in Figure 2.23. The source material is heated up by passing current in the coils to activate the thermionic emission of electrons. These electrons, energized by high accelerating voltage, are then targeted to the metal target. They hit the target to remove electrons from its lower orbit, and the electron in the upper shell tends to fill the empty orbitals in the lower state, emitting a photon of energy ranging from to 10 to 100 keV, which falls in the X-ray spectrum. These rays are then focused on the sample at different angles by the movement of the X-ray generator arm.

- **Instrument:**
  A diffractometer is a measuring instrument used for analyzing the structure of the material from the scattering pattern produced when a beam of

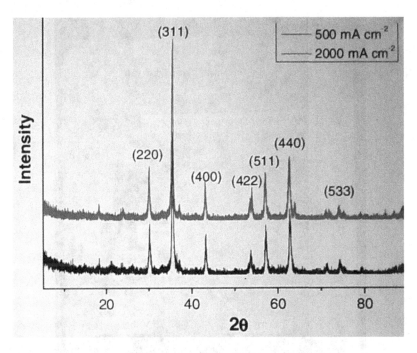

FIGURE 2.22   XRD patterns of maghemite nanoparticles deposited at different current densities (i.e. 500 and 2000 mA/Cm²). XRD, X-ray diffraction. (Adapted from ResearchGate.)

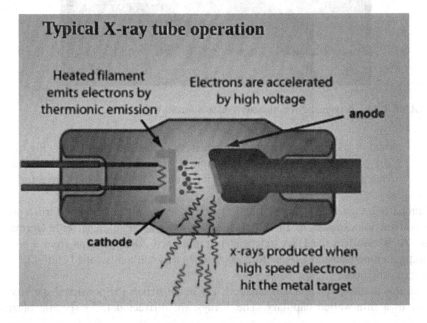

FIGURE 2.23   A schematic view of the components of the X-ray generators. (Adapted from Design World.)

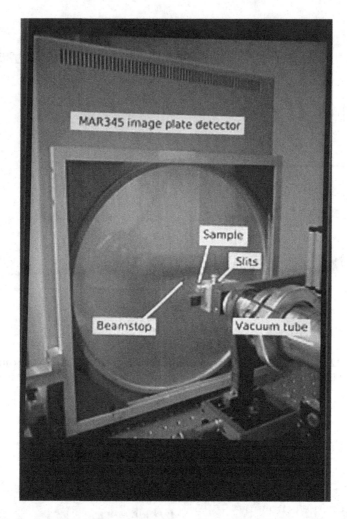

FIGURE 2.24   A typical diffractometer. (Adapted from Wikipedia.)

radiation such as X-rays interact with it. A typical diffractometer (Figure 2.24) consists of a source of radiation, a monochromator to choose the wavelength, slits to adjust the shape of the beam, a sample, and a detector.

Nanoparticles are not very good single crystals. Hence, one needs to perform a powder diffraction experiment. In 1916, Peter J.W. Debye and Paul Scherrer in Germany and A. W. Hull in the USA independently obtained XRD patterns from a single sample in a powder form. Therefore, sometimes this method is called Hull–Debye–Scherrer method.

In this method, the monochromatic X-ray beam strikes the powdered specimens kept in a thin-walled capillary. The X-rays are diffracted from specific planes,

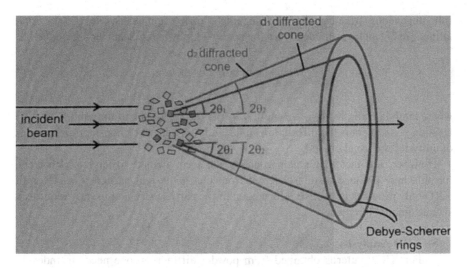

FIGURE 2.25 X-ray powder diffraction. (Adapted from pd.chem.ucl.uk)

making an angle theta with the beam (obeying Bragg's law), as shown in Figure 2.25. The diffracted rays generate cone, concentric with the incident beam. The total opening angle of the diffraction cone 4 theta is determined by measuring the distance S between two diffraction lines corresponding to a particular plane and is related to Bragg's angle by

$$4 \text{ theta} = S/R \times \text{radius}$$

where R is the specimen-to-film distance, usually the radius of the camera housing the film. A list of theta values can be obtained from the measured value of S.

Since the wavelength is known, substitution of theta and lambda in Bragg's equation $2d \sin \text{theta} = n \text{ lambda}$ gives the spacing d between neighboring planes, and n is the integer.

By measuring the distance between symmetrical lines and line intensities, the following information is obtained.

1. A crystalline substance can be distinguished from an amorphous one because the former produces lines on the photographic plate and the latter does not.
2. With the aid of the standard photograph of known chemical compounds, their presence in a mixture can be detected by comparison of the photographs.
3. The size of unit cell and type of lattice of the crystalline solid can be determined. Diffraction peak positions are accurately measured with XRD, which makes it the best method for characterizing homogeneous and inhomogeneous strains.

Inhomogeneous strain causes the broadening of diffraction peaks, where in the crystallite size D can be estimated from the peak width by using Scherrer's formula:

$$D = K \text{ lambda}/B \text{ Cos theta}$$

where lambda is the X-ray wavelength, B is the full width at half maximum of a diffraction peak, theta is the diffraction angle, and K is Scherrer's constant of the order of unity for the usual crystal.

However, the fact that nanoparticles often form twined structures. Scherrer's formula may produce results different from that of its real particle size. Overall, XRD only provides collective information of the particle size and usually requires a sizable amount of powder.

- **XRD Analysis:**
  For XRD patterns obtained from powder diffraction, one needs to index peaks based on combination of Bragg's law and the properties of the seven systems into which all crystals can be classified. This involves determination of the number of atoms per unit cell and the density of a given material. Thus, its chemical composition can be determined. The assignment of atoms to the respective peaks can be found by using the information contained in the intensity of the respective scattering peaks, depending on the scattering power and absolute and relative intensities, which may vary. This variation helps identify the type of atom in a given lattice site. In most cases, XRD data obtained for a given compound are compared to the International Crystal Structure Database, which is known as the Rietveld method. This method refines structural data by using correction parameters with respect intensity, lattice, line shape crystal structure, and possible zero shifts of the detection system.

2.3.2 RAMAN SPECTROSCOPY

Raman Effect:

In 1928, Sir C. V. Raman, while studying the scattering of light by liquids, found that when a beam of monochromatic light was passed through organic liquids such as benzene and toluene, the scattered light contained other frequencies in addition to the incident light. Most of the scattering is elastic in nature. This is called Rayleigh scattering, and the scattered light has some energy, wavelength, and frequency as that of the incident light. The smaller portion of the light is scattered inelastically, which will have higher or lower frequency and, therefore, energy different from that of the incident light, as shown in Figure 2.25; this is called Raman scattering or Raman effect. This makes up the Raman spectrum as, shown in Figure 2.26. The wavelength-shifted photons are called Raman scatter. For this discovery, Sir C.V. Raman was awarded Noble Prize in physics in 1930, and this effect is named after him.

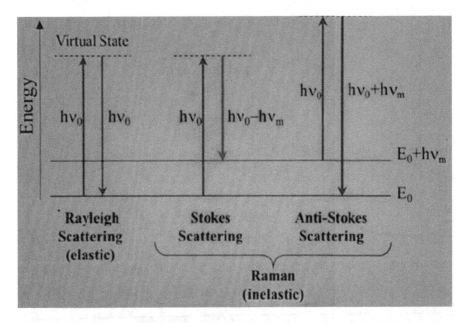

FIGURE 2.26 Energy level diagram of Rayleigh scattering, Stokes Raman scattering, and anti-Stokes Raman scattering. (Adapted from Research gate.)

- **Principle:**
  In Rayleigh scattering, the electron decays back to the same level from which it started. In Stokes and anti-Stokes Raman scattering, the electron decays to a level different from where it started. Stokes scattering is much more common than anti-Stokes scattering because at any given time, an electron in the most common temperature range is most likely to be in its lowest energy state (Figure 2.27).

  A Raman spectrum is a plot of the intensity of Raman-scattered radiation as a function of its frequency difference from the incident radiation (usually in units of wave number cm^{-1}). The difference is called the Raman shift. The additional peaks in the Raman spectrum arise due to the shift in the frequency of the scattered light. The elastic scattered light (Rayleigh scattering) dominates the spectrum with an intense peak at the energy of incoming photons.

  Raman scattering is commonly used in spectroscopy. Raman spectroscopy has been a very powerful tool for finding information about the molecular vibration of materials. There are two different types of Raman spectroscopy: surface-enhanced Raman spectroscopy (SERS) and resonance Raman spectroscopy. Both of these processes enhance the weak signal of the Raman spectra.

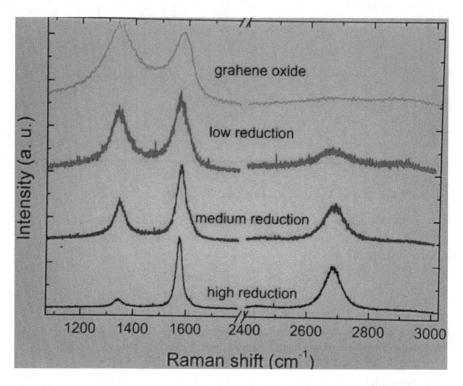

FIGURE 2.27 Raman spectra from RGO at high, medium, and low reduction and from GO. (Adapted from MDPI.)

- **Construction and working of the Raman Spectrometer:**
  A typical Raman spectrometer has three primary components, namely, an excitation source, a sampling apparatus, and a detector. While these three components have come in varying forms over the years, modern Raman instrumentation has developed around using a laser as the excitation source, a spectrometer as the detector, and either a microscope or a fiber optic probe as the sample apparatus. It is essential to use a clean and narrow-bandwidth laser because the quality of Raman peaks is directly affected by the sharpness and stability of the excitation light source. A probe is a collection device that collects the scattered photons, filters out the Rayleigh scatter and any background signal from the fiber optic cables, and sends the Raman scatter to the spectrograph. Here, scattered photons are passed through a transmission grating to separate them by wavelength and passed to a detector, which records the intensity of the Raman signal at each wavelength. These data are plotted as a Raman spectrum (Figure 2.28).

- **Raman scattering microscopy:**
  Raman scattering microscopy is an analytical microscopy and is used as a tool for analyzing advanced nanomaterials such as biomolecules in a live

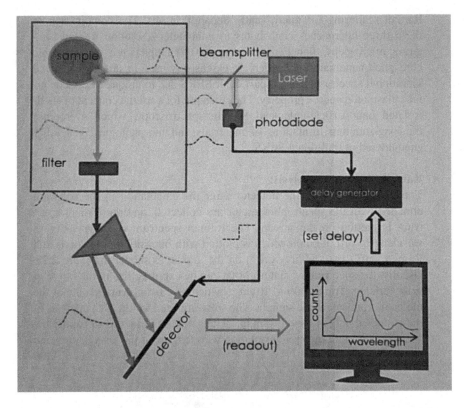

FIGURE 2.28   A schematic overview of a Raman spectrometer. (Adapted from ResearchGate.)

cell for the study of cellular dynamics, semiconductor devices for characterizing strain distribution and contamination and nanocarbons and nano-2D materials.

A Raman microscope begins with a standard optical microscope and adds an excitation laser, a monochromator, and a sensitive detector (such as a CCD or a photomultiplier tube, MT. Nano-Raman spectroscopy yields chemical information on nanoscale without damaging the samples. When used it in conjunction with AFM, it can also produce a topographic image to correlate with the Raman spectra. In order to obtain spectral information with a much spatial resolution, researchers turned to near-field scanning optical microscopy establishing the field of nano-Raman spectroscopy. In near-field scanning optical microscopy, the transmitted light has x, y, and z components, which enable the investigation of a variety of polarization configurations. Satoshi Kawata et al. (2017) have come forward with three approaches for the enhancement of scattering efficiency, and they have shown that the scattering enhancement synergistically increases the spatial resolution. They have also discussed the mechanisms of tip enhanced Raman scattering, UV resonant Raman scattering, and coherent nonlinear

Raman scattering for micro- and nanoapplications. The combination of these three approaches contributes to nanometer-resolution Raman scattering microscopy. Jian Feng Li et al. (2017) emphasized on plasmonic core–shell nanomaterials, which are extensively used in surface-enhanced vibrational spectroscopies, in particular SERS due to unique localized surface plasmon resource property. They also put forward the concept of shell-isolated nanoparticle-enhanced Raman spectroscopy, which overcomes the long-standing limitations of materials and morphological generality encountered in traditional SERS.

- **Raman Spectrum Analysis:**
  The data acquired by the detector after the calculation in the software attached with the spectrophotometer are collected, and the Raman spectrum is plotted on the screen. The Raman spectrum is an intensity vs wavelength shift, a graph which is plotted with intensity on the y-axis and Raman shift (wave number cm⁻¹) on the x-axis, as shown in Figure 2.29. Usually, the spectrum of material that can be captured lies in the range of wave numbers from 4000 to 10 cm⁻¹, which reveals information about the molecules present in the sample. The software used for the data analysis of the collected data is "invia Raman." The software allows us to handle the spectrometer with different excitation wavelengths of the laser source and many controls.

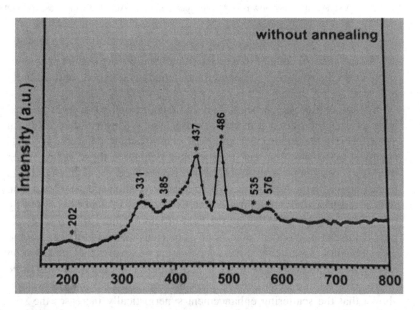

FIGURE 2.29 A typical Raman spectrum recorded for ZnO and plotted with intensity vs wavelength.

2.3.3 XPS Technique: X-Ray Photoelectron Spectroscopy

As the demand for high-performance materials has increased, so has the importance for surface engineering. The material surface is its point of interaction with the external environment and other materials, and this influences factors such as corrosion rates, catalytic activity, adhesive properties, wettability, contact potential, and failure mechanisms.

Surface modifications can be used to alter or improve these characteristics; therefore, surface analysis is used to understand surface chemistry. X-ray photoelectron spectroscopy is one of the standard tools for surface characterization. It finds use from non-stick cookware coatings to thin film electronics and bioactive surfaces.

X-ray photoelectron spectroscopy is also known as electron spectroscopy for chemical analysis. It is a technique for analyzing material surface chemistry. XPS can measure the elemental composition as well as the chemical and electronic states of the atoms within a material. XPS spectra are obtained by irradiating the material surface with a beam of X-rays and measuring the kinetic energy of electrons that are emitted from the top 1 to 10 nm of the material. The sample to be analyzed can be a solid, a liquid, or a gas.

A photoelectron spectrum is recorded by counting ejected electrons over a range of kinetic energies. The energies and intensities of the photoelectron peaks enable the identification and quantification of all surface elements.

Additionally, the change in bond energy caused by the electrons surrounding the atoms to be analyzed, such as atomic valence changes and interatomic distances, tends to be greater than the chemical shift observed in Advanced Encryption Standard (meant for electronic data established by the U.S. National Institute of Standards and Technology in 2001). This makes it relatively easy to identify the state of chemical bonds, which is yet another advantage of XPS (Figure 2.30).

- **Instrumentation:**
  The main components of an XPS system are the source of X-ray, an UHV chamber with mu-metal magnetic shielding, an electron collection lens, an electron energy analyzer, an electron detector system, a sample introduction chamber, sample mounts, a sample stage with ability to heat or cool, a sample, and a set of stage manipulators (Figures 2.31 and 2.32).

- **Excitation Source (scanning microfocus X-ray source):**
  A scanning microfocus X-ray source is an X-ray source that can scan a focused monochrome Al ka beam on sample. In general, characteristics of X-rays such as Al ka rays are widely used as excitation sources for photoelectrons. The X-ray beam diameter can be set between several um zeta to several hundred um zeta, and the scan range can be changed arbitrarily, enabling the measurement of the most appropriate analysis area for the sample. Secondary electron image observation (scanning X-ray image) based on this feature also allows accurate analysis and location. Besides, it supports various analyses, including multipoint simultaneous analysis, large area measurement line analysis, and area analysis (Figure 2.33).

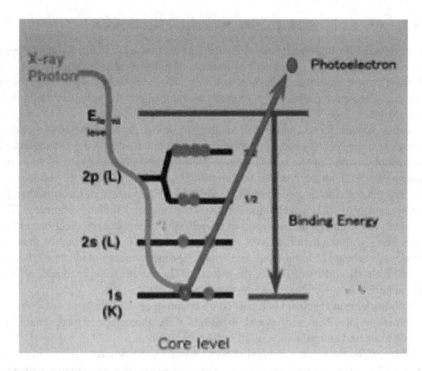

FIGURE 2.30 Electron being ejected from the core level by penetration of X-rays. (Adapted from ULVAC-Phi.)

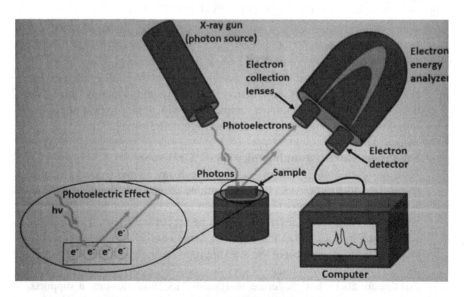

FIGURE 2.31 Diagram and main components of XPS system. (Adapted from Research Gate.)

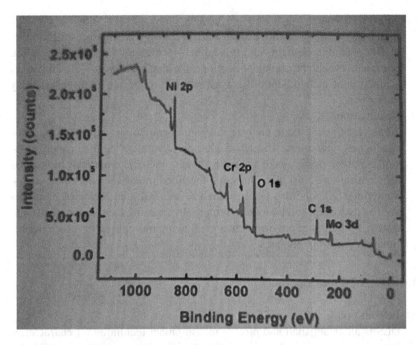

FIGURE 2.32   XPS survey spectrum, on C22 after polarization at oV(+) in a PH = 7. (Adapted from XPS fitting.com)

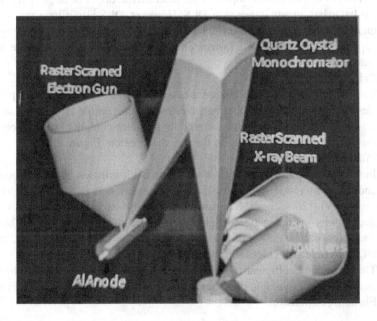

FIGURE 2.33   Scanning microfocus X-ray. (Adapted from ULVAC-Phi.)

- **Analysis Time:**
  It takes typically 1–20 minutes for a broad survey that measures the amount of all detectable elements and typically 1–15 minutes for a high-resolution scan that reveals the chemical depth profile that measures 4–5 elements as a function of etched depth (this time range can vary as many factors play their role).
- **Applications:**
  XPS is routinely used to analyze inorganic compounds, metal alloys, semiconductors, polymers, catalysts, glasses, ceramics, paints, paper, inks, woods, plant parts, make up teeth, bones, medical implants, biomaterials, coatings, viscous oils, glues, ion-modified materials, and many others. Sometimes XPS is also used to analyze the hydrated forms of materials such as hydrogels and biological samples by freezing them in their hydrated state in an ultrapure environment and allowing multiple layers of ice to sublime away prior to analysis.

2.3.4 DLS Technique

- **Introduction:**
  DLS is an established and precise measurement technique for characterizing particle sizes in suspensions and emulsions. It is particularly suitable for nanomaterials. The measurement range for DLS is from 0.3 to 10 nm. This range largely overlaps with the laser diffraction, which starts from 10 nm up to the millimeter range. With much decreasing particle size, the method of DLS becomes a better option than laser diffraction. For larger particles, the latter has an advantage over DLS. However, DLS also allows measurement in a wide concentration range from a few ppm to 40 vol % (sample-dependent), and many DLS instruments offer the possibility to measure zeta potential.
- **Principle:**
  The DLS measurement technique is based on the Brownian motion of particles, which states that smaller particles move faster, while larger ones move slower in a liquid. The light scattered by particles contains information on the diffusion speed and thus on the size distribution (Figure 2.34).
- **Working of DLS:**
  In particle analysis using DLS, the sample is illuminated by a laser beam, and the scattered light is recorded at one detection angle over a period of usually 30–120 seconds. The movement of the particles causes intensity fluctuations in the scattered light. From these fluctuations, the diffusion coefficient can be determined and thus also the particle size (Figure 2.35).
- **DLS particle size calculation formula**
  The diffusion coefficient (D) of particles are inversely proportional to the size (dp– hydrodynamic diameter) of the particle according to the Stokes–Einstein relationship.

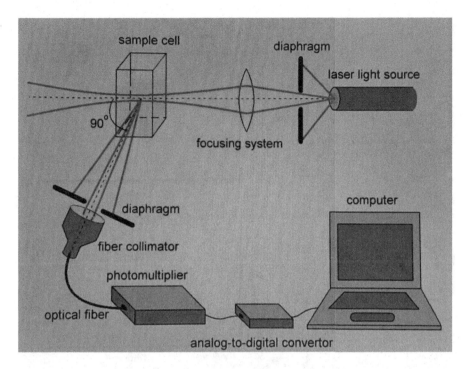

sample cell
diaphragm
laser light source
90°
focusing system
diaphragm
computer
fiber collimator
photomultiplier
optical fiber
analog-to-digital convertor

FIGURE 2.34 Schematic diagram of DLS spectrometer. (Adapted from Research Gate.) DLS, dynamic light scattering.

$$D = KT/3 \text{ pi } n \, dp$$

where K is Boltzmann constant, T is temperature, and n is vicosity; to determine the particle size accurately, the precise value of T and n of the liquid must be known.

- **Technicality involved:**
  Moreover, the temporal fluctuation of the scattered light signal is important because it contains information about the movement of the particles and also contains slight frequency shifts caused by the time-dependent position or velocity of particles. These frequency shifts can be determined by comparison with a coherent optical reference and are on the scale of 1 Hz–100 KHz.

Two approaches exist for optical reference:

1. Homodyne detection (also called "self-beating" or "self-reference" and
2. Heterodyne detection ("reference beating" or controlled reference) (Figure 2.36)

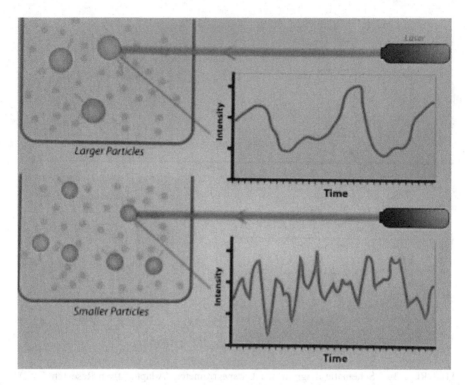

FIGURE 2.35 DLS showing larger particles move slower with longer wavelength and smaller particles move fast with shorter wavelength. (Adapted from Firefly Sci cuvette shop.) DLS, dynamic light scattering.

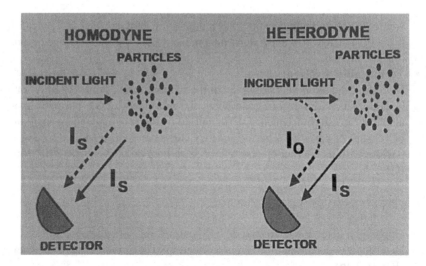

FIGURE 2.36 Scheme of homodyne and heterodyne detection. (Adapted from microtac. com)

Of the two approaches, the heterodyne mode with controlled reference offers many advantages over the homodyne setup in a DLS analyzer. The most important of these is signal intensity, which is proportional to Is2, i.e., the square of mean of the scattered light in the homodyne measurement, while as in case of heterodyne measurement, it is proportional to Is x Io, i.e., the product of the scattered intensity and intensity of the reference. This results in a much stronger measurement.

Signal allows the use of laser diodes as a light source and silicon photodiodes as a detector. The improved signal strength facilitates the measurement of very small low scattering particles down to the lower nanometer range.

- **Evaluation of the DLS signal:**
  The DLS signal can be evaluated in different ways, via a time-dependent auto-correlation function or a frequency power spectrum (FPS). The FPS method is the most reliable and superior method in terms of sensitivity, accuracy, and resolution. The DLS signal from the detector is mathematically transformed into FPS by the fast Fourier transform, and after iterative error minimization, it provides a direct indication of the size distribution.

  The FPS takes the form of a Lorentzian function. The characteristic frequency w is inversely proportional to the particle size, as shown in Figure 2.37.

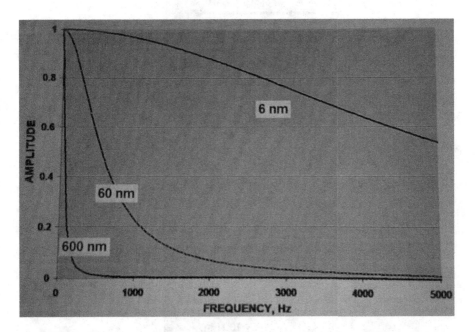

FIGURE 2.37 DLS graph plotted between amplitude vs frequency. (Adapted from microtrac.com)

2.4 CLASSIFICATION OF NANOMATERIALS

Based on their dimensionality, nanomaterials are classified into four types as follows:

0 (zero) dimensional or 0-D
1 (one) dimensional or 1-D
2 (two) dimensional or 2-D
3 (three) dimensional or 3-D

2.4.1 ZERO DIMENSIONAL (0-D) NANOMATERIALS

Nanomaterials with external dimensions at the nanoscale, i.e., between 1 and 100 nm, may be classified as zero dimensional (0-D). This class of nanomaterials includes quantum dots and various types of nanoparticles such as nanospheres, dendrimers, and fullerenes (Figure 2.38).

- **Properties and Applications:**
  Graphene quantum dots (GQDs), carbon quantum dots, inorganic quantum dots, magnetic nanoparticles noble metal nanoparticles upconversion nanoparticles, and polymer dots have attracted extensive research interest due to their optical stability, wavelength-dependent photoluminescence

FIGURE 2.38   0-D nanostructures. (Adapted from Research Gate.) 0-D, zero-dimensional.

(PL) chemical inertness, cellular permeability, and biocompatibility. These 0-D nanomaterials offer great adaptability to biomedical applications such as nanomedicine, cosmetics, bioelectronics, biosensors, and biochips.

Due to their low mass production, low intrinsic toxicity, and multifunctional surface functionalization, 0-D carbon dots are widely studied materials in the field of nanotechnology.

One of the outstanding features of GQDs is their (PL, and their emission wavelength can be changed by adjusting their dimension, morphology, or dopant. The tunable PL property enables GQD use in bioimaging and biosensing. Besides, there are many oxygen-rich functional groups at the edge of GQDs that contribute to their good water solidities and biocompatibility. Furthermore, GQDs have good conductivity and antibleaching properties. Nowadays, GQD-based sensors are widely used in the detection of various ions and biomarkers as well as in the diagnoses of major diseases.

2.4.2   One-Dimensional Nanomaterials

Nanomaterials with two external dimensions at the nanoscale and the third dimension being usually at the microscale are known as one-dimensional (1-D) nanomaterials. This class of nanomaterial include nanofillers, nanotubes, nanowires, nanorods, and nanofilaments (Figure 2.39).

- **Properties and Applications:** Electrons are confined within two dimensions, indicating that electrons cannot move freely. Moreover, 1-D nanomaterials can be amorphous or crystalline, single or polycrystalline; metallic, ceramic, or polymeric similar to 0-D materials. However, 1-D nanomaterials can be pure or impure (such as in doped semiconductors). Furthermore, 1-D nanostructured metal oxide materials exhibit exceptional properties and potential applications in many areas such as electronics, miniatured devices, and catalysis.

1-D/fibrous nanomaterials, due to their high aspect ratio and often good cohesion to the matrix (providing efficient stress transfer between the matrix and the fillers), are frequently used to improve the mechanical properties in composites. Carbon nanotubes are popular components for this purpose because of their high stiffness with Young's modulus of 1TPa; protein nanofillers display a lower modulus but are more hydrophilic than carbon nanomaterials, which makes them easier to disperse in aqueous solution. Remarkably, the use of protein nanofillers from hen egg white lysozyme improved the stiffness of a silicone elastomer to a level twice that obtained with the use of carbon nanotubes (Xinchen Ye et al 2019).

Boron nitride nanotubes (BNNTs) possess many extraordinary physical properties, many of which are comparable or even superior to those of CNTs. For instance, Young's modulus of up to 1.3 TPa and the tensile strength of up to 33 GPa have been reported for BNNTs. They are also resistant to oxidation at high temperatures and

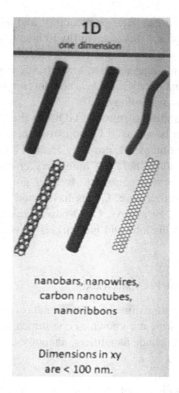

FIGURE 2.39   1-D nanostructure. (Adapted from Research Gate.) 1-D, one-dimensional.

inert to almost all harsh chemicals. BNNTs also possess a large band gap (5–6 eV) and therefore are excellent electrical insulators. They also are excellent neutron absorber materials. They are biocompatible and exhibit no apparent toxicity in interaction with protein and liver cells. They find use in biomedicines, nanocomposites, protective shields, and insulators.

2.4.3   Two-Dimensional Nanomaterials

Nanomaterials with one external dimension at the nanoscale and other two dimensions outside the nanoscale are known as two-dimensional (2-D) nanomaterials. This class of nanomaterials include thin nanofilms, nanolayers, nanocoatings, and nanoplates (Figure 2.40).

2.4.3.1   Properties and Applications

Two-dimensional nanomaterials are a special class of nanomaterials that are ultrathin and have a single layer of atoms. Because of their large surface area, 2-D layered nanomaterials exhibit peculiar electronic properties such as graphene. Other than graphene, 2-D nanomaterials are those of hexagonal boron nitride, different transition metal chalcogenides, germanane, silicone, and phosphorene. The layered

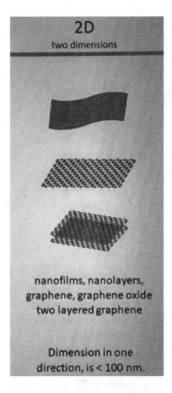

FIGURE 2.40   2-D nanostructures. (Adapted from ResearchGate.) 2-D, two-dimensional.

structures of 2-D materials have properties of lubricants. These materials show a range of electronic properties and can be used in nanoelectronics, optoelectronics, catalysis, sensors, energy storage, and other flexible devices. Moreover, 2-D metal nanomaterials have also found applications in surface enhanced Raman scattering, bioimaging, solar cells, and photothermal therapy (Ye Cheng, Zhanxi Fan et al. 2018). Furthermore, 2-D nanomaterials such as metal nanosheets, graphene-based materials, and transition metal oxides/dichalcogenides provide enhanced physical and chemical functions owing to their ultrathin structures, high surface-to-volume ratios, and surface charges. They have also been found to have high catalytic activities in terms of natural enzymes such as peroxidase, oxidase, catalase, and superoxide dismutase. (Shuangfei Cai and Rong Yag 2020).

2.4.4   THREE-DIMENSIONAL NANOMATERIALS

Nanomaterials that are not confined to the nanoscale in any dimension but display internal nanoscale features are known as three-dimensional (3-D) nanomaterials. This class of nanomaterials includes nanocomposites, dispersions of nanoparticles, bundles of nanowires, and nanotubes as well as multi-nanolayers, nanopowders, and nanostructured materials (Figure 2.41).

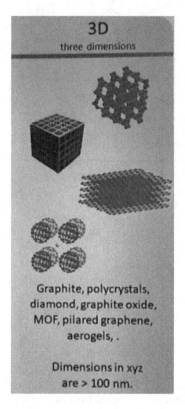

FIGURE 2.41 3-D nanostructures. (Adapted from Research Gate.) 3-D, three-dimensional.

- **Properties and Applications**: Hierarchical nanostructures are 3-D molecules that gradually grow from one parent structure into a more complex form. These nanostructures are synthesized by used building blocks such as biomolecules and organic molecules. They exhibit catalytic chemiluminescence properties, superhydrophobicity, optical properties, magnetic properties, catalytic activity, photolytic activity, flammability, and applications in lithium-ion batteries, Ni_MH batteries, and so on.

  For example, it is proved that solvent-assisted 3-D ZnS hierarchical nanostructures have good photocatalytic and electrochemical performance. All samples possess high photocatalytic efficiency for degrading eosin Y dye, i.e., more than 94%. Samples ZA2 and ZA3 possess superior characteristics, i.e., highest specific capacitance, as it has smallest crystallite size, maximum pore diameter, and high diffusion coefficient compared with other samples (Medha Bhushan, Ranjana Jha 2020).
- Cellulose nanocrystals (CNCs) have been used as a reinforcement agent in composite materials or hydrogels to improve their mechanical properties, thermal stability, and/or water absorption properties.

- Hydrogels of hemicellulose and CNCs showed improved toughness, good recovery behavior, and acceptable swelling and mechanical properties. They have found to have load-bearing biomedical applications such as AC replacements.
- CNCs possess a large surface area, size, crystallinity, and hydrophilic characteristic and have demonstrated a positive impact on food packaging performance.

2.5 FUNCTIONALLY GRADED MATERIALS

2.5.1 INTRODUCTION

Functionally graded materials (FGMs) are multifunctional materials that contain a spatial variation in composition and/or microstructure for the specific purpose of controlling variations in thermal, structural, or functional properties. FGMs have been developed as ultrahigh-resistant materials for aircrafts, space vehicles, and other engineering specifications. Most of the FGMs are particle-reinforced, and their compositions depend on position.

Since 1990, an international congress on FGMs is organized by the international Advisory Committee of AFGMs every 2 years. The proceedings of these congresses give a lot of information about research on FGMs. FGMs is a broad research area and has attracted tremendous attention today in the material science and engineering society. In recent years, FGMs have experienced remarkable developments in manufacturing methods. FGMs can be produced using several well-known processing techniques from conventional to advanced. Research made in this field for the past 30 years provides an overview of manufacturing methods for FGMs and can be considered as a milestone for their future production and analysis investigations and beneficial for researchers, designers, and manufacturers in this regard. A better modified version of composite materials can build a structure that is higher in weight and yet harder than its base materials. The applications and uses of FGMs are immense from aerospace industries, engineering structures, automobile industries to paints, coatings, and oil industries. The advances in technology have added impetus to the development and usage of FGMs. 3D printing can now be used for the fabrication of FGMs. New coating techniques and surface modifications have added extra value to the marketing and processing of matrixes and composites.

Traditionally, four potential methods were used for fabrication of FGMs; powder metallurgy, CVD, self-propagating high-temperature synthesis, and plasma spraying. Some of the recently developed methods are the cast–decant–cast process, friction stir processing, and laser engineered net shaping, which are usually cost-effective and used to make changes in properties. FGMs are also available in nature, in the form of culms of bamboo and barley, in the bones, seashells, and teeth. These naturally occurring FGMs completely match advanced FGMs in terms of desired properties and function at the desired surface.

The FGM is one of the promising candidate for a variety of applications, such as biomaterial implants, thermal barriers, materials for energy conversion, metal cutting and rock drilling tools, mechanical element as gears, and optical and optoelectronic applications. FGMs possess a good chance of reducing mechanical and thermal stress concentrations in many structural materials that can be developed for specific applications. Mathematical modeling and numerical simulations are extremely helpful tools for the design and investigation of FGMs. With the help of numerical simulations, chemical composition and design optimization may be achieved, which increases the performance of FGMs significantly. For instance, an optimized composition gradient of a metal cutting tool may improve the tool life and wear resistance. Similarly, in dental implants, optimized composition gradients and gradient thickness result in improvement of implant life and its biocompatibility.

2.5.2 FGM Types

FGMs can be categorized based on their uses, gradation, composition, and material combination, as shown in the table below:

Classification of FGMs based on type		
Area of application	Biomaterial FGM	HA-Ti
	Piezoelectric FGM	PZT-Ag
	Heat-resistant FGM	PSZ-Steel
	Energy conversion FGM	NiO-YSZ
Type of gradient	Continuous graded type	Al-AlZr
	Stepwise graded type	WC-Co
Changing composition	Functionally gradient type	WC-Co
	Functionally gradient coating type	WC-Co-Fe
Material combination	Metal/ceramic	PSZ-Steel
	Ceramic/ceramic	Al_2O_3-ZrO_2
	Ceramic/glass	$CaO/ZrO_2/SiO_2$

2.5.3 FGM Applications

FGMs are versatile in nature with a broad area of applications, some of which are listed below:

- Durable thermal barrier coating is needed for space shuttle materials because when it enters into space, it has to face extremely high temperature on the outer side (the temperature difference between the outside and the inside is 1000°C). In such conditions, FGMs sustain very effectively with the combination of a ceramic and a metal of a low thermal expansion coefficient, e.g., partially stabilized zirconia ceramic/austenitic steel. A similar kind of combination is used in cryogenics, nuclear fusion, nuclear fission (where high-energy radiation hits a cold container), and laser and plasma technology.

- FGMs can be used as an effective material for metal cutting and rock drilling tools to attain economic cutting condition, optimum cutting speed, and maximized tool life, e.g., diamond–Co and WC–Co are combination materials for these purposes.
- Corrosion of metallic and nonmetallic components at high temperatures can be prevented effectively by FGM coating. Gas turbine blades and areas subjected to thermomechanical stresses or fatigue exposure are protected by FGM coating like MeCrALY.
- Gears and other functional components of machines can be manufactured with greater economy and good performance as compared to conventional costly alloy steel.
- FGMs have been successfully fabricated for energy conversion materials. There is noticeable improvement in the performance of photoelectric, thermoelectric, thermionic, and nuclear energy conversion.
- Piezoelectricity and ferroelectricity are other fields of application of FGM techniques. Piezoelectric FGMs have the ability to reduce stress concentration, improve residual stress distribution, and reduce delamination and have good thermal properties and greater fracture toughness.
- In the human body, the concept of gradation is present in the teeth and bone. Keeping this in mind, it is possible to manufacture FGMs as an implant material. Thus, FGMs can be used in the healing of fractures as a bone plate, dental implant, biomaterial disc in the lumbar spine, femoral component for knee replacement, and hip prothesis coating. Metallic alloys with functionally graded porosity can be successfully prepared for bioengineering. In order to prepare the HAP/Ti dental implant, cold isostatic pressing and sintering process were successfully adopted.
- Optical and optoelectronics are other fields of application of FGMs in which glass fiber can be made with a variation of refractive index and lenses with varying optical wave transmissions and absorption properties.

2.6 CONCLUSION

Nanoscience and nanotechnology have dramatically influenced almost every major sphere of life, be it health care, environment, electronics, energy conversions, optoelectronics, agriculture, veterinary sciences, engineering sciences, etc. Besides it has to show its more, and much needed involvement in pollution control, defense and missions for space endeavors. A lot of developments of nanotechnology are arises in terms of fabrication of nanomaterials, which the researchers are improving upon day by day with new discoveries, and novel designs that are further refining the existing techniques. The method of compound/element characterization is gaining importance with new methods and models of instrumentations, with each equipment working in its own capacity with much more precision and accuracy than its respective previous version. The latest techniques light DLS has taken nanotechnology to a next level. The innovation of FGMs in 1990s has added one more laurel to the techniques of nanotechnology, with compositions of various materials mixed in such proportions, which have enhanced and added to the dynamic and excellent properties of the substances that are serving their end use to desired maxima. There is always scope for further improvements and developments which will always persist to be there so long the human pursuits are constantly persevering with their exploring and curious minds.

BIBLIOGRAPHY

Inigo Agote and Miguel Angel Lagos. Fundamentals and applications of field-assisted sintering techniques (FAST). *Encyclopedia of Materials: Metals and Alloys*, 272–280, 2022.

Sajid Ali Ansari and Moo Hwan Cho. Highly visible light responsive narrow band gap TiO2 nanoparticles modified by elemental red phosphorus for photocatalysis and petrochemical applications. *Scientific Reports*, 6, 25405, 2016.

Anton Paar. Principles of dynamic light scattering. https://wiki.anton-paar.com/in-en/multiple-detection-angles-in-dynamic-light-scattering-analysis/

Valmik Bhavan, Prakash Kattire, Sandeep Thakare, Sachin Patil, and R.K.P. Singh. A review on functionally gradient materials (FGMs) and their applications. In *IOP Conference Series: Materials Science and Engineering*, Vol. 229, No. 1, p. 012021. IOP Publishing, 2017.

Medha Bhushan and Ranjana Jha. Surface activity correlations of mesoporous 3-D hierarchical ZnS nanostructures for enhanced photo and electro catalytic performance. *Applied Surface Science*, 2020. DOI:10.1016/j.apsusc.2020.146988

Britannica. Molecular beam epitaxy – (Material Science).

Andrei A. Bunaciu, Elena Gabriela Udristioice, and Hassan Y. Aboul Enain. X-ray diffraction: Instrumentation and applications. *Reviews in Analytical Chemistry*, 45(4): 289–299, 2015.

Shuangfei Cai and Rong Yang. Two dimensional nanomaterials with enzyme-like properties for biomedical applications. *Frontiers in Chemistry*, 2020. DOI:10.3389/fchem.2020.565940

Shixuan Chen, Bing Liu, Mark A. Carlson, Adrian F. Gombart, Debra A. Reilley, and Jingwei Xie. Recent advances in electrospun nanofibers for wound healing. *Nanomedicine (London)*, 12(11), 1335–1352, 2017.

Ye Chen, Zhanxi Fan, Zhicheng Zhang, Wenxin Niu, Cuiling Li, Nailiang Yang, Bo Chen, and Hua Zhang. Two dimensional metal nano-materials: Synthesis, properties and applications. *Chemical Reviews*, 118(13), 6409–6455, 2018.

Anthony Dichiara, Jin-kai Yuan, Sheng-Hong Yao, and Alain Sylvestre. Chemical vapour deposition synthesis of carbon nanotube-graphene nanosheet hybrids and their application in polymer composites. *Journal of Nanoscience and Nanotechnology*, 12, 6935–6940. DOI:10.1166/jnn.2012.6573.2012.

Dynamic light scattering – An overview. Science Direct Topics.

Yong X. Gan, Ahalapitya H. Jayatissa, Zhen Yu, Xi Chen, and Mingheng Li. Hydrothermal synthesis of nanoparticles. *Journal of Nanomaterials*, 2020. DOI:10.1155/2020/8917013

HORIBA. Dynamic light scattering (DLS) particle size distribution analysis. https://www.horiba.com/cze/scientific/technologies/dynamic-light-scattering-dls-particle-size-distribution-analysis/dynamic-light-scattering-dls-particle-size-distribution-analysis/

Satoshi Kawata, Taro Ichimura, Atsushi Taguchi, and Yasuaki Ku Mamoto. Nano-Raman scattering microscopy: Resolution and enhancement. *Chemical reviews*, 117(7), 4983–5001, 2017.

Ibrahim Khan and Idrees Khan. Nanoparticles: Properties, applications and toxicities. *Arabian Journal of Chemistry*, 12(7), 908–931, 2017.

Sulabha K. KulKarni. *Nanotechnology: Principles and practices*. Springer eBooks, 2015.

Anuj Kumar, Yun Kuang, Zheng Liang, and Xiamomin Sun. Microwave chemistry, recent advancements, and eco-friendly microwave-assisted synthesis of nanoarchitectures and their applications: A review. *Materials Today Nano*, 11, 100076, 2020.

S. Kumar. Advances in additive manufacturing and tooling. *Comprehensive Materials Processing*, 10, 303–344, 2014.

Jian-Feng Li, Yue-Jiao Zhang, Song-Yuan Ding, Rajapandiyan Panneerselvam, and Zhong-Qun Tian. Core–shell nanoparticle-enhanced Raman spectroscopy. *Chemical Reviews*, 117(7), 5002–5069, 2017.

Guodong Liu, Zhengbiao Gu, Yan Hong, Li Chang, and Caiming Li. Electrospum starch nanofibers: Recent advances, challenges and strong strategies for potential pharmaceutical applications. *Journal of Controlled Release*, 28, 95–107, 2017.

Yehia M. Manawi, Ihsanullah, and Muataz A. Atieh. A review of carbon nanomaterials, synthesis via the chemical vapor deposition CVD method. *Journal of Materials*, 11, 822, 2018.

S. Maruyama, R. Kojima, Y. Miyauchi, S. Chiashi, and M. Kohno. Low-temperature synthesis of high-purity single-walled carbon nanotubes from alcohol. *Chemical Physics Letters*, 360(3–4), 229–234, 2002.

Adnan Memic, Tuerdimaimait Abudula, Halimates S. Mohammad, Kassturia Josh Navare, Thibault Colombani, and Sids A. Bencherif. Latest progress in electrospum nanofibers, for wound healing applications. *ACS Applied Bio Materials*, 2, 952–969, 2019.

Microtrac. Dynamic light scattering: DLS particle analyzer. https://www.microtrac.com/products/particle-size-shape-analysis/dynamic-light-scattering/

Molecular beam Epitaxy: An overview. Science Direct.

M. Naguib, Y. Gogotsi, and M. W. Barsoum. Mxenes: A new family of two-dimensional materials and its application as electrodes for Li and Na-ion batteries. In *Electrochemical Society Meeting Abstracts 227*. The Electrochemical Society, Inc., April 2015, No. 9, p. 849.

Kannan Badri Narayanam, Gyu Tae Park, and Sung Soo Han. Electrospun poly (vinyl alcohol)/reduced graphene oxide nanofibrous scaffolds for skin tissue engineering. *Colloids and Surfaces B: Biointerfaces*, 191, 110994, 2020.

Seiichi Normua and Donna M. Sheahen. Micromechanical approach to the thermomechanical analysis of FGMs. Proceedings of the 4th International Symposium on Functionally Graded Materials, AIST Tsukuba Research Center, Tsukuba, Japan, October 21–24, 1996.

Steffen Oswald. *X-ray photoelectron spectroscopy in analysis of surfaces*. Wiley, 2013.

Rityuj Singh Parihar, Srinivasu Gang Setti, and Rajkumar Sahu. Recent advances in the manufacturing processes of functionally graded materials a review. *Science and Engineering of Composite Materials*, 25(2), 309–336, 2016.

T. Pradeep. *Nano: The essentials. Understanding nanoscience and nanotechnology*. McGraw-Hill, 2009.

M. Virginia Roldan, Nora Pellegri, and Oscar de Sanctis. Electrochemical method for Ag-PEG nanoparticle synthesis. *Journal of Nanoparticles*, April 2013. DOI:10.1155/2013/524150.

Micheal B. Ross, Chad A. Mirkin, and George C. Schatz. Optical properties of one, two and three dimensional arrays of plasmonic nanostructures. *The Journal of Physical Chemistry C*, 120(2), 816–830, 2015.

Amir Reza Sadrolhosseini, Mohd Adzir Mahdi, Farideh Alizadeh, and Suraya Abdul Rashid. Laser ablation technique for synthesis of metal nanoparticle in liquid. In Yufei Ma (ed.) *Laser Technology and Its Applications*, pp. 63–83. IntechOpen, 2018.

Charlene M. Schaldach, Graham Bench, James J. Deyoreo, Tony Esposito, David P. Fergenson, James Ferreira, … & Stephan P. Velsko. State of the art in characterizing threats: Non-DNA methods for biological signatures. In S. Schutzer, R.G. Breeze, B. Budowle (eds.) *Microbial Forensics*, pp. 251–294. Academic Press, 2005.

M. A. Shah and K. A. Shah. *Nanotechnology: The science of small*. Wiley, 2019.

R. Singaravelan and S. Bangaru Sudarsan Alwan. Electrochemical synthesis, characterization and phytogenic properties for silver nanoparticles. *Applied Nanoscience*, 5(8), 983, 2015.

Sourabh Soltani, Nasrin Khanian, Thomas Shean Yaw Choong, and Umer Rashid. Recent progresses in design and synthesis of nanofibers over diverse synthetic methodologies, characterization and potential applications. *New Journal of Chemistry*, 2000. DOI:10.1039/d0nj01071e

Yuta Suzuki, Hideitsu Hino, and Kauta Ono. Symmetry Prediction and knowledge discovery from x-ray diffraction patterns using an interpretable machine learning approach. *Scientific Reports*, 10, 21790, 2020.

Faris B. Sweidan and Ho Jin Ryu. One step functionally graded materials fabrication using ultra large temperature gradients obtained through finite element analysis of field assisted sintering technique. *Materials & Design*, 192, 108714, 2020.

Thermo Fischer Scientific. X-ray photoelectron spectroscopy, XPS surface analysis of materials ranging from metals to polymers. https://www.thermofisher.com/in/en/home/materials-science/xps-technology.html#:~:text=X%2Dray%20photoelectron%20spectroscopy%20(XPS,the%20atoms%20within%20a%20material

T. Varghese and K. M Balakrishna. *Nanotechnology: An introduction to synthesis properties and applications of nanomaterials*. Atlantic Publishers & Distributors, 2012.

Jurina M. Wessels, Heinz Geog Nothofer, William E. Ford, Florian von Wrochem, Frank Scholz, Tobias Vossmeyer, Andrea Schroedter, Horst Weller, and Akio Yasuda. Optical and electrical properties of three-dimensional interlinked gold nanoparticle assemblies. *Journal of the American Chemical Society: JACS*, 126, 3349–3356, 2004.

Wikipedia. X-ray photoelectron spectroscopy. https://en.wikipedia.org/wiki/X-ray_photoelectron_spectroscopy

Chris Woodford. Molecular beam epitaxy. March 2021.

Xiaoting Ye, Liyan Chen, Liyang Liu, and Yan Bai. Electrochemical synthesis of selenium nanoparticles and formation of sea urchin-like selenium nanoparticles by electrostatic assembly. *Materials Letters*, 196, 381–384, 2017.

M. Zdrojek, J. Sobieski, A. Duzynska, and J. Judek. Synthesis of carbon nanotubes from propane ÃÃ. *Chemical Vapor Deposition*, 21, 1–5, 2015.

Zhigang Zhang, G. Yao, X. Zhang, J. Ma, and Hao Lin. Synthesis and characterization of nickel ferrite nanoparticles via planetary ball milling assisted solid state reaction. *Ceramics International*, 41(3), 4523–4530, 2015.

3 Mechanical, Thermal, Electrical, Optical, and Magnetic Properties of Organic Nanofillers

Tousif Reza, Rokib Uddin,
Mohammad Fahim Tazwar,
and Md Enamul Hoque
Military Institute of Science and Technology (MIST)

Yashdi Saif Autul
Worcester Polytechnic Institute (WPI)

CONTENTS

3.1 Introduction 72
3.2 Properties of Organic Nanofillers 74
    3.2.1 Cellulose Nanocrystal (CNC) 74
        3.2.1.1 Mechanical Properties of CNC 74
        3.2.1.2 Thermal Properties of CNC 75
        3.2.1.3 Electrical Properties of CNC 75
        3.2.1.4 Optical Properties of CNC 78
        3.2.1.5 Magnetic Properties of CNC 79
    3.2.2 Cellulose Nanofiber (CNF) 79
        3.2.2.1 Mechanical Properties of CNF 80
        3.2.2.2 Electrical Properties of CNF 81
        3.2.2.3 Thermal Properties of CNF 81
        3.2.2.4 Optical Properties of CNF 82
        3.2.2.5 Magnetic Properties of CNF 82
    3.2.3 Bacterial Nanocellulose (BNC) 83
        3.2.3.1 Mechanical Properties of BNC 83
        3.2.3.2 Electrical Properties of BNC 84
        3.2.3.3 Thermal Properties of BNC 84
        3.2.3.4 Optical Properties of BNC 84
        3.2.3.5 Magnetic Properties of BNC 85
    3.2.4 Chitin Nanocrystal (ChNC) 86
        3.2.4.1 Mechanical Properties of ChNC 86

DOI: 10.1201/9781003279372-3

        3.2.4.2   Thermal Properties of ChNC 88
        3.2.4.3   Electrical Properties of ChNC 89
        3.2.4.4   Optical Properties of ChNC...89
        3.2.4.5   Magnetic Properties of ChNC 90
    3.2.5   Chitin Nanofiber (ChNF)...90
        3.2.5.1   Mechanical Properties of ChNF 90
        3.2.5.2   Electrical Properties of ChNF...92
        3.2.5.3   Thermal Properties of ChNF..92
        3.2.5.4   Optical Properties of ChNF 92
        3.2.5.5   Magnetic Properties of ChNF..93
    3.2.6   Starch Nanocrystal (SNC) 93
        3.2.6.1   Mechanical Properties of SNC 94
        3.2.6.2   Thermal Properties of SNC..95
        3.2.6.3   Electrical Properties of SNC..96
        3.2.6.4   Optical Properties of SNC 96
        3.2.6.5   Magnetic Properties of SNC..96
3.3   Conclusion 98
List of Abbreviations...98
Nomenclature... 100
References... 100

3.1   INTRODUCTION

Recent years have seen a surge in curiosity in the design and production of light poly-mer composites with improved mechanical and thermal properties (Ravichandran et al., 2022). Thus, biopolymers have garnered considerable attention due to their critical role in daily life and their unique tunable properties, which make them desir-able for an expansive range of applications (Biswas et al., 2022). Nanofillers are used to create nanocomposites because of the application's key issue. It is not necessary to modify the density of composites when adding nanofillers to increase their mechani-cal, electrical, thermal, magnetic, and optical characteristics (Lamouroux & Fort, 2016). The procedures for nanocomposite synthesis and dispersion modification have been highlighted, and also a detailed examination of the effect of several nanofillers on the physicomechanical characteristics of nanocomposites in comparison to those of natural bone (Michael et al., 2016). Typically, nanofillers aggregate and create a filler cluster. This aggregation inhibits their dispersion within the polymer network and has an unfavorable effect on the nanocomposite characteristics (Sagadevan et al., 2019). As a result of this extensive research and analysis of nanofillers, a wide range of nanocomposites are now possible. Most nanocomposite material's mass comprises nanofiller particles, which can be made up of either inorganic or organic compo-nents or even a combination of both inorganic and organic components (Saba et al., 2014). At least one dimension of a filler must be between 1 and 100nm for it to be referred to as a nanofiller and for composites to be referred to as nanocomposites. One family of nanomaterials is the group of biocomposites, where a nano-object is spread in a matrix or a fraction of a more considerable substance (Lamouroux & Fort, 2016). Organic, carbon-based nanomaterials, inorganic, and clays are the most common types of nanofillers. A carbon nanostructure is a carbon nanofiller that includes fullerenes, carbon nanotubes (CNTs), and different nanofibers. Chitin, starch, and cellulose are examples of organic nanofillers. The use of organic nanofillers in polymer composites improves their mechanical characteristics (tensile, flexural, impact, and tribological) and, most notably, their self-healing property, which results in their widespread application. For several technical applications, the usage of anti-bacterial nanocomposite coatings has increased dramatically in recent years. The interaction between organic nanofillers and eco-friendly biopolymers leads to the better functioning of nanocomposite materials. CNC, an organic nanofiller, is rapidly being used as a reinforcement in biodegradable polymers because of its beneficial properties. It is possible to use CNFs to improve the performance of solar cells and other electronic equipment. Colorful optical behaviors in self-assembling films may be achieved by using ChNCs to construct the chiral nematic crystalline state in suspension. An organic nanofiller known as BNC can be used for biomedical applications. The block diagram of the applications of organic nanofillers can be seen in Figure 3.1.

We shall examine the properties of organic nanofillers in detail in this chapter. We will discuss six different organic nanofillers composed of cellulose, starch, and chitin, and outline those materials' mechanical, thermal, electrical, optical, and magnetic properties. The six nanofillers include (i) cellulose nanocrystal (CNC), (ii) cellulose nanofiber (CNF), (iii) bacterial nanocellulose (BNC), (iv) chitin nanocrystal (ChNC), (v) chitin nanofiber (ChNF), and (vi) starch nanocrystal (SNC). The block diagram of six types of organic nanofillers is shown in Figure 3.2.

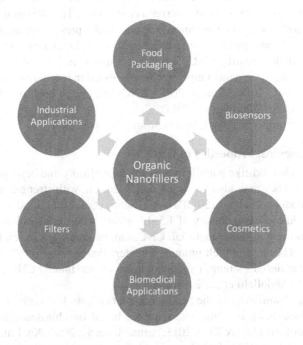

FIGURE 3.1 Block diagram of the application of organic nanofillers.

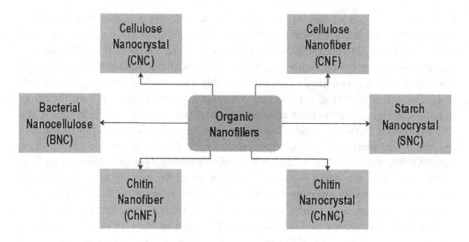

FIGURE 3.2   Block diagram of various organic nanofillers.

3.2   PROPERTIES OF ORGANIC NANOFILLERS

3.2.1   Cellulose Nanocrystal (CNC)

Nanocellulose is a lightweight, cellulose-based material with a high aspect ratio composed of nanosized cellulose fibers (Tusnim et al., 2020). Biodegradable polymers increasingly use CNCs as reinforcement because of their favorable characteristics and a wide variety of potential uses (Ferreira et al., 2018). This development is due to creating nanocomposites with enhanced features while preserving the matrix's biodegradability. Solvent casting, in situ polymerization, and melt mixing have been the primary methods for creating CNC-based nanocomposites. Several recent advancements in producing these nanocomposites are discussed in recent review publications (Kargarzadeh et al., 2018; Mondal, 2018). CNCs are low in cost, readily available, high in strength, etc. (Khalil et al., 2014).

The different properties that CNCs bear are discussed later.

3.2.1.1   Mechanical Properties of CNC

CNCs are extended rod-like particles with high crystallinity and bending strength of around 10 GPa. They may also be strong under tension, with strength up to 7.5 GPa and Young's modulus of about 150 GPa (Azizi Samir et al., 2004; Jonoobi et al., 2015). Crystalline cellulose has a density of 1.5–1.6 g/cm^3, whereas glass has a density of 2.5 g/cm^3, making it a lighter material. CNCs can be employed as reinforcement in nanocomposites because of their unique characteristics (Azeredo et al., 2017). CNCs often raise the material's strength and modulus to a "maximum" CNC content, typically about 5% (Abdollahi et al., 2013).

The structural similarity of the two biopolymers allows for combining each component's characteristics in composite matrices based on chitosan reinforced with CNCs as nanofiller (Celebi & Kurt, 2015; Fernandes et al., 2009; Xu, Liu, et al., 2018). When 0.5% CNC was added to chitosan-reinforced nanocomposites (CH-CNC1h,

CH-CNC2h, and CH-CNC3h), the mechanical properties improved in the case of CH-CNC2h films which can be seen from Figure 3.3a. Similarly, the change in stress due to the incorporation of CNC in the chitosan-based films can be viewed in Figure 3.3b. The stress value of unreinforced chitosan films was determined to be 47.3 MPa, but reinforced films reinforced with 0.75%, 0.5%, 0.25%, and 0.1% CNC enhanced the stress values to 60.6, 68.2, 53.9, and 55 MPa, respectively (less than or equal to 0.05). The increase amounted to a 28.2%, 44.1%, 14%, and 16.4% increase in strength, respectively, when related to the CH matrix (Marín-Silva et al., 2019).

Additionally, by and large, starch films have excellent barrier characteristics against oxygen, lipids, and CO_2. However, they exhibit poor mechanical attributes compared to typical synthetic polymer films (El Miri et al., 2015). If CNCs are put through a carboxymethylation reaction, carboxymethyl cellulose can be obtained via a slurry method (Hebeish et al., 2010). From Figure 3.4, we can see the elongation and tensile strength of the cassava starch-based nanocomposite films. Tensile strength of the N-CMCs/cassava starch film with N-CMCs was 5.6 and 2.8 times stronger than the cassava starch film without N-CMCs and the CMCs/cassava starch film with CMCs, respectively. A concentration of 0.5 g/100 mL of N-CMCs improved the tensile strength, as shown in Figure 3.4. The tensile strength of the CNC/cassava starch was reduced due to a rise in the concentration of CNCs from 0.4 to 0.5 g/100 mL. This phenomenon is because as CNC aggregates easily, it has poorer dispersion when compared to N-CMCs due to its higher crystallinity index (Ma et al., 2017).

3.2.1.2 Thermal Properties of CNC

The melting temperature (T_m), crystallization temperature (T_c), and degree of crystallinity (X_c) of a polymer are all augmented by CNCs as nucleating agents. Polymer crystalline lamellae with thicker lamellae can increase T_m (Lipparelli Morelli et al., 2015). As the nucleating agent, the T_c of the nanocomposite is generally pushed to a little higher temperature than the T_c of the neat polymer, which is directly connected to a higher number of disparate nuclei for crystallization (Ferreira et al., 2017). A reduction in T_c may also be observed (J. Chen et al., 2016).

Additionally, the hydrogen bonding between the filler and polyvinyl alcohol (PVA) matrix was strengthened by adding CNC, making nanocomposites much more thermally stable (Mandal & Chakrabarty, 2014). Moreover, the degradation temperature was raised by 7°C–9°C due to the addition of CNC to the gelatin matrix (Echegaray et al., 2016). The addition of lignin-coated CNC (L-CNC) increased the thermal stability at a higher temperature in the case of a 3D printed nanocomposite (Alam et al., 2014).

3.2.1.3 Electrical Properties of CNC

When it comes to electrical and electronics applications, CNCs are a potential material because of their insulating characteristics and high resistivity compared to paper (Kang et al., 2012; Le Bras et al., 2015; Nair et al., 2016). A dielectric paper might be replaced with cellulosic nanocrystals in the fabrication of batteries, supercapacitors, and electric motors. Many electrical devices use the dielectric form, which degrades or absorbs humidity over time, causing them to lose their quality or collapse and financial losses for both the vendor and the customer. Dielectric paper is utilized

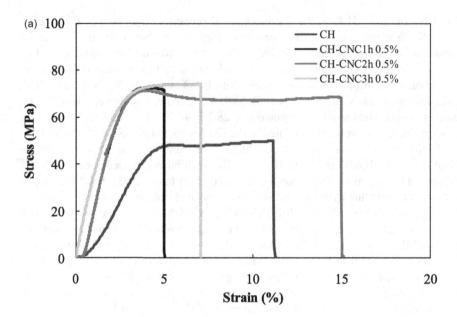

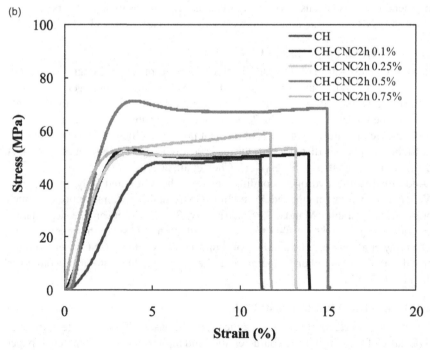

FIGURE 3.3 Behavior of tensile tension and strain during (a) CNC-0.5% nanocomposites and CH films and (b) different CNC concentrations used to create CH films and nanocomposites (Marín-Silva et al., 2019).

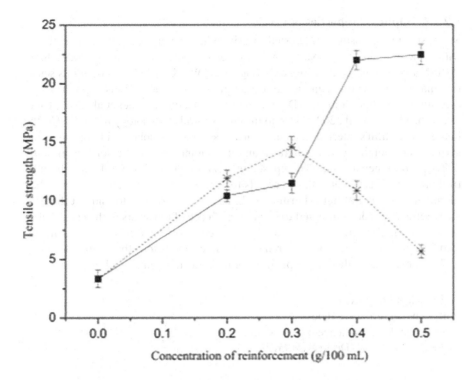

FIGURE 3.4 Variations in tensile strength by the concentration of N-CMCs and CNCs (X. Ma et al., 2017).

in many electrical products. It would be wonderful to replace them with products that have better qualities. CNCs might be used in the power industry since they are readily available materials with fascinating features (Hernández-Flores et al., 2020).

Nowadays, piezoresistive sensors with extensive response bands and excellent sensitivity are in high demand as the flexible electronic sector proliferates (J. Huang et al., 2019; W. Huang et al., 2017; H. Liu et al., 2018; Xiao-peng et al., 2019; Zhai et al., 2019; Zhang et al., 2019). CNCs were added to a WPU/CNT composite foam in order to uplift its piezoresistive performance. As CNC loading increased, the gauge factor (GF) of WCT foam rose, and the GF jumped to 2.5 times the GF of composite foam without CNC. $GF = (\Delta R/R_0)/\varepsilon$, ε represents the applied compression strain, which is the ratio of resistance change/unit and length change/unit. That is all good for the sensor's use as wearable electronics that can detect a wide range of human movements with varying levels of compressive strain (Zhang et al., 2020).

Additionally, various PVDF/CNC nanocomposites were electrospun with different CNC contents (0%–5%), and their piezoelectric properties were studied. The usage of CNC drastically improved the piezoelectric properties of the nanocomposite. It was found that a sample with 3% or 5% CNC could quickly charge a 33-μF capacitor over a voltage of 6. The model was further able to light up a LED for a period of over half a minute (Fashandi et al., 2016).

3.2.1.4 Optical Properties of CNC

Scientists are increasingly interested in using photonic crystals to create materials with structural shades (Baba, 2008; Drake & Genack, 1989; Huang, 2020; John, 1991; Marcelo et al., 2021; Wang & Zhang, 2013; Wen et al., 2020). Materials scientists may tune the chiral nematic nanostructure of CNCs-based films to produce colored films, and this exhibits a 1D photonic crystal feature (Bardet et al., 2015; Gu et al., 2016; Khattab et al., 2020). Due to the chiral spiral morphology of the CNCs, the liquid crystal film's internal structure generates circularly polarized light reflection, giving it outstanding optical qualities and its vibrant structural color (Duan et al., 2021). A chiral nematic liquid composed of cellulose photonic crystals, which selectively reflect visible light, causes crystals to reveal stunning structural colors when illuminated by natural light (Furumi et al., 2003). It is possible to find out the final mean refractive index through the orientation of each of the layers of the crystal molecules. Bragg diffraction occurs when light reflected off the film's surface is irradiated from a source with a different relative refractive index than that of the film itself.

Three parameters dictate the pitch (P) of the reflected light, which is

λ = reflected wavelength,
n_{avg} = mean refractive index,
$\sin\theta$ = angle of incidence of the light.
$\lambda = n_{avg} \times P \times \sin\theta$ (De Vries, 1951)

Because of its customized optical characteristics, cellulose chiral liquid crystal film may be used to create stimuli-responsive displays.

Cellulose chiral nematic liquid crystals bear some other properties also that include strong birefringence, and CNCs tend to manifest a high birefringence phenomenon (Dumanli et al., 2014). Optical rotation is also quite fascinating in the case of chiral nematic crystals. The vibration plane of the light incident of the liquid crystal undergoes a specific deviation, and there will be a deflection angle between the vibration plane of the incident and the transmitted light, which can be identified through a circular polarizing filter. Thus, the crystals are able to express more vibrant colors on the left-hand side polarizing filter. However, the color on the right-hand side filter will be dim in comparison to the left-hand side. With further scrutinization, the optical properties of the reflecting left-handed polarized light and the transmitting right-handed transmitting light can be corroborated (Tran et al., 2020). Again, in light of the eminent properties of CNCs to exhibit liquid-crystalline activity in the water, one notable application is the fabrication of iridescent products. As a visual phenomenon, iridescence has attracted inventors, artists, and scientists for ages, including Newton, Aristotle, and Darwin (Meadows et al., 2009). Additionally, nanopapers obtained from CNCs showed an exorbitant visible light transmittance which was over 90% and possessed a wavelength of around 300–800 nm. This high value of transmittance is beneficial for nanopapers because it allows light to pass through them without causing a significant amount of light scattering (Yao et al., 2017).

3.2.1.5   Magnetic Properties of CNC

Linear chains of hydrogen-bonded β-D glucopyranose are packed together into nanorods with diameters of 10–20 nm and lengths of 200–400 nm in CNCs. CNCs are suitable template elements for nanomaterials because of their negative charge and hydroxyl (OH) groups (e.g., Au, 16). In the last few years, silica–CNCs composites have attracted much attention and have been utilized to create chiral nematic carbon needles and mesoporous carbon. Silica functioned as robust nanoreactors and frameworks accordingly (Asefa, 2012; Shopsowitz et al., 2011; Silva et al., 2012). To support an iron oxide nanoparticle, silica and CNCs were used as the coating material and supporting template. Thus, a CNC@Fe_3O_4@SiO_2 hybrid nanomaterial was formed (Chen et al., 2014). A vibrating sample magnetometer was used to investigate the magnetic characteristics of CNC@Fe_3O_4@SiO_2 hybrids with varied silica shell depths at 300 K. Coercivity was shown to be low in the Fe_3O_4 and CNC@Fe_3O_4 hybrids, confirming the superparamagnetism of the hybrid systems through hysteresis loop analysis. Superparamagnetic features of the hybrids led to a rapid reaction to external magnetic fields and a brisk redistribution under mild shaking when the magnetic field was withdrawn. Many applications benefit from the reversibility of this procedure.

Again, in another work, Fe_3O_4 nanoparticles (MGCNCs) were incorporated into CNCs produced from bamboo utilizing an in situ coprecipitation method. After being disseminated in polylactic acid (PLA), a directional magnetic field was used to align the MGCNCs in the nanocomposites after being disseminated in PLA. Magnetically susceptible MGCNCs may be readily aligned in a polymeric matrix because of their rod-like form when the magnetic field is modest. In PLA/MGCNC nanocomposites, better scattering of CNCs in the polymeric matrix is envisaged as a consequence of the synergistic effects of two separate nanofillers (Fe_3O_4 nanoparticles and CNCs). Nanofillers have also been shown to increase the mechanical, thermal, electrical, and magnetic characteristics of nanocomposite films when aligned in a particular direction in a magnetic field (Dhar et al., 2016).

In addition, iron oxides (IONCs) were mixed with CNCs, upraising the filler's magnetic properties. It was realized through the movement and agglomeration of particles in the direction of an external magnet without precipitation. Thus, it can be implemented in many applications, i.e., externally generated magnetic fields might be used to align them in various matrices and augment their mechanical characteristics in a particular direction (L. Chen et al., 2020). A magnified image of a CNC filler mixed with IONC can be viewed in Figure 3.5.

3.2.2   Cellulose Nanofiber (CNF)

Due to the enormous specific surface area and interfacial properties, CNFs have significantly enhanced features such as mechanical, crystallinity, resistance, barrier, and flexibility qualities (H. Lee et al., 2014; Pires et al., 2019). It is possible to mechanically remove CNFs, also known as "nanofibrillated cellulose," from the fibrous cellulose fibers (Moohan et al., 2019). Wood cellulose fibers floating in water may be mechanically degraded into CNFs with 10–50 nm diameter by high-intensity

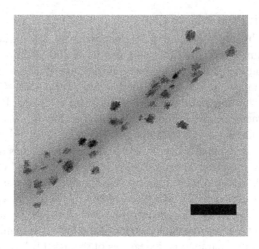

FIGURE 3.5   The IONPs and their distribution on the CNCs surface under a transmission electron microscope (L. Chen et al., 2020).

homogenization. However, the manufacturing of such mechanically disintegrated CNFs consumes a tremendous amount of energy (>200 kWh/kg), making CNF production expensive (Isogai, 2013; Isogai et al., 2010; Klemm et al., 2011).

The different properties that CNFs bear are later.

3.2.2.1   Mechanical Properties of CNF

CNFs are capable of augmenting the mechanical attributes of nanocomposites considerably. When CNF was included in a PVA film, intriguing results were obtained. PVA nanocomposite films tensile strength and modulus increased by 74% and 90%, respectively, when 15% CNF loading was added (Kassab et al., 2019). Additionally, the PVA film tensile strength and modulus were significantly increased by 44% and 210%, respectively, when 15% of CNF was introduced (D. Liu et al., 2013). Similarly, when CNFs (0.5% $w_{/w}$) were added to starch films, the mechanical strength of the biofilm rose by 3.4% and 35%, which was related to the way of processing each of the components (Karimi, 2014).

Again, lignin was combined with CNFs and LCNFs in another research by Yang et al. (2020). When LCNFs were introduced to PVA composites, tensile strength and Young's modulus were elevated considerably.

The inclusion of 5% CNFs with PLA nanocomposites increased the material's modulus from around 3 to 3.6 GPa (Jonoobi et al., 2010). Furthermore, adding CNF nanofillers to polyion complex (PIC) gels elevated their mechanical properties (Yataka et al., 2020). The mechanical properties of the CNF-reinforced films were inspected in both dry and wet conditions. In the dried state, the modulus increased to around 3300 MPa, which exhibited an enhancement of around 65% compared to the original film. However, the improvement in Young's modulus was significantly more significant in the wet state, which was a mammoth 1300% enhancement compared to the original film.

3.2.2.2 Electrical Properties of CNF

Because CNF is innately insulating, novel approaches must be devised to transmit electrical characteristics to it (Lay et al., 2016). CNFs and conductive polymers (CPs) can therefore be integrated to enhance the functionality of supercapacitors, solar cells, and other electrical gadgets that use CNFs (Luo et al., 2014; Nyholm et al., 2011; Tammela et al., 2015; Z. Wang et al., 2015; Zheng et al., 2013; Zhu et al., 2013). Since CNFs are naturally insulating, the addition of merely 8% PPy filler resulted in a semiconductor product because of the consistently linked conducting networks formed on the CNF's surface during drying and filtering (Koga et al., 2014). Nanopapers contain up to 20% PPy, and the electrical conductivity was 5.2×10^{-2} S/cm, which is three times larger than the conventional semiconductor-like silicon (1.5×10^{-5} S/cm) (Lay et al., 2016).

Again, in a work by Hou et al. (n.d.), a CNF layer is placed between two reduced graphene oxide (RGO) layers. Thus, a highly conductive network was formed with significantly higher electrical conductivity to around 4382 Sm^{-1} than the base composite paper. Insights for biobased flexible electronic devices were provided by this successful technique offered at an affordable price. Additionally, supercapacitor electrodes made from CNFs have the most full-electrode-normalized volumetric (122 F/cm^3) and gravimetric (127 F/g) capacitances at high current densities (300 mA/cm² 33 A/g) to date (Z. Wang et al., 2015). Moreover, dielectric constant (k) is a fundamental property for high-touch-sensitivity applications—ultralong metal nanofibers embedded in CNF films improve k significantly. Thus, different electronic gadgets have been established based on this principle (Ji et al., 2017).

3.2.2.3 Thermal Properties of CNF

Many researchers have investigated starch-based bioplastics (Abral et al., 2017; Panchagnula & Kuppan, 2019; Syafri et al., 2017). However, starch possesses low thermal stability, which can be overcome by using organic nanofillers. Biomass derived from CNFs can enhance bioplastic's thermal characteristics. The loss sample was reduced due to an addition in the CNF content in the case of a CNF/PCC nanocomposite. As the number of nanofillers was elevated, the loss was further reduced (Syafri et al., 2018). Raised filler content improves thermal stability because of the strong bond between fibers and matrix (Prachayawarakorn et al., 2013). Additionally, when CNFs were added to kenaf/epoxy composite, due to the cross-linked CNF filler's better heat receptivity and the epoxy matrix's limitation of molecular mobility, CNF/kenaf/epoxy hybrid nanocomposite demonstrated exceptional thermal stability (S. H. Lee et al., 2011; Saba et al., 2017).

Thermoplastic starch (TPS) is one of the most popular natural polymers due to its inexpensive cost and good physical attributes (Balakrishnan et al., 2017). On the other hand, TPS has several downsides, such as lower heat stability than other thermoplastic polymers (Glenn et al., 2011). When CNFs were added to TPSs, the thermal results showed that 1.5% CNF provided better resistivity toward heat. Moreover, the TPS-foamed composites' Tg was raised by the addition of CNFs (Ghanbari et al., 2018).

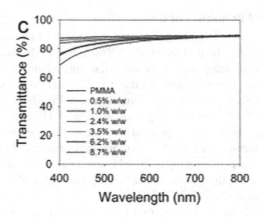

FIGURE 3.6   Transmittance spectra of PMMA/CNF films (D. W. Kim et al., 2020).

3.2.2.4   Optical Properties of CNF

CNFs possess high optical properties. PMMA nanocomposites reinforced with CNFs were developed in the research by D. W. Kim et al. (2020). The optical transparency of the composite films in the range of CNF concentration was equivalent to that of the pristine PMMA films, with a transmittance of around 87% at 600 nm which we can see in Figure 3.6. This value was similar to the *N,N*-dimethylformamide solvent cast PMMA/CNF nanocomposite film (Fujisawa et al., 2018).

In another work, a simple water solution method was used to create poly(ethylene oxide)/CNF nanocomposite (Safdari et al., 2017). When CNFs were mixed, the optical characteristics of PEO were unaffected by the addition of CNFs, indicating that the scattering of CNFs in PEO was excellent (Petersson & Oksman, 2006; Safdari et al., 2016). Only a 5% drop in transmittance values was recorded by inserting 5 wt% CNFs into PLA film. Optoelectronic and optical use was possible for this very reason (Janardhnan & Sain, 2011; Kalia et al., 2014; Safdari et al., 2016; Siro & Plackett, 2010; Tercjak et al., 2015; X. Xu et al., 2013).

Furthermore, CNFs made by electrospinning cellulose acetate from bamboo cellulose and then deacetylating it were employed as supports to create optically transparent composites. Even though the final composite film had a large percentage of fibers, it nevertheless manifested an excellent transmittance of light (Cai et al., 2016).

3.2.2.5   Magnetic Properties of CNF

Nanocellulose, a biodegradable, low-density, superior mechanical and elevated surface area substance, has substantial advantages over other materials employed as templates for the manufacture of nanomaterials (Moon et al., 2011). Crystalline packing, vital physical characteristics, and nanomaterial's formation/condensation are all controlled by H-bonding between cellulose's OH groups on the surface, which also provides a nanofiber network for the formation of nanoparticles. The next generation of renewable nanomaterials for high-performance, long-term-use materials is being studied, and nanocellulose is one (Walther et al., 2011). In a work by Amiralian et al., assembling a relatively homogeneous flexible nanocomposite membrane with high levels of magnetic nanoparticles by synthesizing nanomaterials on the surfaces of colloidal suspensions of colloidal fibers was exhibited. CNFs were effectively employed as a layout to produce magnetic nanoparticles with diameters below 20 nm and limited size distribution. A densely packed iron oxide nanoparticle structure on the surfaces of nanofibers, which contributes to the high magnetic characteristics, was shown to form due to the nucleation and development of nanomaterials on nanofiber surfaces first and subsequently on Fe_3O_4 nanomaterials grafted onto CNF (Amiralian et al., 2020). Similarly, CNFs were mixed with magnetite nanoparticles, and a membrane was created for loudspeakers (Tarrés et al., 2017). The present loudspeakers, which are usually electrodynamic and rely on the motion of a magnet owing to the induction of a magnetic field when a coil is stimulated electrically, might benefit from the employment of these nanoparticles on membranes.

3.2.3 BACTERIAL NANOCELLULOSE (BNC)

Due to its remarkable physical and biological properties, BNC has arisen as a natural biopolymer of extraordinary value in several technical fields. In the form of an exopolysaccharide of β-D glucopyranose, bacteria can produce the promising natural biopolymer known as BNC. BNC is well suited for adhesion and cellular immobilization due to its nanostructured shape and water-holding capability, which is comparable to that of collagen, an extracellular matrix protein. As a result of its numerous unique features, BNC falls under the group of items generally regarded as safe (El-Hoseny et al., 2015; Sharma & Bhardwaj, 2019). Young's modulus of BNC in a dry state is around 15–35 GPa. Its tensile strength is around 300 MPa (Stanisławska, 2016).

The different properties that BNCs bear are discussed later.

3.2.3.1 Mechanical Properties of BNC

Studies in the papermaking sector have shown that BNC may be used for biomedical purposes. Inter-fiber interaction is enhanced by its nanoscale fibers and vast free OH groups. As a result, BNC is an excellent reinforcing material for recycled paper and nonwoody cellulosic paper with high mechanical strength (Skočaj, 2019). Moreover, when BNCs were added to cement particles, the mechanical features of the cement increased. Mechanical characteristics are improved when BNC between 0.05% and 0.20% was added to the cement. This improvement might also be due to increased hydration, which boosts the substance's strength (Barría et al., 2021).

Again, it was found that by combining polyethylene glycol diacrylate (PEGDA) and BNC gels using polyethylene glycol as a solvent (BNC/PEG gel), the gel's mechanical strength was increased, resulting in a composite structure (Numata et al., 2015). Mechanical qualities like tear resistance and form retention of natural BNC are superior to synthetic materials. BNC treated into a film or sheet has excellent mechanical toughness when combined with organic layers, like cellophane or polypropylene (Stanisławska, 2016).

3.2.3.2   Electrical Properties of BNC

The usage of BNC nanocomposites in flexible electrodes, supercapacitors, or biosensors has recently been researched (X. Chen et al., 2016; Qiu & Netravali, 2014). After being pyrolyzed in nanocomposites, BNC can be utilized as an electrical conductor (L.-F. Chen et al., 2013; Liang et al., 2012; B. Wang et al., 2013; Z.-Y. Wu et al., 2013). Nanocomposites may also be made using BNC as an ideal matrix for conducting nanocomposites with other materials (Gutierrez et al., 2012; Kang et al., 2012; J. Shah & Brown, 2005; Wan et al., 2018). A bio-triboelectric-based nanogenerator (Bio-TENG) based on BNC (an eco-friendly and plentiful biomaterial) was described for the first time in a research carried out by H.-J. Kim et al., (2017, p. 130–137). At a load resistance of $1\,M\Omega$, the BNC bio-accumulated TENG's charges and maximum power density may reach around 8 $\mu C/m^2$ and 5 mW/m^2, respectively.

Thanks to BNC films, flexible optoelectronic devices and systems may now use an entirely new application area. Disposable/biocompatible electronics, including paper RFID, smart packaging, and point-of-care systems in bio-applications, may be made using BNC because of its compatibility with large-scale/extensive area deposition processes (Fortunato, 2016).

3.2.3.3   Thermal Properties of BNC

BNC's low thermal conductivity makes it an excellent support material for interface solar steam production because it effectively transports water and manages heat (Q. Jiang et al., 2016). Layers of BNC and PDA were used to create a biodegradable, scalable, and flexible solar steam generator (Q. Jiang et al., 2017). The device showed excellent photothermal conversion and localization of heat. Additionally, the layers of PDA/BNC illustrated remarkable steam generation capability under the sun. The efficacy was around the 78% mark. Again, polydopamine (PDA) particulates and BNC were used to create a bilayer photothermal membrane with a permeate flow of 1 kg/m^2/h under single solar irradiation, and the efficiency of solar energy to collected water was shown to be as high as 68% (X. Wu et al., 2021). It is an excellent solution for reducing nondegradable polymer pollutants, as both BNC and PDA are biodegradable materials compared to other polymer-based membranes (Q. Jiang et al., 2017).

A mass loss of 53.57% on average and a residual mass of 46.43% were found by thermogravimetric analysis (TGA) in the thermal degradation of BNC. The activation energy ($E_a = 59.39$ kJ/mol) and preexponential factor ($k_o = 1.62 \times 10^{10}\,min^{-1}$) for BNC were calculated using kinetic modeling, suggesting that the compound is highly reactive, possessing a very high thermal reactivity (Nyakuma et al., 2021).

3.2.3.4   Optical Properties of BNC

Excellent features such as high flexibility and porosity make BNC an ideal choice for use as a substrate to create optical (bio)sensors. It was possible to create a portable, efficient, inexpensive assay kit for HSA monitoring by integrating the optical characteristics of curcumin with the intriguing qualities of BNC nanopaper as a promising sensing substrate (Naghdi et al., 2019). Moreover, a technique for making nanocomposites using BNC and metal nanoparticles that self-assemble has been discovered for the first time. An increase in absorption wavelengths may be achieved by compressing BNC–AuNP composites, which results in regulated near-field interaction

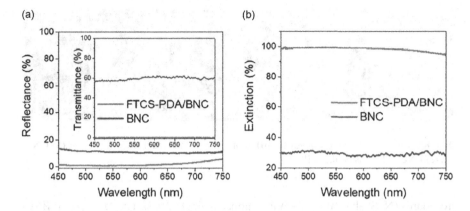

FIGURE 3.7 (a) Transmittance and reflectance of light of FTCS-PDA/BNC and neat BNC membranes, (b) extinction of light of FTCS-PDA/BNC and neat BNC membranes (X. Wu et al., 2021).

between the two nanoparticles. For colorimetric biosensor development, it is critical that biomolecule adsorption on NPs can thoroughly screen their near-field contact upon being compressed (Eskilson et al., 2020).

Furthermore, BNC and PDA (polydopamine)/BNC membranes were analyzed at wavelengths ranging from 450 to 700 nm for optical transmission and reflection. BNC's light transmittance was 59%, while its reflectance was just 11%, resulting in a 30% light extinction. However, FTCS-PDA/BNC showed exceedingly minimal light transmittance (0%) and reflectance (2%), which led to a light extinction (98%) in the visible range after polydopamine particles were loaded. The FTCS-PDA/BNC membrane's good light-to-heat conversion may be ascribed to the PDA particles' strong light extinction in the BNC network, which is responsible for the membrane's high light extinction (X. Wu et al., 2021). We can view the comparison between the reflectance, the transmittance, and the extinction in Figure 3.7a and b.

For water purification, the development of flexible, porous, and self-supported BNC/MoS_2 hybrid aerogel films with photocatalytic and absorptive capabilities was reported. For in-flow photo-assisted removal of organic model pollutants (MB dye) and hazardous heavy metal compounds (Cr(VI)), the material's efficacy in a specially built membrane photoreactor was examined. MoS_2 nanostructures mixed with BNC membranes almost perfectly removed pollutants by boosting the membrane's adsorption ability and providing visible light photocatalytic properties (Ferreira-Neto et al., 2020).

3.2.3.5 Magnetic Properties of BNC

BNC lacks magnetic properties, but ferrites (Fe_3O_4) can be mixed to give it a magnetic character. BNC/Fe_3O_4 nanocomposite films may be prepared using the in situ synthesis approach, which involves two or three processes. This method can be executed at room temperature as well. For the nanocomposites, it was crucial to boost the saturation magnetization (Ms) dependent on nanoparticle dispersion while using simple

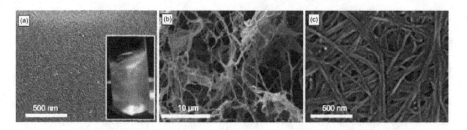

FIGURE 3.8 Microscopic images of different kinds of nanocellulose: (a) CNCs, (b) CNF, and (c) BNC (Thomas et al., 2018).

procedures (N. Shah et al., 2013; Yingkamhaeng et al., 2018). The M_S value of BNC/Fe$_3$O$_4$ CNFs was around 40.57 emu/g, which is a high number. It was clear from the FE-SEM and TGA data that Fe$_3$O$_4$ NPs may be found embedded in BNC nanofibrils, with an exceptional distribution of nanoparticles in the BNC matrix. BNCs act as a template for various metals capable of providing magnetic properties (Yingkamhaeng et al., 2018). Again, BNC nanoribbons were used as a template in preparing cobalt ferrite nanotubes (NTs). At ambient temperature, the BNC-templated CoFe$_2$O$_4$ nanotubes displayed magnetic characteristics. Nanomaterials in the superparamagnetic condition were found to alter the magnetic behavior (Menchaca-Nal et al., 2016).

Furthermore, coprecipitation of natural IONPs in situ resulted in a magnetically synthesized BNC. The BNC fiber framework and IONPs formed a strong connection, and the IONPs were distributed uniformly. It was also shown to have a paramagnetic reaction suggesting it might be an excellent choice for coating an aneurysm-neck stent. Thus, this MBNC will help implement a magnetic field (external) for the treatment of PBI injuries in a unique magnetic approach (Arias et al., 2016).

The different properties related to CNCs, CNFs, and BNCs have been illustrated above. The different types of CNCs can be seen in Figure 3.8, and the molecular structure of cellulose can be viewed in Figure 3.9.

3.2.4 Chitin Nanocrystal (ChNC)

ChNCs, crystalline rod-like nanomaterials obtained from natural sources, have a high aspect ratio, a high modulus, and excellent reinforcing capacity toward polymers (Y. Liu et al., 2018). It is common for ChNCs to have widths of around 10–50 nm and lengths of around 150–200 nm, which leads to a high ratio. Polymers benefit from CNCs' high elastic modulus of 150 GPa, which provides a significant amount of reinforcement (M. Liu et al., 2015). Physical cross-linking via H-bonding with polymer chains may be achieved using ChNC reinforcements for polymer hydrogels (Y. Huang et al., 2015; M. Liu et al., 2015).

The different properties that ChNCs bear are discussed later.

3.2.4.1 Mechanical Properties of ChNC

In a research, poly(lactic acid) was employed as the core and polyacrylonitrile/ChNCs as the shell for core–shell nanofibers. Coaxial membranes with nanocrystals saw a

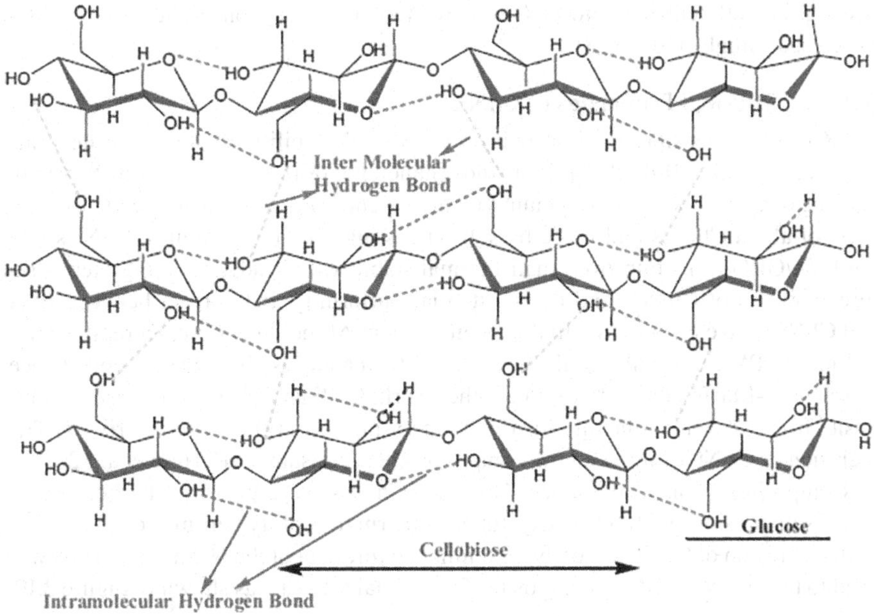

FIGURE 3.9 H-bonding among several cellulosic units (Rana et al., 2021).

rise in tensile strength and elastic modulus compared to coaxial membranes that did not. The most extraordinary mechanical characteristics were found at 15 wt% ChNC. ChNC-coated coaxial fibers were very hydrophilic in nature (Jalvo et al., 2017). A novel epoxidized natural rubber (ENR)/ChNCs composites were successfully manu-factured without standard cross-linking agents in another study. ChNC OH groups interacted with epoxy groups in ENR, resulting in hydrogen bonds between the two molecules. This interaction allowed ChNCs to develop supramolecular networks, enhancing mechanical characteristics while also providing ENR/CNC composites with improved self-healing capabilities. There has been an almost two-fold increase in ENR/CNC composite's mechanical strength, including 20% ChNCs compared to the clean ENR (1.1 MPa) (Nie et al., 2019).

ChNCs, when used as nanofillers in a cross-linked styrene-butadiene rubber (S-xSBR) matrix, enhance their mechanical properties substantially. Static and dynamic tensile strength and modulus may both be improved greatly by including CNCs. For S-xSBR/CNC/CNC composites, the tensile strength and the tensile mod-ulus, when 4 wt% ChNCs are added, are 97% and 62.5% greater than neat S-xSBR (L. Ma et al., 2016). ChNCs are also responsible for elevating the tensile strength of NR/CNC amalgams. 5.75 MPa is the tensile strength of the NR/CNCs nanocom-posite comprising 5% ChNCs, which is around six times more than neat NR. The uniform distribution of CNCs and the excellent interfacial contacts between rubber and ChNC molecular chains are responsible for the improved NR characteristics (Y. Liu et al., 2018). The tensile strength of PLA nanocomposite is augmented by about 30 MPa from 41 MPa when orientated using ChNCs. The prolongation at break was significantly enhanced to 55% more than the neat nanocomposite when ChNCs were used (Singh et al., 2018).

3.2.4.2 Thermal Properties of ChNC

ChNCs are responsible for improving the thermal stability in biocomposite films (Salaberria et al., 2014, 2015). In a study, nanocomposite films based on PVA with varying concentrations of Origanum vulgare essential oil were reinforced with 0.5% (w/v) of alpha ChNCs and made by solvent casting. The integration of ChNCs into the PVA/OEO films improved their thermal stability. Degradation at high temperatures (380°C) of ChNCs and the substantial uniformity and contact between PVA and ChNCs have been linked to this phenomenon. At the same time, increased crystallinity of PVA molecules in the presence of the nanocrystals is also a good chance (Fernández-Marín et al., 2020). In another work, CNF and ChNC self-bonded composite films were manufactured using hot pressing to test their characteristics. The inclusion of ChNCs improved the composite's heat resistance (Herrera et al., 2016).

Composites of natural rubber (NR) and ChNCs were prepared by combining ChNCs dispersions with NR latex. Rubber's thermal stability was improved because of the inclusion of ChNCs equally disseminated throughout the NR matrix. TGA was used to test the NR/ChNCs composites for thermal stability, as shown in Figure 3.10.

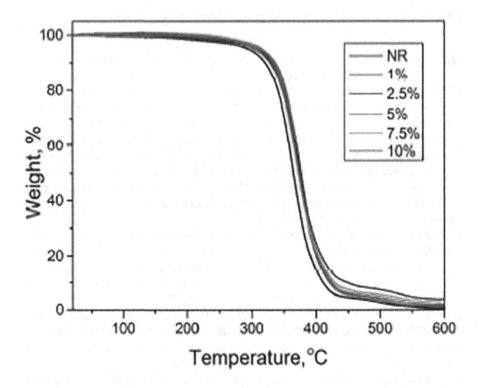

FIGURE 3.10 Composite NR/CNCs with varying ChNCs contents as determined by TA (Y. Liu et al., 2018).

In the whole temperature range, the weight loss curve of plain NR is lower than the curves of NR/ChNC compounds. The thermal stability of NR seems to be enhanced by using ChNCs. At 600°C, ChNC-infused NR/ChNC composites had a higher residue than NR/ChNCs without CNCs (Y. Liu et al., 2018). ChNCs depicted a positive impact on the thermomechanical properties of PLA composites. The tan delta peak shifted to an elevated temperature with rising ChNC concentration in the material. Both the ChNC in the nanoparticles and the arrangement of the ChNC with the polymer chains caused by solid-state drawing are predicted to have synergistic impacts on these improvements in thermal characteristics (Singh et al., 2018).

3.2.4.3   Electrical Properties of ChNC

By using chitosan-based nanocomposite films reinforced with ChNCs, several dielectric and conductivity tests were performed in a research performed by Salaberría et al. (2018). Under isothermal circumstances at a temperature of 298 K, the complex conductivity was measured at a frequency of 10^{-2} to 10^7 Hz. At lower frequencies, direct current (DC) showed a significant impact, but at higher frequencies, alternant current (AC) had a significant impact. As chitin concentration in the bionanocomposites grew, the AC conductivity of the bionanocomposites marginally improved until the CH/ChNC had the same chitin and chitosan fractions. However, a drop in DC conductivity was seen for chitin content in bionanocomposites as its value rose minimally. This pattern continued until the DC conductivity was the same for both CH and ChNC. The conductivity rocketed for chitin levels greater than around 50%, and the overall functioning of the heterogeneous system was impacted by the dispersal of the ChNCs in the CH matrix. These findings matched up with those from examining the dielectric relaxation spectra. Due to a phase transition break, the ChNCs clogged the chitosan matrix's conductivity pathways.

3.2.4.4   Optical Properties of ChNC

For self-assembling substances (films) to display colorful optical behaviors that originate from reflected wavelengths, ChNC can produce the chiral nematic crystalline phase in suspensions (Y. Wang et al., 2019). ChNCs were used in an investigation by Zhong et al. to see if they could be used in tandem with a current pliable packaging surface coating. Oxidized ChNCs were used in a water-based acrylic resin (WBAR). Flexible films of biaxially oriented polypropylene that contained ChNCs, either as an extra to the matrix or as a continuous network in a lamina, did not reduce the original packaging's optical transparency but instead preserved it. Contemporary packaging materials seek transparency as a desirable quality (Zhong et al., 2019). Additionally, in a work, acrylated epoxy soybean oil (AESO) dispersions with ChNCs as the only amalgamate were successfully created employing ChNC as a biobased aqueous coalescence. Increased toughness may be achieved by keeping optical transparency intact by including ChNCs in the treated layers (Cheikh et al., 2021).

The properties shown by ChNCs in a chitosan/O-ChNC matrix were found by Wu et al. ChNC nanofillers build up in the chitosan matrix-induced light diffusion or reduced optical transmission when ChNCs were incorporated into the chitosan matrix. However, the film obtained showed high optical transparency (C. Wu et al., 2019).

3.2.4.5 Magnetic Properties of ChNC

Very insufficient research has been performed so far regarding the magnetic properties of ChNCs. In a study, in a strong magnetic field of around 5 Tesla, a ChNC/polyacrylic acid composite film was created. The liquid crystal form of ChNC nanoparticles was stabilized via free-radical polymerization, which was photo-initiated. The long axis of these rod-shaped nanoparticles was found to form a 90° angle with the magnetic field (Morganti et al., 2006).

3.2.5 CHITIN NANOFIBER (CHNF)

Chitin has recently gained increased attention as a promising material for enzyme immobilization due to its biodegradability, biocompatibility, renewability, sustainability, high affinity for enzymes, accessibility of reactive groups for chemical treatment, and mechanical stability. Nanofibers of chitin have diameters of 15 nm and up to several micrometers in length (Gopalan Nair & Dufresne, 2003; J. Wu et al., 2014). Because of the protonation of amine units under suitably acidic circumstances, ChNFs have a total ionic charge on their surfaces which is positive in nature (Junkasem et al., 2006).

The different properties that ChNFs bear are discussed later.

3.2.5.1 Mechanical Properties of ChNF

The viscoelastic reaction of the ChNF network is rate-dependent so that displacement decreases with a higher applied load for a constant load. Although surface quality and inherent inhomogeneity are likely to account for the minor dispersion in the stress-strain data, the average result of the material beneath the punch diameter is sufficiently repeatable (Hassanzadeh et al., 2014). The axially symmetric flattened punch nano-indentation of viscoelastic layers and the extraction of material properties, most notably via creep tests, have been explored experimentally and theoretically (L. Cheng et al., 2000; Y.-T. Cheng & Yang, 2009). Chitin has a strain-dependent reaction, and when compressed to 40%, the chitin layer exhibits a functional elasticity in the region of around 4–5 GPa. The small strain modulus (about 2 GPa) is consistent with the bulk tensile tests mentioned previously, while the significant strain modulus is consistent with the single nanofiber modulus computed previously. A set of 600 s creep trials (all performed at different sites) with constant loads of one, two, and four mN demonstrate the characteristic steady rise in strain with time associated with a creeping process. Within each load step, the dispersion in data is caused by somewhat varied beginning strains and variable drift rates, which typically decrease with each consecutive indentation as the device stabilizes (Hassanzadeh et al., 2014). Again, in another work, SEM, phase diagram, and textural and rheological investigations were performed to determine the mechanical characteristics of ChNF-SPI complex gel. ChNFs aided in the development of gel by forming an initial framework. The electrostatic interactions made the SPI gel stiffer. The textural features of SPI-ChNF gels were evaluated using massive deformation rheology. Gels of varied protein concentrations were tested for hardness, chewiness, cohesiveness, gumminess, and resilience. The gel's hardness is augmented with the addition of ChNF at

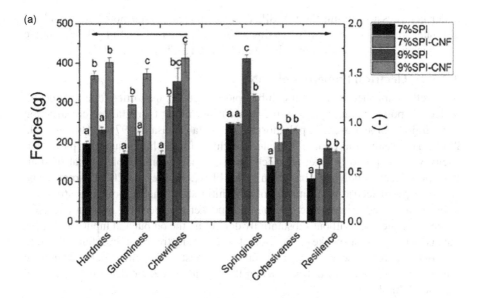

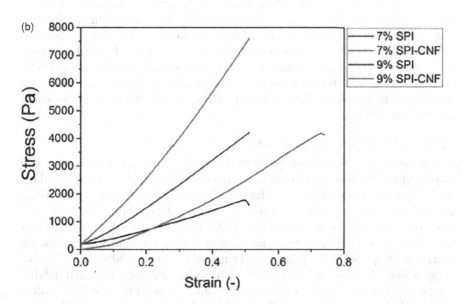

FIGURE 3.11 (a) Texture profile analysis (TPA) and (b) stress-strain curve of ChNF-SPI gels as a function of protein and ChNF concentration (Q.-H. Chen et al., 2021).

both protein concentrations, which is in line with the minor deformation rheological findings. There was a similar pattern in the findings of gumminess and chewiness. We can see the change in properties in Figure 3.11a. As a result of the ChNFs, fracture strain and compressive stress have increased. After combining with ChNFs at about 7% protein concentration, the fracture strain rose from around 0.5 to 0.7. Even under compression strain, the 9% SPI-ChNF complex gel did not shatter, and the ultimate stress value was nearly two times higher than that of the SPI gel. We can see the changes in stress and strain in Figure 3.11b (Q.-H. Chen et al., 2021).

3.2.5.2 Electrical Properties of ChNF

It is well established that chitin exhibits piezoelectric capabilities (due to inherent molecular polarization caused by the noncentrosymmetric crystalline construction of both β-chitin and α-chitin polymorphs) (Fukada & Sasaki, 1975; J. Huang et al., 2017). The easily controllable ferroelectric chitin film exhibits superior piezoelectricity when exposed to external mechanical pressure, comparable to that of standard piezoelectric polymers based on stand Fl (K. Kim et al., 2018). Notably, the adequate piezoelectric performance of the chitin sheet enables to demonstrate not only an increased printed microphone and speaker that perform over an extensive range of frequencies without substantially degrading the output and input resonances but also a transparent speaker composed of AgNWs electrodes on freestanding chitin film that also reproduces the original sound. The sustainable chitin polymer may be effectively destroyed within 8 days using the chitinase enzyme, leaving no harmful residues behind.

Jin et al. described the manufacturing and characterization of a transparent ChNF paper and illustrated its promise as a base material for foldable sustainable electronics by building an OLED gadget on the ChNF paper. The OLED device created on the ChNF substrate surface performed satisfactorily; with time, it beats a benchmark device produced on a more excellent artificial plastic film. We envision using ChNF translucent sheet as a structural foundation for foldable green electronics due to its reassuring macroscopic features, biofriendly traits, and abundance of the raw material (Jin et al., 2016).

3.2.5.3 Thermal Properties of ChNF

A simple blending procedure was performed to manufacture the nanocomposite with varying ChNF and BACNF concentrations (Hai et al., 2020). Hai et al. explored the nanocomposite's morphologies, mechanical characteristics, chemical interactions, thermal characteristics, water contact angles, and biodegradability. The nanocomposite is more thermally stable than pure BACNF. TGA was used to study the heat breakdown of the chitin-cellulose nanocomposite. Evaporation of water molecules in the nanocomposites is found at the first temperature range of 30°C–100°C. ChNF begins to degrade at approximately 230°C. The ChNF has superior thermal stability when contrasted to the BACNF (Hai et al., 2020). Compared to plain PVA, adding α-chitin whiskers resulted in a significant delay in the heat degradation of PVA nanocomposites (Sriupayo et al., 2005).

3.2.5.4 Optical Properties of ChNF

Naghdi et al. took benefit of the excellent features of CNF paper to construct bioactive, productive, transparent, flexible, and shrunken optical sensing bioplatforms by incorporating numerous plasmonic nanomaterials (silver and gold nanomaterials), photoluminescent nanoparticles, and colorimetric reagents (dithizone, curcumin, etc.) in ChNF paper's nanonetwork framework. Mobile phones were used in conjunction with the established ChNF paper-based sensing bioplatforms by Naghdi et al. to digitize the pictures of the acquired colorimetric/fluorometric readings. Indeed, by linking ChNF paper-based sensors to phone technologies and designing an IoNT model, these new biosensing stages can be used as economical, consumer, and transportable sensing or monitoring devices, particularly in evolving smart municipalities (Naghdi et al., 2020). Because of their unique nanostructure and superior physicochemical parameters, ChNF films are considered to be suitable for chiral separation membranes (Ifuku & Saimoto, 2012).

3.2.5.5 Magnetic Properties of ChNF

In W.-C Huang et al.'s work, a novel biodegradable magnetic ChNF composite (MCNC) was developed as a platform for enzyme immobilization.

Chymotrypsin (CT) was utilized as a prototype enzyme, and its immobilization in the MCNCs was successful. The stability of the immobilized CTs was much greater than that of the free CTs. Cross-linking of CTs significantly increased the load-bearing capacity of the MCNCs. Additionally, the synthetic biocatalytic system demonstrated high recyclability. These findings indicated that the MCNC has considerable potential to be used as a substrate for the immobilization of a variety of enzymes in biomedical applications (W.-C. Huang et al., 2018).

3.2.6 STARCH NANOCRYSTAL (SNC)

Starch is a biodegradable and renewable polymer that many plants manufacture for energy storage. Human beings and their forefathers have always consumed starchy foods. Starch grains were recently (2010) detected in grinding stones from 30,000-year-old Europe (Italy, the Czech Republic, and Russia) (Healy & Healy, 1999). Beans, maize, potatoes, rice, wheat, and sago are just a few of the numerous crops that provide starch (Hoque et al., 2013). Cereal kernels account for most of the composition, at about 60%. The fundamental advantage of starch is that it can be readily isolated from other parts and the granules have varying forms, dimensions, topologies, and chemical compositions based on their various botanical sources (Saha et al., 2020). The practical utilization of starch derivatives (i.e., nonfood applications) began when Egyptians glued papyrus strips together with a wheat starch adhesive (Revedin et al., 2010).

Nowadays, the main applications of starch have remained relatively constant, with approximately 60% being utilized for foods and 40% for commercial applications (Huber & BeMiller, 2010). SNCs are the crystalline structures that remain after the amorphous structures of starch granules are destroyed by acid hydrolysis. SNCs, starch crystallites, microcrystalline starch, and hydrolyzed starches are all terms that refer to the crystalline portion of starch that is formed during hydrolysis. They differ in the amount to which they have been hydrolyzed (from the most to the least). They must be differentiated from starch nanoparticles, of which they are subtypes and capable of becoming amorphous (Le Corre & Angellier-Coussy, 2014). The following review focuses on SNC, a kind of SNP possessing crystalline properties and a platelet-like shape. SNC can be extracted from various starch botanical sources, resulting in a wide variety of amylose concentrations, form, suspension viscosity, surface reactivity, and thermal resistance. So far, the most frequently used method for obtaining SNC is moderate acid hydrolysis of native granular starch's amorphous components. Until now, alternative procedures have resulted in significantly lower yields. Since the inception of SNC research, the primary objective has been to utilize them as reinforcement for polymer matrices. Due to the high reactivity of starch, the surface of SNCs can be changed via grafting or cross-linking, resulting in a more dispersible material in the polymer matrix (Le Corre & Angellier-Coussy, 2014).

The different properties that SNCs bear are discussed later.

3.2.6.1 Mechanical Properties of SNC

Interfacial interactions between the polymer and the filler were crucial in increasing the mechanical properties of nanocomposites due to the nanoparticle's unusually high surface areas (N.-J. Huang et al., 2017). When nanoclay is added to nanocomposite layers, the tensile characteristics of the material and the elastic modulus rise, while the elasticity at break decreases in comparison to the starch/PVA mixture. These improvements in mechanical qualities corroborated prior findings (N.-J. Huang et al., 2017). They demonstrated that increasing the amount of Na-MMT in TPS/Na-MMT nanocomposites from 0% to 30% by weight increased the tensile strength to 27.34 MPa from around 6 MPa and decreased the elongation at break from around 85% to 18%. Additionally, Wetzel et al. established that for starch/PVA/Na-MMT nanostructures to have enhanced mechanical characteristics, H-bonds among the nanoparticles and polymer chains must be formed first (Wetzel et al., 2003). Additionally, Ali et al. demonstrated that adding Na-MMT with a proportion of 10–100 enhanced the interfacial contacts among nanoclays and polymer chains through hydrogen bondings (Ali et al., 2011); besides the physical intertwining of polymer chains generated by nanoclays, the important hydrogen bonding contact among nanoclays and polymer matrix increased the mechanical properties. Compared to neat PLA, with the addition of SNC and Ac-SNC (acetylated SNC), all nanocomposites tensile properties and elastic modulus were enhanced. Tensile strength measures a material's ability to withstand stress (Yu et al., 2008). As the proportion of nanofiller was increased to 3% by weight (Yan et al., 2013), the reinforcement effect became more pronounced. The PLA-Ac-SNC-3 wt% had the highest tensile strength, which may result from the boosted interfacial adhesion and dissemination of Ac-SNC inside the PLA matrix (Xu, Chen, et al., 2018; Yin et al., 2015; Lin et al., 2011; Arrieta et al., 2014). When SNC and Ac-SNC were added to neat PLA, the pattern in elongation at break values shifted in the other direction, with values dropping. The reduction of value was more noticeable in PLA-SNC nanocomposite; this indicated that the nanocomposite had increased brittleness due to the existence of rigid SNC (Fortunati et al., 2015; Robles et al., 2015). However, there was a growth in PLA-Ac-SNC nanocomposites compared to PLA-SNC nanocomposites. PLA nanocomposite containing untreated and treated CNCs (surfactant) yielded similar results (Fortunati et al., 2015). This phenomenon highlighted the beneficial effect of acetylation on nanocomposites. Tensile testing validated the plasticizing effect induced by surface treatment of SNC nanocomposites, and the results supplemented the nanocomposite's morphology (Fortunati et al., 2015). Overall, this suggested that depending on the concentration and alteration of the surface of SNC, the ultimate properties of nanocomposites may be altered to meet the specifications of the intended application. As a result, biodegradable nanocomposites are formed. Additionally, all packed nanocomposites outperformed their unfilled counterparts or tidy PLA (Bondeson & Oksman, 2007).

3.2.6.2 Thermal Properties of SNC

The addition of nanoparticles to the mixture augmented the thermal stability of nanostructured materials in general (Sessini et al., 2019). Nanoclay platelets worked as gas-permeable barriers, and the polymeric matrix's thermal stability improved the quality of this barrier effect alongside its inorganic nature. Additionally, physico-chemical dalliances and significant hydrogen bonding between nanoclay and poly-mer chain layers hindered polymer chain mobility (Leszczynska & Pielichowski, 2008). Consequently, the temperature at which particular nanocomposites began to degrade was more significant than the temperature at which the pure mix began to degrade. These findings agreed with those of Wang et al., who discovered that com-bining cellulose nanowhiskers and SNCs in an aqueous polyurethane matrix could improve the thermal stability of nanocomposites. However, the BSN5 nanocomposite with 5 % weight SNCs had the lowest thermal stability due to SNC agglomeration. These findings were consistent with the deficient mechanical properties obtained for BSN5. At 500°C, in general, the addition of nanofillers boosted residual mass. The residual mass for samples BCL5, BCL3, and BCL1 was around 17%, 15%, and 14.7%, respectively. For the blend that was pure, the amount was close to 15%. In gen-eral, the addition of nanofillers increased residual mass. Two variables most likely contributed to this increase in residual mass: (i) nanofillers mass and (ii) an elevation in the interfacial area between fillers and polymers as a consequence of nanofillers being uniformly dispersed throughout the polymer matrix leading to more effective interactions between them.

From the work of Takkalkar et al., we can see the TGA profiles of unprocessed waxed maize starch and SNC. Both raw waxy maize starch and SNC went through three significant stages of degradation (Takkalkar et al., 2019). Between 30°C and 110°C, the first stage is evident, which can be attributable to moisture loss through dehydration/evaporation. As shown by a lower weight loss % for SNC, SNC appears to have a lower moisture content than refined waxed maize starch. It is predicted that loose nanoparticle powder is formed when moisture is removed from the SNC dur-ing freeze-drying. At 254°C and 275°C, SNC and raw waxy maize starch undergo thermal breakdown, causing carbohydrate polymer chain breakdown. The difference in breakdown behavior between SNC and raw waxy maize starch can be associated with the presence of sulfate anion of SNC, which acts as a catalyst for polymer chain degradation. Once all sulfate ions have been destroyed at 200°C–300°C, the rest of the unprocessed waxy maize starch trend is the same (D. D. Jiang et al., 1999). LeCorre et al. detected a similar trend in the depolymerization of unprocessed waxed maize starch and SNC (LeCorre et al., 2012).

Furthermore, T. Jiang et al. contended that the humidity content of the samples during decomposition does not affect the degradation temperature. The temperature stays unaffected because the first phase of degradation eliminates all moisture con-tents before reaching the second stage. The third stage is the total burning of carbo-hydrates to ash, which happens at temperatures between 380°C and 400°C. Inside the experimental margins of error, both SNC and unprocessed waxed maize starch have the same combustion temperature (T. Jiang et al., 2020).

3.2.6.3 Electrical Properties of SNC

The abrasion-resistant properties of rubber composites can be significantly improved by adding benzyl SNCs to silica/NR composites (C. Wang et al., 2014). It can also reduce rubber tire rolling resistance because of its excellent bonding strength and other essential characteristics. Finally, the low substitution benzyl SNCs were anticipated to be used as an active biopolymer ingredient in polymer composites.

3.2.6.4 Optical Properties of SNC

By measuring transmittance from 400 to 800 nm, the optical transparency of the nanocomposite film was determined (Bel Haaj et al., 2016). When measured in the visible light spectrum (400–800 nm), the neat PBMA matrix had a light transmittance of over 90%. Adding starch nanofillers results in a loss of transparency. SNCs, on the other hand, saw a more significant decline in transmittance than SNPs. In contrast, even in the presence of 2% nanofiller, a significant decrease in transmittance was observed for SNCs. Because of the particle's light scattering, which is highly dependent on particle size, there will be a decrease in transparency when nanoparticles are incorporated into a polymer matrix. To prevent Rayleigh scattering from reducing the transmitted light's intensity, the maximum diameter of a nanoparticle is typically 40–50 nm (Althues et al., 2007). The nanofiller's dispersion will strongly influence a nanocomposite film's translucency in the host matrix. Nanoparticle accumulation during film processing increases scattering and reduces film transparency by definition. Since SNCs cluster into a dense network throughout the film development process, the reduced transmittance observed in the existence of SNCs is perhaps most probably attributable to hydrogen-bonding-mediated self-interaction. Another reason for the reduced opacity of SNC-based nanocomposite is the size distribution discrepancy between SNC and SNP. As indicated in Figure 3.12, substantial light scattering was created by adding nanoparticles greater than 60 nm to SNC.

3.2.6.5 Magnetic Properties of SNC

SNC hydrolyzed in sulfuric acid and phosphorylated SNC have more ordered regions and thus exhibit greater crystallinities than SNC treated enzymatically (Zhao et al., 2021). By graft copolymerizing SNCs with styrene, an amphiphilic SNC copolymer was formed (Song et al., 2008). The amorphous portion of starch was hydrolyzed and eliminated during the hydrolysis process, producing crystalline microparticles of 50 nm termed SNCs. The SEM pattern indicated that the amphiphilic SNCs had particle dimensions of around 80–100 nm. The amphiphilic SNCs spread efficiently in polar and nonpolar liquids (Song et al., 2008). Its exceptional amphiphilic capabilities are due to the hydrophobic polystyrene side chains and the hydrophilic starch core displaying different conformational changes in nonpolar and polar liquids. The amphiphilic starch-g-polystyrene NCs synthesized in Song et al.'s experiment can be employed as fillers in polymer matrixes, using NR, PLA, and polycaprolactone to make nanocomposites with a variety of exclusive properties (Song et al., 2008).

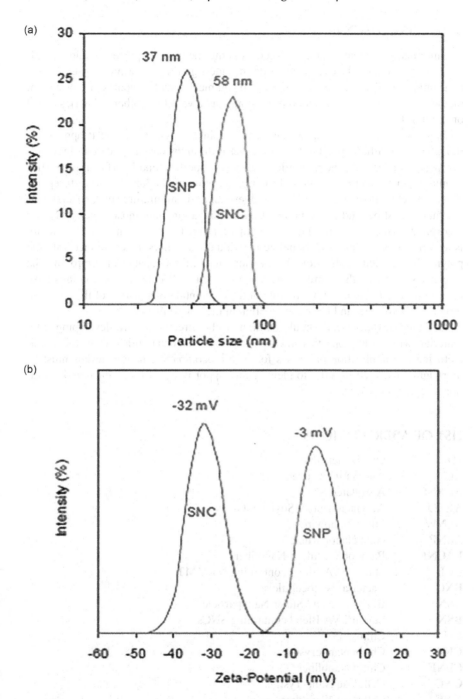

FIGURE 3.12 (a) Dimensions of a given particle and (b) ζ-potential distribution of SNP and SNC suspension from waxy maize (Haj & Thielemans, 2016).

3.3 CONCLUSION

Polymer nanocomposites have emerged as being the most empowering and appealing family of materials science in recent decades, gaining widespread attention for their unique ability to improve the physical and mechanical attributes of fabrication, medical sciences, cosmetics, food packaging, and a variety of other industries based on composite.

Nanocomposites are nanoparticles encased in a polymer matrix. It appears as though the nanofillers specific structure governs the envisioned applications for nanocomposites. Therefore, the nanofiller's shapes, properties, and interfacial properties should be tuned via modification of their preparation methodologies or postpreparation treatments. If someone wishes to achieve an optimal mixture state, special consideration must be paid to the preparation, modification, and, most importantly, the incorporation techniques used in the polymer matrix. The remaining major difficulty is to carefully select and functionalize individual nanofillers in accordance with the polymer's intended application. The various nanofillers capable of improving the properties of materials or mitigating some of their well-known shortcomings have been introduced, emphasizing the role of the elemental composition of the nanomaterials, their volume, and their composition in the materials.

However, advancements should continue to be directed toward developing safer materials with novel properties and superior functional capabilities. Thus, the manufacturing and application processes for such biocompatible nanomaterials must be thoroughly analyzed in order to identify any opportunities for technical and operational segment optimization.

LIST OF ABBREVIATIONS

1D	One Dimensional
AC	Alternating Current
Ac-SNC	Acetylated SNC
AESO	Acrylated Epoxy Soybean Oil
AgNW	Silver Nanowire
AuNP	Gold Nanoparticle
BACNF	Bamboo Cellulose Nanofiber
BCL	Starch/PVA Blend containing Na-MMT
BNC	Bacterial Nanocellulose
BSN	Biosynthesized Silver Nanoparticle
BSN	Starch/PVA Blend containing SNCs
CH	Chitosan
ChNC	Chitin Nanocrystal
ChNF	Chitin Nanofiber
CNC	Cellulose Nanocrystals
CNF	Cellulose Nanofiber
CP	Conductive Polymer
CT	Chymotrypsin
DC	Direct Current

ENR	Epoxidized Natural Rubber
FE-SEM	Field Emission Scanning Electron Microscopes
FTCS	(tridecafluoro-1,1,2,2-tetrahydrooctyl)-trichlorosilane
HSA	Human Serum Albumin
IONC	Iron Oxide
IONP	Iron Nanoparticle
IoNT	Internet of Nanothings
L-CNC	Lignin-Coated CNC
LCNF	Lignin-Coated Cellulose Nanofiber
LED	Light-Emitting diode
MB	Methylene Blue
MBNC	Magnetic Bacterial Nanocellulose
MCNC	Magnetic Chitin Nanofiber Composite
MGCNC	Magnetic Cellulose Nanocrystal
MS	Saturation Magnetization
Na-MMT	Sodium Montmorillonite
NC	Nanocellulose
N-CMC	Carboxymethyl Cellulose
nm	Nanometer
NP	Nanoparticle
NR	Natural Rubber
NT	Nanotube
O-CNC	Oxidized Chitin Nanocrystal
OEO	Oregano Essential Oil
OLED	Organic Light-Emitting Diode
PBI	Penetrating Brain Injury
PBMA	Polybutyl Methacrylate
PCC	Precipitated Calcium Carbonate
PDA	Polydopamine
PEGDA	Polyethylene glycol diacrylate
PEO	Polyethylene Oxide
PIC	Polyion Complex
PLA	Polylactic Acid
PMMA	Polymethyl Methacrylate
PPy	Polypyrrole
PVA	Polyvinyl Alcohol
PVDF	Polyvinylidene Fluoride
PVOH	Polyvinyl Alcohol
RFID	Radio Frequency Identification
RGO	Reduced Graphene Oxide
SNC	Starch Nanocrystal
SNP	Starch Nanoparticle
SPI	Soy Protein Isolate
TEM	Transmission Electron Microscope
TENG	Triboelectric-Based Nanogenerator
TGA	Thermogravimetric Analysis

TPA	Texture Profile Analysis
TPS	Thermoplastic Starch
WBAR	Water-Based Acrylic Resin
WPU/CNT	Waterborne Polyurethane/Carbon Nanotube
xSBR	Cross-linked Styrene-Butadiene Rubber

NOMENCLATURE

Au	Gold
CO_2	Carbon Dioxide
$CoFe_2O_4$	Cobalt Ferrite
Cr	Chromium
Fe_3O_4	Magnetite
Fl	Fluorine
H	Hydrogen
MoS_2	Molybdenum Disulfide
OH	Hydroxyl
SiO_2	Silicon dioxide

REFERENCES

Abdollahi, M., Alboofetileh, M., Behrooz, R., Rezaei, M., & Miraki, R. (2013). Reducing water sensitivity of alginate bio-nanocomposite film using cellulose nanoparticles. *International Journal of Biological Macromolecules*, *54*, 166–173. https://doi.org/10.1016/j.ijbiomac.2012.12.016

Abral, H., Putra, G., Asrofi, M., Park, J.-W., & Kim, H.-J. (2017). Effect of vibration duration of high ultrasound applied to bio-composite while gelatinized on its properties. *Ultrasonics Sonochemistry*, *40*. https://doi.org/10.1016/j.ultsonch.2017.08.019

Alam, M. K., Islam, M. T., Mina, M. F., & Gafur, M. A. (2014). Structural, mechanical, thermal, and electrical properties of carbon black reinforced polyester resin composites. *Journal of Applied Polymer Science*, *131*(13), n/a-n/a. https://doi.org/10.1002/app.40421

Ali, S. S., Tang, X., Alavi, S., & Faubion, J. (2011). Structure and physical properties of starch/poly vinyl alcohol/sodium montmorillonite nanocomposite films. *Journal of Agricultural and Food Chemistry*, *59*(23), 12384–12395. https://doi.org/10.1021/jf201119v

Althues, H., Henle, J., & Kaskel, S. (2007). Functional inorganic nanofillers for transparent polymers. *Chemical Society Reviews*, *36*(9), 1454–1465. https://doi.org/10.1039/B608177K

Amiralian, N., Mustapic, M., Hossain, M. S. A., Wang, C., Konarova, M., Tang, J., Na, J., Khan, A., & Rowan, A. (2020). Magnetic nanocellulose: A potential material for removal of dye from water. *Journal of Hazardous Materials*, *394*, 122571. https://doi.org/10.1016/j.jhazmat.2020.122571

Arias, S. L., Shetty, A. R., Senpan, A., Echeverry-Rendón, M., Reece, L. M., & Allain, J. P. (2016). Fabrication of a functionalized magnetic bacterial nanocellulose with iron oxide nanoparticles. *Journal of Visualized Experiments*, *111*, 52951. https://doi.org/10.3791/52951

Arrieta, M. P., Fortunati, E., Dominici, F., Rayón, E., López, J., & Kenny, J. M. (2014). PLA-PHB/cellulose based films: Mechanical, barrier and disintegration properties. *Polymer Degradation and Stability*, *107*, 139–149. https://doi.org/10.1016/j.polymdegradstab.2014.05.010

Asefa, T. (2012). Chiral nematic mesoporous carbons from self-assembled nanocrystalline cellulose. *Angewandte Chemie (International Ed. in English)*, *51*, 2008–2010. https://doi.org/10.1002/anie.201107332

Azeredo, H. M. C., Rosa, M. F., & Mattoso, L. H. C. (2017). Nanocellulose in bio-based food packaging applications. *Industrial Crops and Products*, *97*, 664–671. https://doi.org/10.1016/j.indcrop.2016.03.013

Azizi Samir, M. A. S., Alloin, F., Sanchez, J.-Y., El Kissi, N., & Dufresne, A. (2004). Preparation of cellulose whiskers reinforced nanocomposites from an organic medium suspension. *Macromolecules*, *37*(4), 1386–1393. https://doi.org/10.1021/ma030532a

Baba, T. (2008). Slow light in photonic crystals. *Nature Photonics*, *2*(8), 465–473. https://doi.org/10.1038/nphoton.2008.146

Balakrishnan, P., Sreekala, M. S., Kunaver, M., Huskić, M., & Thomas, S. (2017). Morphology, transport characteristics and viscoelastic polymer chain confinement in nanocomposites based on thermoplastic potato starch and cellulose nanofibers from pineapple leaf. *Carbohydrate Polymers*, *169*. https://doi.org/10.1016/j.carbpol.2017.04.017

Bardet, R., Belgacem, N., & Bras, J. (2015). Flexibility and color monitoring of cellulose nanocrystal iridescent solid films using anionic or neutral polymers. *ACS Applied Materials & Interfaces*, *7*(7), 4010–4018. https://doi.org/10.1021/am506786t

Barría, J. C., Vázquez, A., Pereira, J.-M., & Manzanal, D. (2021). Effect of bacterial nanocellulose on the fresh and hardened states of oil well cement. *Journal of Petroleum Science and Engineering*, *199*, 108259. https://doi.org/10.1016/j.petrol.2020.108259

Bel Haaj, S., Thielemans, W., Magnin, A., & Boufi, S. (2016). Starch nanocrystals and starch nanoparticles from waxy maize as nanoreinforcement: A comparative study. *Carbohydrate Polymers*, *143*, 310–317. https://doi.org/10.1016/j.carbpol.2016.01.061

Biswas, M. C., Jony, B., Nandy, P. K., Chowdhury, R. A., Halder, S., Kumar, D., Ramakrishna, S., Hassan, M., Ahsan, M. A., Hoque, M. E., & Imam, M. A. (2022). Recent advancement of biopolymers and their potential biomedical applications. *Journal of Polymers and the Environment*, *30*(1), 51–74. https://doi.org/10.1007/s10924-021-02199-y

Bondeson, D., & Oksman, K. (2007). Polylactic acid/cellulose whisker nanocomposites modified by polyvinyl alcohol. *Composites Part A: Applied Science and Manufacturing*, *38*(12), 2486–2492. https://doi.org/10.1016/j.compositesa.2007.08.001

Cai, J., Chen, J., Zhang, Q., Lei, M., He, J., Xiao, A., Ma, C., Li, S., & Xiong, H. (2016). Well-aligned cellulose nanofiber-reinforced polyvinyl alcohol composite film: Mechanical and optical properties. *Carbohydrate Polymers*, *140*, 238–245. https://doi.org/10.1016/j.carbpol.2015.12.039

Celebi, H., & Kurt, A. (2015). Effects of processing on the properties of chitosan/cellulose nanocrystal films. *Carbohydrate Polymers*, *133*, 284–293. https://doi.org/10.1016/j.carbpol.2015.07.007

Cheikh, F., Ben Mabrouk, A., Lancelon-Pin, C., & Putaux, J.-L. (2021). Honeycomb organization of chitin nanocrystals (ChNCs) in nanocomposite films of UV-cured waterborne acrylated epoxidized soybean oil emulsified with ChNCs. *Biomacromolecules*, *22*. https://doi.org/10.1021/acs.biomac.1c00612

Chen, J., Defeng, W., Tam, K., Pan, K., & Zheng, Z. (2016). Effect of surface modification of cellulose nanocrystal on nonisothermal crystallization of poly(β-hydroxybutyrate) composites. *Carbohydrate Polymers*, *157*. https://doi.org/10.1016/j.carbpol.2016.11.071

Chen, L., Berry, R., & Tam, K. (2014). Synthesis of β-cyclodextrin-modified cellulose nanocrystals (CNCs)@Fe$_3$O$_4$@SiO$_2$ superparamagnetic nanorods. *ACS Sustainable Chemistry & Engineering*, *2*, 951–958. https://doi.org/10.1021/sc400540f

Chen, L., Sharma, S., Darienzo, R. E., & Tannenbaum, R. (2020). Decoration of cellulose nanocrystals with iron oxide nanoparticles. *Materials Research Express*, *7*(5), 055003. https://doi.org/10.1088/2053-1591/ab8a82

Chen, L.-F., Huang, Z.-H., Liang, H.-W., Guan, Q.-F., & Yu, S.-H. (2013). Bacterial-cellulose-derived carbon nanofiber@MnO2 and nitrogen-doped carbon nanofiber electrode materials: An asymmetric supercapacitor with high energy and power density. *Advanced Materials (Deerfield Beach, Fla.)*, *25*. https://doi.org/10.1002/adma.201204949

Chen, Q.-H., Li, X.-Y., Huang, C.-L., Liu, P., Zeng, Q.-Z., Yang, X.-Q., & Yuan, Y. (2021). Development and mechanical properties of soy protein isolate-chitin nanofibers complex gel: The role of high-pressure homogenization. *LWT*, *150*, 112090. https://doi.org/10.1016/j.lwt.2021.112090

Chen, X., Yuan, F., Zhang, H., Huang, Y., Yang, J., & Sun, D. (2016). Recent approaches and future prospects of bacterial cellulose-based electroconductive materials. *Journal of Materials Science*, *51*(12), 5573–5588. https://doi.org/10.1007/s10853-016-9899-2

Cheng, L., Xia, X., Yu, W., Scriven, L. E., & Gerberich, W. W. (2000). Flat-punch indentation of viscoelastic material. *Journal of Polymer Science Part B: Polymer Physics*, *38*(1), 10–22. https://doi.org/10.1002/(SICI)1099-0488(20000101)38:1<10::AID-POLB2>3.0.CO;2-6

Cheng, Y.-T., & Yang, F. (2009). Obtaining shear relaxation modulus and creep compliance of linear viscoelastic materials from instrumented indentation using axisymmetric indenters of power-law profiles. *Journal of Materials Research*, *24*(10), 3013–3017. https://doi.org/10.1557/jmr.2009.0365

De Vries, Hl. (1951). Rotatory power and other optical properties of certain liquid crystals. *Acta Crystallographica*, *4*(3), 219–226. https://doi.org/10.1107/S0365110X51000751

Dhar, P., Kumar, A., & Katiyar, V. (2016). Magnetic cellulose nanocrystal based anisotropic polylactic acid nanocomposite films: Influence on electrical, magnetic, thermal, and mechanical properties. *ACS Applied Materials & Interfaces*, *8*(28), 18393–18409. https://doi.org/10.1021/acsami.6b02828

Drake, J. M., & Genack, A. Z. (1989). Observation of nonclassical optical diffusion. *Physical Review Letters*, *63*(3), 259–262. https://doi.org/10.1103/PhysRevLett.63.259

Duan, C., Cheng, Z., Wang, B., Zeng, J., Xu, J., Li, J., Gao, W., & Chen, K. (2021). Chiral photonic liquid crystal films derived from cellulose nanocrystals. *Small*, *17*(30), 2007306. https://doi.org/10.1002/smll.202007306

Dumanli, A. G., van der Kooij, H. M., Kamita, G., Reisner, E., Baumberg, J. J., Steiner, U., & Vignolini, S. (2014). Digital color in cellulose nanocrystal films. *ACS Applied Materials & Interfaces*, *6*(15), 12302–12306. https://doi.org/10.1021/am501995e

Echegaray, M., Mondragon, G., Martin, L., González, A., Peña, C., & Arbelaiz, A. (2016). Physicochemical and mechanical properties of gelatin reinforced with nanocellulose and montmorillonite. *Journal of Renewable Materials*, *4*. https://doi.org/10.7569/JRM.2016.634106

El Miri, N., Abdelouahdi, K., Barakat, A., Zahouily, M., Fihri, A., Solhy, A., & El Achaby, M. (2015). Bio-nanocomposite films reinforced with cellulose nanocrystals: Rheology of film-forming solutions, transparency, water vapor barrier and tensile properties of films. *Carbohydrate Polymers*, *129*, 156–167. https://doi.org/10.1016/j.carbpol.2015.04.051

El-Hoseny, S., Basmaji, P., Molina de Olyveira, G., Costa, L. M., Abdulwahid, M., Oliveira, J., & Francozo, G. (2015). Natural ECM-bacterial cellulose wound healing—Dubai study. *Journal of Biomaterials and Nanobiotechnology*, *06*, 237–246. https://doi.org/10.4236/jbnb.2015.64022

Eskilson, O., Lindström, S. B., Sepulveda, B., Shahjamali, M. M., Güell-Grau, P., Sivlér, P., Skog, M., Aronsson, C., Björk, E. M., Nyberg, N., Khalaf, H., Bengtsson, T., James, J., Ericson, M. B., Martinsson, E., Selegård, R., & Aili, D. (2020). Self-assembly of mechanoplasmonic bacterial cellulose–metal nanoparticle composites. *Advanced Functional Materials*, *30*(40), 2004766. https://doi.org/10.1002/adfm.202004766

Fashandi, H., Abolhasani, M. M., Sandoghdar, P., Zohdi, N., Li, Q., & Naebe, M. (2016). Morphological changes towards enhancing piezoelectric properties of PVDF electrical generators using cellulose nanocrystals. *Cellulose*, *23*(6), 3625–3637. https://doi.org/10.1007/s10570-016-1070-3

Fernandes, S. C. M., Oliveira, L., Freire, C. S. R., Silvestre, A. J. D., Neto, C. P., Gandini, A., & Desbriéres, J. (2009). Novel transparent nanocomposite films based on chitosan and bacterial cellulose. *Green Chemistry, 11*(12), 2023. https://doi.org/10.1039/b919112g

Fernández-Marín, R., Labidi, J., Andrés, Ma. M., & Fernandes, S. C. M. (2020). Using α-chitin nanocrystals to improve the final properties of poly (vinyl alcohol) films with Origanum vulgare essential oil. *Polymer Degradation and Stability, 179*, 109227. https://doi.org/10.1016/j.polymdegradstab.2020.109227

Ferreira, F., Franceschi, W., Menezes, B., Brito, F. S., Lozano, K., Coutinho, A. R., Cividanes, L., & Thim, G. (2017). Dodecylamine functionalization of carbon nanotubes to improve dispersion, thermal and mechanical properties of polyethylene based nanocomposites. *Applied Surface Science, 410*. https://doi.org/10.1016/j.apsusc.2017.03.098

Ferreira, F. V., Dufresne, A., Pinheiro, I. F., Souza, D. H. S., Gouveia, R. F., Mei, L. H. I., & Lona, L. M. F. (2018). How do cellulose nanocrystals affect the overall properties of biodegradable polymer nanocomposites: A comprehensive review. *European Polymer Journal, 108*, 274–285. https://doi.org/10.1016/j.eurpolymj.2018.08.045

Ferreira-Neto, E., Ullah, S., Silva, T., Domeneguetti, R., Perissinotto, A., De Vicente, F., Rodrigues Filho, U. P., & Ribeiro, S. (2020). Bacterial nanocellulose/MoS2 hybrid aerogels as bifunctional adsorbent/photocatalyst membranes for in-flow water decontamination. *ACS Applied Materials & Interfaces, XXXX*. https://doi.org/10.1021/acsami.0c14137

Fortunati, E., Luzi, F., Puglia, D., Petrucci, R., Kenny, J. M., & Torre, L. (2015). Processing of PLA nanocomposites with cellulose nanocrystals extracted from Posidonia oceanica waste: Innovative reuse of coastal plant. *Industrial Crops and Products, 67*, 439–447. https://doi.org/10.1016/j.indcrop.2015.01.075

Fortunato, E., Gaspar, D., Duarte, P., Pereira, L., Águas, H., Vicente, A., ... & Martins, R. (2016). Optoelectronic devices from bacterial nanocellulose. In *Bacterial Nanocellulose* (pp. 179–197). Elsevier.

Fujisawa, S., Togawa, E., & Kimura, S. (2018). Large specific surface area and rigid network of nanocellulose govern the thermal stability of polymers: Mechanisms of enhanced thermomechanical properties for nanocellulose/PMMA nanocomposite. *Materials Today Communications, 16*. https://doi.org/10.1016/j.mtcomm.2018.05.002

Fukada, E., & Sasaki, S. (1975). Piezoelectricity of α-chitin. *Journal of Polymer Science: Polymer Physics Edition, 13*(9), 1845–1847. https://doi.org/10.1002/pol.1975.180130916

Furumi, S., Yokoyama, S., Otomo, A., & Mashiko, S. (2003). Electrical control of the structure and lasing in chiral photonic band-gap liquid crystals. *Applied Physics Letters, 82*(1), 16–18. https://doi.org/10.1063/1.1534613

Ghanbari, A., Tabarsa, T., Ashori, A., Shakeri, A., & Mashkour, M. (2018). Thermoplastic starch foamed composites reinforced with cellulose nanofibers: Thermal and mechanical properties. *Carbohydrate Polymers, 197*, 305–311. https://doi.org/10.1016/j.carbpol.2018.06.017

Glenn, G., Imam, S., & Orts, W. J. (2011). Starch-based foam composite materials: Processing and bioproducts. *MRS Bulletin, 36*, 696–702. https://doi.org/10.1557/mrs.2011.205

Gopalan Nair, K., & Dufresne, A. (2003). Crab shell chitin whisker reinforced natural rubber nanocomposites. 1. Processing and swelling behavior. *Biomacromolecules, 4*(3), 657–665. https://doi.org/10.1021/bm020127b

Gu, M., Jiang, C., Liu, D., Prempeh, N., & Smalyukh, I. I. (2016). Cellulose nanocrystal/poly(ethylene glycol) composite as an iridescent coating on polymer substrates: Structure-color and interface adhesion. *ACS Applied Materials & Interfaces, 8*(47), 32565–32573. https://doi.org/10.1021/acsami.6b12044

Gutierrez, J., Tercjak, A., Algar, I., Retegi, A., & Mondragon, I. (2012). Conductive properties of TiO2/bacterial cellulose hybrid fibres. *Journal of Colloid and Interface Science, 377*, 88–93. https://doi.org/10.1016/j.jcis.2012.03.075

Hai, L., Choi, E. S., Zhai, L., Panicker, P. S., & Kim, J. (2020). Green nanocomposite made with chitin and bamboo nanofibers and its mechanical, thermal and biodegradable properties for food packaging. *International Journal of Biological Macromolecules, 144*, 491–499. https://doi.org/10.1016/j.ijbiomac.2019.12.124

Haj, S., & Thielemans, W. (2016). Starch nanocrystals and starch nanoparticles from waxy maize as nanoreinforcement: A comparative study. *Carbohydrate Polymers, 143*. https://doi.org/10.1016/j.carbpol.2016.01.061

Hassanzadeh, P., Sun, W., de Silva, J. P., Jin, J., Makhnejia, K., Cross, G. L. W., & Rolandi, M. (2014). Mechanical properties of self-assembled chitin nanofiber networks. *Journal of Materials Chemistry B, 2*(17), 2461–2466. https://doi.org/10.1039/C3TB21550D

Healy, J. F. (1999). *Pliny the Elder on Science and Technology*. Oxford University Press.

Hebeish, A., El-Rafie, M., Abdel-Mohdy, F., Abdel-Halim, E., & Emam, H. (2010). Carboxymethyl cellulose for green synthesis and stabilization nanoparticles. *Carbohydrate Polymers, 82*, 933–941. https://doi.org/10.1016/j.carbpol.2010.06.020

Hernández-Flores, J. A., Morales-Cepeda, A. B., Castro-Guerrero, C. F., Delgado-Arroyo, F., Díaz-Guillén, M. R., de la Cruz-Soto, J., Magallón-Cacho, L., & León-Silva, U. (2020). Morphological and electrical properties of nanocellulose compounds and its application on capacitor assembly. *International Journal of Polymer Science, 2020*, 1–14. https://doi.org/10.1155/2020/1891064

Herrera, N., Salaberria, A. M., Mathew, A. P., & Oksman, K. (2016). Plasticized polylactic acid nanocomposite films with cellulose and chitin nanocrystals prepared using extrusion and compression molding with two cooling rates: Effects on mechanical, thermal and optical properties. *Composites Part A: Applied Science and Manufacturing, 83*, 89–97. https://doi.org/10.1016/j.compositesa.2015.05.024

Hoque, M. E., Ye, T. J., Yong, L. C., & Mohd Dahlan, K. (2013). Sago starch-mixed low-density polyethylene biodegradable polymer: Synthesis and characterization. *Journal of Materials, 2013*, 1–7. https://doi.org/10.1155/2013/365380

Hou, M., Xu, M., & Li, B. (n.d.). *Enhanced electrical conductivity of cellulose nanofiber/graphene composite paper with a sandwich structure*, 25.

Huang, C.-L. (2020). A study of the optical properties and fabrication of coatings made of three-dimensional photonic glass. *Coatings, 10*(8), 781. https://doi.org/10.3390/coatings10080781

Huang, J., Li, D., Zhao, M., Ke, H., Mensah, A., Tian, X., & Wei, Q. (2019). Flexible electrically conductive biomass-based aerogels for piezoresistive pressure/strain sensors. *The Chemical Engineering Journal*. https://doi.org/10.1016/j.cej.2019.05.136

Huang, J., Zhong, Y., Zhang, L., & Cai, J. (2017). Extremely strong and transparent chitin films: A high-efficiency, energy-saving, and "green" route using an aqueous KOH/Urea solution. *Advanced Functional Materials, 27*(26), 1701100. https://doi.org/10.1002/adfm.201701100

Huang, N.-J., Zang, J., Zhang, G.-D., Guan, L.-Z., Li, S.-N., Zhao, L., & Tang, L.-C. (2017). Efficient interfacial interaction for improving mechanical properties of polydimethyl-siloxane nanocomposites filled with low content of graphene oxide nanoribbons. *RSC Advances, 7*(36), 22045–22053. https://doi.org/10.1039/C7RA02439H

Huang, W., Dai, K., Zhai, Y., Liu, H., Zhan, P., Gao, J., Zheng, G.-Q., Liu, C., & Shen, C. (2017). Flexible and lightweight pressure sensor based on carbon nanotube/thermoplastic polyurethane aligned conductive foam with superior compressibility and stability. *ACS Applied Materials & Interfaces, 9*. https://doi.org/10.1021/acsami.7b16975

Huang, W.-C., Wang, W., Xue, C., & Mao, X. (2018). Effective enzyme immobilization onto a magnetic chitin nanofiber composite. *ACS Sustainable Chemistry & Engineering, 6*(7), 8118–8124. https://doi.org/10.1021/acssuschemeng.8b01150

Huang, Y., Yao, M., Zheng, X., Liang, X., Su, X., Zhang, Y., Lu, A., & Zhang, L. (2015). Effects of chitin whiskers on physical properties and osteoblast culture of alginate based nanocomposite hydrogels. *Biomacromolecules, 16*(11), 3499–3507. https://doi.org/10.1021/acs.biomac.5b00928

Huber, K. C., & BeMiller, J. N. (2010). *Modified starch: Chemistry and properties. En: Starches: Characterization, properties and applications, (AC Bertolini ed.)* Pp. 145–204. CRC Press.

Ifuku, S., & Saimoto, H. (2012). Chitin nanofibers: Preparations, modifications, and applications. *Nanoscale, 4*(11), 3308–3318. https://doi.org/10.1039/C2NR30383C

Isogai, A. (2013). Wood nanocelluloses: Fundamentals and applications as new bio-based nanomaterials. *Journal of Wood Science, 59.* https://doi.org/10.1007/s10086-013-1365-z

Isogai, A., Saito, T., & Fukuzumi, H. (2010). TEMPO-oxidized cellulose nanofibers. *Nanoscale, 3,* 71–85. https://doi.org/10.1039/c0nr00583e

Jalvo, B., Mathew, A. P., & Rosal, R. (2017). Coaxial poly(lactic acid) electrospun composite membranes incorporating cellulose and chitin nanocrystals. *Journal of Membrane Science, 544,* 261–271. https://doi.org/10.1016/j.memsci.2017.09.033

Janardhnan, S., & Sain, M. (2011). Bio-treatment of natural fibers in isolation of cellulose nanofibres: Impact of pre-refining of fibers on bio-treatment efficiency and nanofiber yield. *Journal of Polymers and the Environment, 19,* 615–621. https://doi.org/10.1007/s10924-011-0312-6

Ji, S., Jang, J., Cho, E., Kim, S.-H., Kang, E.-S., Kim, J., Kim, H.-K., Kong, H., Kim, S.-K., Kim, J.-Y., & Park, J.-U. (2017). High dielectric performances of flexible and transparent cellulose hybrid films controlled by multidimensional metal nanostructures. *Advanced Materials, 29*(24), 1700538. https://doi.org/10.1002/adma.201700538

Jiang, D. D., Yao, Q., McKinney, M. A., & Wilkie, C. A. (1999). TGA/FTIR studies on the thermal degradation of some polymeric sulfonic and phosphonic acids and their sodium salts. *Polymer Degradation and Stability, 63*(3), 423–434. https://doi.org/10.1016/S0141-3910(98)00123-2

Jiang, Q., Gholami Derami, H., Ghim, D., Cao, S., Jun, Y.-S., & Singamaneni, S. (2017). Polydopamine-filled bacterial nanocellulose as a biodegradable interfacial photothermal evaporator for highly efficient solar steam generation. *Journal of Materials Chemistry A, 5*(35), 18397–18402. https://doi.org/10.1039/C7TA04834C

Jiang, Q., Tian, L., Liu, K.-K., Tadepalli, S., Raliya, R., Biswas, P., Naik, R. R., & Singamaneni, S. (2016). Bilayered biofoam for highly efficient solar steam generation. *Advanced Materials, 28*(42), 9400–9407. https://doi.org/10.1002/adma.201601819

Jiang, T., Duan, Q., Zhu, J., Liu, H., & Yu, L. (2020). Starch-based biodegradable materials: Challenges and opportunities. *Advanced Industrial and Engineering Polymer Research, 3*(1), 8–18. https://doi.org/10.1016/j.aiepr.2019.11.003

Jin, J., Lee, D., Im, H.-G., Han, Y. C., Jeong, E. G., Rolandi, M., Choi, K. C., & Bae, B.-S. (2016). Chitin nanofiber transparent paper for flexible green electronics. *Advanced Materials, 28*(26), 5169–5175. https://doi.org/10.1002/adma.201600336

John, S. (1991). Localization of light. *Physics Today, 44*(5), 32–40. https://doi.org/10.1063/1.881300

Jonoobi, M., Harun, J., Mathew, A. P., & Oksman, K. (2010). Mechanical properties of cellulose nanofiber (CNF) reinforced polylactic acid (PLA) prepared by twin screw extrusion. *Composites Science and Technology, 70*(12), 1742–1747. https://doi.org/10.1016/j.compscitech.2010.07.005

Jonoobi, M., Oladi, R., Davoudpour, Y., Oksman, K., Dufresne, A., Hamzeh, Y., & Davoodi, R. (2015). Different preparation methods and properties of nanostructured cellulose from various natural resources and residues: A review. *Cellulose, 22*(2), 935–969. https://doi.org/10.1007/s10570-015-0551-0

Junkasem, J., Rujiravanit, R., & Supaphol, P. (2006). Fabrication of α-chitin whisker-reinforced poly(vinyl alcohol) nanocomposite nanofibres by electrospinning. *Nanotechnology, 17*(17), 4519–4528. https://doi.org/10.1088/0957-4484/17/17/039

Kalia, S., Celli, A., & Kango, S. (2014). Nanofibrillated cellulose: Surface modification and potential applications. *Colloid and Polymer Science, 292.* https://doi.org/10.1007/s00396-013-3112-9

Kang, Y., Chun, S.-J., Lee, S.-S., Kim, B.-Y., Kim, J., Chung, H., Lee, S., & Kim, W. (2012). All-solid-state flexible supercapacitors fabricated with bacterial nanocellulose papers, carbon nanotubes, and triblock-copolymer ion gels. *ACS Nano, 6*, 6400–6406. https://doi.org/10.1021/nn301971r

Kargarzadeh, H., Mariano, M., Gopakumar, D., Ahmad, I., Thomas, S., Dufresne, A., Huang, J., & Lin, N. (2018). Advances in cellulose nanomaterials. *Cellulose, 25*(4), 2151–2189. https://doi.org/10.1007/s10570-018-1723-5

Karimi, S. (2014). A comparative study on characteristics of nanocellulose reinforced thermoplastic starch biofilms prepared with different techniques. *Nordic Pulp and Paper Research Journal, 29*, 041–045. https://doi.org/10.3183/NPPRJ-2014-29-01-p041-045

Kassab, Z., Boujemaoui, A., Ben Youcef, H., Hajlane, A., Hannache, H., & El Achaby, M. (2019). Production of cellulose nanofibrils from alfa fibers and its nanoreinforcement potential in polymer nanocomposites. *Cellulose, 26*. https://doi.org/10.1007/s10570-019-02767-5

Khalil, H. P. S. A., Davoudpour, Y., Aprilia, N. A. S., Mustapha, A., Hossain, S., Islam, N., & Dungani, R. (2014). Nanocellulose-based polymer nanocomposite: Isolation, characterization and applications. In V. K. Thakur (Ed.), *Nanocellulose Polymer Nanocomposites* (pp. 273–309). John Wiley & Sons, Inc. https://doi.org/10.1002/9781118872246.ch11

Khattab, T. A., Abdelrahman, M. S., Ahmed, H. B., & Emam, H. E. (2020). Molecularly imprinted cellulose sensor strips for selective determination of phenols in aqueous environment. *Fibers and Polymers, 21*(10), 2195–2203. https://doi.org/10.1007/s12221-020-1325-3

Kim, D. W., Shin, J., & Choi, S. Q. (2020). Nano-dispersed cellulose nanofibrils-PMMA composite from pickering emulsion with tunable interfacial tensions. *Carbohydrate Polymers, 247*, 116762. https://doi.org/10.1016/j.carbpol.2020.116762

Kim, H.-J., Yim, E.-C., Kim, J.-H., Kim, S.-J., Park, J.-Y., & Oh, I.-K. (2017). Bacterial nano-cellulose triboelectric nanogenerator. *Nano Energy, 33*, 130–137. https://doi.org/10.1016/j.nanoen.2017.01.035

Kim, K., Ha, M., Choi, B., Joo, S. H., Kang, H. S., Park, J. H., Gu, B., Park, C., Park, C., Kim, J., Kwak, S. K., Ko, H., Jin, J., & Kang, S. J. (2018). Biodegradable, electro-active chitin nanofiber films for flexible piezoelectric transducers. *Nano Energy, 48*, 275–283. https://doi.org/10.1016/j.nanoen.2018.03.056

Klemm, D., Kramer, F., Moritz, S., Lindström, T., Ankerfors, M., Gray, D., & Dorris, A. (2011). Nanocelluloses: A new family of nature-based materials. *Angewandte Chemie (International Ed. in English), 50*, 5438–5466. https://doi.org/10.1002/anie.201001273

Koga, H., Nogi, M., Komoda, N., Nge, T., Sugahara, T., & Suganuma, K. (2014). Uniformly connected conductive networks on cellulose nanofiber paper for transparent paper electronics. *NPG Asia Materials, 6*, e93. https://doi.org/10.1038/am.2014.9

Lamouroux, E., & Fort, Y. (2016). An overview of nanocomposite nanofillers and their functionalization. In S. Thomas, D. Rouxel, & D. Ponnamma (Eds.), *Spectroscopy of Polymer Nanocomposites* (pp. 15–64). Elsevier. https://doi.org/10.1016/B978-0-323-40183-8.00002-1

Lay, M., Méndez, J. A., Delgado-Aguilar, M., Bun, K. N., & Vilaseca, F. (2016). Strong and electrically conductive nanopaper from cellulose nanofibers and polypyrrole. *Carbohydrate Polymers, 152*, 361–369. https://doi.org/10.1016/j.carbpol.2016.06.102

Le Bras, D., Strømme, M., & Mihranyan, A. (2015). Characterization of dielectric properties of nanocellulose from wood and algae for electrical insulator applications. *The Journal of Physical Chemistry B, 119*(18), 5911–5917. https://doi.org/10.1021/acs.jpcb.5b00715

LeCorre, D., & Angellier-Coussy, H. (2014). Preparation and application of starch nanoparticles for nanocomposites: A review. *Reactive and Functional Polymers, 85*, 97–120. https://doi.org/10.1016/j.reactfunctpolym.2014.09.020

LeCorre, D., Bras, J., & Dufresne, A. (2012). Influence of native starch's properties on starch nanocrystals thermal properties. *Carbohydrate Polymers*, *87*(1), 658–666. https://doi. org/10.1016/j.carbpol.2011.08.042

Lee, H., Abd Hamid, S. B., & Zain, S. (2014). Conversion of lignocellulosic biomass to nanocellulose: Structure and chemical process. *TheScientificWorldJournal*, *2014*, 631013. https://doi.org/10.1155/2014/631013

Lee, S. H., Teramoto, Y., & Endo, T. (2011). Cellulose nanofiber-reinforced polycaprolactone/polypropylene hybrid nanocomposite. *Composites Part A: Applied Science and Manufacturing*, *42*, 151–156. https://doi.org/10.1016/j.compositesa.2010.10.014

Leszczynska, A., & Pielichowski, K. (2008). Application of thermal analysis methods for characterization of polymer/montmorillonite nanocomposites. *Journal of Thermal Analysis and Calorimetry*, *93*(3), 677–687. https://doi.org/10.1007/s10973-008-9128-6

Liang, H.-W., Guan, Q.-F., Zhu, Z., Song, L.-T., Yao, H.-B., Lei, X., & Yu, S.-H. (2012). Highly conductive and stretchable conductors fabricated from bacterial cellulose. *NPG Asia Materials*, *4*. https://doi.org/10.1038/am.2012.34

Lin, N., Huang, J., Chang, P. R., Feng, J., & Yu, J. (2011). Surface acetylation of cellulose nanocrystal and its reinforcing function in poly(lactic acid). *Carbohydrate Polymers*, *83*(4), 1834–1842. https://doi.org/10.1016/j.carbpol.2010.10.047

Lipparelli Morelli, C., Belgacem, N., Branciforti, M., Bretas, R., Crisci, A., & Bras, J. (2015). Supramolecular aromatic interactions to enhance biodegradable film properties through incorporation of functionalized cellulose nanocrystals. *Composites Part A: Applied Science and Manufacturing*, *83*. https://doi.org/10.1016/j.compositesa.2015.10.038

Liu, D., Sun, X., Tian, H., Maiti, S., & Ma, Z. (2013). Effects of cellulose nanofibrils on the structure and properties on PVA nanocomposites. *Cellulose*, *20*. https://doi.org/10.1007/s10570-013-0073-6

Liu, H., Li, Q., Zhang, S., Yin, R., Liu, X., He, Y., Dai, K., Shan, C.-X., Guo, J., Liu, C., Shen, C., Wang, X., Wang, N., Wang, Z., Wei, R., & Guo, Z. (2018). Electrically conductive polymer composites for smart flexible strain sensors: A critical review. *Journal of Materials Chemistry C*, *6*. https://doi.org/10.1039/C8TC04079F

Liu, M., Huang, J., Luo, B., & Zhou, C. (2015). Tough and highly stretchable polyacrylamide nanocomposite hydrogels with chitin nanocrystals. *International Journal of Biological Macromolecules*, *78*, 23–31. https://doi.org/10.1016/j.ijbiomac.2015.03.059

Liu, Y., Liu, M., Yang, S., Luo, B., & Zhou, C. (2018). Liquid crystalline behaviors of chitin nanocrystals and their reinforcing effect on natural rubber. *ACS Sustainable Chemistry & Engineering*, *6*(1), 325–336. https://doi.org/10.1021/acssuschemeng.7b02586

Luo, Y., Zhang, J., Li, X., & Liao, C. (2014). The cellulose nanofibers for optoelectronic conversion and energy storage. *Journal of Nanomaterials*, *2014*. https://doi.org/10.1155/2014/654512

Ma, L., Liu, M., Peng, Q., Liu, Y., Luo, B., & Zhou, C. (2016). Crosslinked carboxylated SBR composites reinforced with chitin nanocrystals. *Journal of Polymer Research*, *23*(7), 134. https://doi.org/10.1007/s10965-016-1025-2

Ma, X., Cheng, Y., Qin, X., Guo, T., Deng, J., & Liu, X. (2017). Hydrophilic modification of cellulose nanocrystals improves the physicochemical properties of cassava starch-based nanocomposite films. *LWT*, *86*, 318–326. https://doi.org/10.1016/j.lwt.2017.08.012

Mandal, A. & Chakrabarty, D. (2014). Studies on the mechanical, thermal, morphological and barrier properties of nanocomposites based on poly(vinyl alcohol) and nanocellulose from sugarcane bagasse. *Journal of Industrial and Engineering Chemistry*, *20*(2), 462–473. https://doi.org/10.1016/j.jiec.2013.05.003

Marcelo, G., López-González, M. del M., Vega, M., & Pecharromán, C. (2021). Colored surfaces made of synthetic eumelanin. *Nanomaterials*, *11*(9), 2320. https://doi.org/10.3390/nano11092320

Marín-Silva, D. A., Rivero, S., & Pinotti, A. (2019). Chitosan-based nanocomposite matrices: Development and characterization. *International Journal of Biological Macromolecules*, *123*, 189–200. https://doi.org/10.1016/j.ijbiomac.2018.11.035

Meadows, M., Butler, M., Morehouse, N., Taylor, L., Toomey, M., Mcgraw, K., & Rutowski, R. (2009). Iridescence: Views from many angles. *Journal of the Royal Society, Interface/the Royal Society*, *6 Suppl 2*, S107–13. https://doi.org/10.1098/rsif.2009.0013.focus

Menchaca-Nal, S., Londoño-Calderón, C. L., Cerrutti, P., Foresti, M. L., Pampillo, L., Bilovol, V., Candal, R., & Martínez-García, R. (2016). Facile synthesis of cobalt ferrite nanotubes using bacterial nanocellulose as template. *Carbohydrate Polymers*, *137*, 726–731. https://doi.org/10.1016/j.carbpol.2015.10.068

Michael, F. M., Khalid, M., Walvekar, R., Ratnam, C. T., Ramarad, S., Siddiqui, H., & Hoque, M. E. (2016). Effect of nanofillers on the physico-mechanical properties of load bearing bone implants. *Materials Science and Engineering: C*, *67*, 792–806. https://doi.org/10.1016/j.msec.2016.05.037

Mondal, S. (2018). Review on nanocellulose polymer nanocomposites. *Polymer-Plastics Technology and Engineering*, *57*(13), 1377–1391. https://doi.org/10.1080/03602559.2017.1381253

Moohan, J., Stewart, S. A., Espinosa, E., Rosal, A., Rodríguez, A., Larrañeta, E., Donnelly, R. F., & Domínguez-Robles, J. (2019). Cellulose nanofibers and other biopolymers for biomedical applications. A review. *Applied Sciences*, *10*(1), 65. https://doi.org/10.3390/app10010065

Moon, R. J., Martini, A., Nairn, J., Simonsen, J., & Youngblood, J. (2011). Cellulose nanomaterials review: Structure, properties and nanocomposites. *Chemical Society Reviews*, *40*(7), 3941–3994. https://doi.org/10.1039/C0CS00108B

Morganti, P., Muzzarelli, R., & Muzzarelli, C. (2006). Multifunctional use of innovative chitin nanofibrils for skin care. *Journal of Applied Cosmetology*, *24*, 105–114.

Naghdi, T., Golmohammadi, H., Vosough, M., Atashi, M., Saeedi, I., & Maghsoudi, M. T. (2019). Lab-on-nanopaper: An optical sensing bioplatform based on curcumin embedded in bacterial nanocellulose as an albumin assay kit. *Analytica Chimica Acta*, *1070*, 104–111. https://doi.org/10.1016/j.aca.2019.04.037

Naghdi, T., Golmohammadi, H., Yousefi, H., Hosseinifard, M., Kostiv, U., Horák, D., & Merkoçi, A. (2020). Chitin nanofiber paper toward optical (bio)sensing applications. *ACS Applied Materials & Interfaces*. https://doi.org/10.1021/acsami.9b23487

Nair, J. R., Bella, F., Angulakshmi, N., Stephan, A. M., & Gerbaldi, C. (2016). Nanocellulose-laden composite polymer electrolytes for high performing lithium–sulphur batteries. *Energy Storage Materials*, *3*, 69–76. https://doi.org/10.1016/j.ensm.2016.01.008

Nie, J., Mou, W., Ding, J., & Chen, Y. (2019). Bio-based epoxidized natural rubber/chitin nanocrystals composites: Self-healing and enhanced mechanical properties. *Composites Part B: Engineering*, *172*, 152–160. https://doi.org/10.1016/j.compositesb.2019.04.035

Numata, Y., Sakata, T., Furukawa, H., & Tajima, K. (2015). Bacterial cellulose gels with high mechanical strength. *Materials Science and Engineering: C*, *47*, 57–62. https://doi.org/10.1016/j.msec.2014.11.026

Nyakuma, B. B., Wong, S., Utume, L. N., Abdullah, T. A. T., Abba, M., Oladokun, O., Ivase, T. J.-P., & Ogunbode, E. B. (2021). Comprehensive characterisation of the morphological, thermal and kinetic degradation properties of *Gluconacetobacter xylinus* synthesised bacterial nanocellulose. *Journal of Natural Fibers*, 1–14. https://doi.org/10.1080/15440478.2021.1907833

Nyholm, L., Nyström, G., Mihranyan, A., & Strømme, M. (2011). Toward flexible polymer and paper-based energy storage devices. *Advanced Materials (Deerfield Beach, Fla.)*, *23*. https://doi.org/10.1002/adma.201004134

Panchagnula, K. K., & Kuppan, P. (2019). Improvement in the mechanical properties of neat GFRPs with multi-walled CNTs. *Journal of Materials Research and Technology, 8*(1), 366–376. https://doi.org/10.1016/j.jmrt.2018.02.009

Petersson, L., & Oksman, K. (2006). Biopolymer based nanocomposites: Comparing layered silicates and microcrystalline cellulose as nanoreinforcement. *Composites Science and Technology, 66*, 2187–2196. https://doi.org/10.1016/j.compscitech.2005.12.010

Pires, J., Souza, V., & Fernando, A. (2019). *Production of Nanocellulose from Lignocellulosic Biomass Wastes: Prospects and Limitations* (pp. 719–725). Springer Verlag. https://doi.org/10.1007/978-3-319-91334-6_98

Prachayawarakorn, J., Chaiwatyothin, S., Mueangta, S., & Hanchana, A. (2013). Effect of jute and kapok fibers on properties of thermoplastic cassava starch composites. *Materials & Design, 47*, 309–315. https://doi.org/10.1016/j.matdes.2012.12.012

Qiu, K., & Netravali, A. N. (2014). A review of fabrication and applications of bacterial cellulose based nanocomposites. *Polymer Reviews, 54*(4), 598–626. https://doi.org/10.1080/15583724.2014.896018

Rana, A. K., Frollini, E., & Thakur, V. K. (2021). Cellulose nanocrystals: Pretreatments, preparation strategies, and surface functionalization. *International Journal of Biological Macromolecules, 182*, 1554–1581. https://doi.org/10.1016/j.ijbiomac.2021.05.119

Ravichandran, S., Sagadevan, S., & Hoque, M. E. (2022). Physical, mechanical, and thermal properties of fiber-reinforced hybrid polymer composites. In Senthilkumar Krishnasamy, Senthil Muthu Kumar Thiagamani, Chandrasekar Muthukumar, Rajini Nagarajan, & Suchart Siengchin (Eds.), *Natural Fiber-Reinforced Composites* (pp. 309–320). John Wiley & Sons, Ltd. https://doi.org/10.1002/9783527831562.ch18

Revedin, A., Aranguren, B., Becattini, R., Longo, L., Marconi, E., Lippi, M. M., Skakun, N., Sinitsyn, A., Spiridonova, E., & Svoboda, J. (2010). Thirty thousand-year-old evidence of plant food processing. *Proceedings of the National Academy of Sciences, 107*(44), 18815–18819. https://doi.org/10.1073/pnas.1006993107

Robles, E., Urruzola, I., Labidi, J., & Serrano, L. (2015). Surface-modified nano-cellulose as reinforcement in poly(lactic acid) to conform new composites. *Industrial Crops and Products, 71*, 44–53. https://doi.org/10.1016/j.indcrop.2015.03.075

Saba, N., Safwan, A., Sanyang, M. L., Mohammad, F., Pervaiz, M., Jawaid, M., Alothman, O. Y., & Sain, M. (2017). Thermal and dynamic mechanical properties of cellulose nanofibers reinforced epoxy composites. *International Journal of Biological Macromolecules, 102*, 822–828. https://doi.org/10.1016/j.ijbiomac.2017.04.074

Saba, N., Tahir, P., & Jawaid, M. (2014). A review on potentiality of nano filler/natural fiber filled polymer hybrid composites. *Polymers, 6*(8), 2247–2273. https://doi.org/10.3390/polym6082247

Safdari, F., Bagheriasl, D., Carreau, P., Heuzey, M.-C., & Kamal, M. (2016). Rheological, mechanical, and thermal properties of polylactide/cellulose nanofiber biocomposites. *Polymer Composites, 39*. https://doi.org/10.1002/pc.24127

Safdari, F., Carreau, P. J., Heuzey, M. C., Kamal, M. R., & Sain, M. M. (2017). Enhanced properties of poly(ethylene oxide)/cellulose nanofiber biocomposites. *Cellulose, 24*(2), 755–767. https://doi.org/10.1007/s10570-016-1137-1

Sagadevan, S., Fareen, A., Hoque, M. E., Chowdhury, Z. Z., Johan, Mohd. R. B., Rafique, R. F., Aziz, F. A., & Lett, J. A. (2019). Chapter 12-Nanostructured polymer biocomposites: Pharmaceutical applications. In S. K. Swain & M. Jawaid (Eds.), *Nanostructured Polymer Composites for Biomedical Applications* (pp. 227–259). Elsevier. https://doi.org/10.1016/B978-0-12-816771-7.00012-0

Saha, T., Hoque, M. E., & Mahbub, T. (2020). Biopolymers for sustainable packaging in food, cosmetics, and pharmaceuticals. In Faris M. Al-Oqla & S. M. Sapuan (Eds.), *Advanced Processing, Properties, and Applications of Starch and Other Bio-Based Polymers* (pp. 197–214). Elsevier. https://doi.org/10.1016/B978-0-12-819661-8.00013-5

Salaberria, A. M., Fernandes, S. C. M., Diaz, R. H., & Labidi, J. (2015). Processing of α-chitin nanofibers by dynamic high pressure homogenization: Characterization and antifungal activity against A. niger. *Carbohydrate Polymers, 116,* 286–291. https://doi.org/10.1016/j.carbpol.2014.04.047

Salaberria, A. M., Labidi, J., & Fernandes, S. C. M. (2014). Chitin nanocrystals and nanofibers as nano-sized fillers into thermoplastic starch-based biocomposites processed by melt-mixing. *Chemical Engineering Journal, 256,* 356–364. https://doi.org/10.1016/j.cej.2014.07.009

Salaberría, A. M., Teruel-Juanes, R., Badia, J. D., Fernandes, S. C. M., Sáenz de Juano-Arbona, V., Labidi, J., & Ribes-Greus, A. (2018). Influence of chitin nanocrystals on the dielectric behaviour and conductivity of chitosan-based bionanocomposites. *Composites Science and Technology, 167,* 323–330. https://doi.org/10.1016/j.compscitech.2018.08.019

Sessini, V., Raquez, J.-M., Kenny, J. M., Dubois, P., & Peponi, L. (2019). Melt-processing of bionanocomposites based on ethylene-co-vinyl acetate and starch nanocrystals. *Carbohydrate Polymers, 208,* 382–390. https://doi.org/10.1016/j.carbpol.2018.12.095

Shah, J., & Brown, R. (2005). Towards electronic paper displays made from microbial cellulose. *Applied Microbiology and Biotechnology, 66,* 352–355. https://doi.org/10.1007/s00253-004-1756-6

Shah, N., Ul-Islam, M., Khattak, W. A., & Park, J. K. (2013). Overview of bacterial cellulose composites: A multipurpose advanced material. *Carbohydrate Polymers, 98*(2), 1585–1598. https://doi.org/10.1016/j.carbpol.2013.08.018

Sharma, C., & Bhardwaj, N. K. (2019). Bacterial nanocellulose: Present status, biomedical applications and future perspectives. *Materials Science and Engineering: C, 104,* 109963. https://doi.org/10.1016/j.msec.2019.109963

Shopsowitz, K. E., Hamad, W. Y., & MacLachlan, M. J. (2011). Chiral nematic mesoporous carbon derived from nanocrystalline cellulose. *Angewandte Chemie International Edition, 50*(46), 10991–10995. https://doi.org/10.1002/anie.201105479

Silva, R., Al-Sharab, J., & Asefa, T. (2012). Edge-plane-rich nitrogen-doped carbon nanoneedles and efficient metal-free electrocatalysts. *Angewandte Chemie International Edition, 51*(29), 7171–7175. https://doi.org/10.1002/anie.201201742

Singh, A. A., Wei, J., Herrera, N., Geng, S., & Oksman, K. (2018). Synergistic effect of chitin nanocrystals and orientations induced by solid-state drawing on PLA-based nanocomposite tapes. *Composites Science and Technology, 162,* 140–145. https://doi.org/10.1016/j.compscitech.2018.04.034

Siro, I., & Plackett, D. (2010). Microfibrillated cellulose and new nanocomposite materials: A review. *Cellulose, 17,* 459–494. https://doi.org/10.1007/s10570-010-9405-y

Skočaj, M. (2019). Bacterial nanocellulose in papermaking. *Cellulose, 26*(11), 6477–6488. https://doi.org/10.1007/s10570-019-02566-y

Song, S., Wang, C., Pan, Z., & Wang, X. (2008). Preparation and characterization of amphiphilic starch nanocrystals. *Journal of Applied Polymer Science, 107*(1), 418–422. https://doi.org/10.1002/app.27076

Sriupayo, J., Supaphol, P., Blackwell, J., & Rujiravanit, R. (2005). Preparation and characterization of α-chitin whisker-reinforced chitosan nanocomposite films with or without heat treatment. *Carbohydrate Polymers, 62*(2), 130–136. https://doi.org/10.1016/j.carbpol.2005.07.013

Stanisławska, A. (2016). Bacterial nanocellulose as a microbiological derived nanomaterial. *Advances in Materials Science, 16*(4), 45–57. https://doi.org/10.1515/adms-2016-0022

Syafri, E., Kasim, A., Abral, H., & Asben, A. (2017). Effect of precipitated calcium carbonate on physical, mechanical and thermal properties of cassava starch bioplastic composites. *International Journal on Advanced Science, Engineering and Information Technology, 7,* 1950. https://doi.org/10.18517/ijaseit.7.5.1292

Syafri, E., Kasim, A., Abral, H., Sudirman, Sulungbudi, G. T., Sanjay, M. R., & Sari, N. H. (2018). Synthesis and characterization of cellulose nanofibers (CNF) ramie reinforced cassava starch hybrid composites. *International Journal of Biological Macromolecules, 120*, 578–586. https://doi.org/10.1016/j.ijbiomac.2018.08.134

Takkalkar, P., Ganapathi, M., Dekiwadia, C., Nizamuddin, S., Griffin, G., & Kao, N. (2019). Preparation of square-shaped starch nanocrystals/polylactic acid based bio-nanocomposites: Morphological, structural, thermal and rheological properties. *Waste and Biomass Valorization, 10*(11), 3197–3211. https://doi.org/10.1007/s12649-018-0372-0

Tammela, P., Wang, Z., Frykstrand, S., Zhang, P., Sintorn, I.-M., Nyholm, L., & Strømme, M. (2015). Asymmetric supercapacitors based on carbon nanofibre and polypyrrole/nanocellulose composite electrodes. *RSC Advances, 5.* https://doi.org/10.1039/C4RA15894F

Tarrés, Q., Deltell, A., Espinach, F. X., Pèlach, M. À., Delgado-Aguilar, M., & Mutjé, P. (2017). Magnetic bionanocomposites from cellulose nanofibers: Fast, simple and effective production method. *International Journal of Biological Macromolecules, 99*, 29–36. https://doi.org/10.1016/j.ijbiomac.2017.02.072

Tercjak, A., Gutierrez, J., Barud, H., Domeneguetti, R., & Ribeiro, S. (2015). Nano- and macroscale structural and mechanical properties of in situ synthesized bacterial cellulose/ PEO- b -PPO- b -PEO biocomposites. *ACS Applied Materials & Interfaces, 7*, 4142–4150. https://doi.org/10.1021/am508273x

Thomas, B., Raj, M., B, A., H, R., Joy, J., Moores, A., Drisko, G., & Sanchez, C. (2018). Nanocellulose, a versatile green platform: From biosources to materials and their applications. *Chemical Reviews, 118.* https://doi.org/10.1021/acs.chemrev.7b00627

Tran, Q. N., Kim, I. T., Hur, J., Kim, J. H., Choi, H. W., & Park, S. J. (2020). Composite of nanocrystalline cellulose with tin dioxide as Lightweight Substrates for high-performance Lithium-ion battery. *Korean Journal of Chemical Engineering, 37*(5), 898–904. https://doi.org/10.1007/s11814-020-0506-5

Tusnim, J., Hoque, M. E., Hossain, S. A., Abdel-Wahab, A., Abdala, A., & Wahab, M. A. (2020). Nanocellulose and nanohydrogels for the development of cleaner energy and future sustainable materials. In F. Mohammad, H. A. Al-Lohedan, & M. Jawaid (Eds.), *Sustainable Nanocellulose and Nanohydrogels from Natural Sources* (pp. 81–113). Elsevier. https:// doi.org/10.1016/B978-0-12-816789-2.00004-3

Walther, A., Timonen, J., Díez, I., Laukkanen, A., & Ikkala, O. (2011). Multifunctional high-performance biofibers based on wet-extrusion of renewable native cellulose nanofibrils. *Advanced Materials (Deerfield Beach, Fla.), 23*, 2924–2928. https://doi.org/10.1002/adma.201100580

Wan, Y., Li, J., Yang, Z., Ao, H., Xiong, L., & Luo, H. (2018). Simultaneously depositing polyaniline onto bacterial cellulose nanofibers and graphene nanosheets toward electrically conductive nanocomposites. *Current Applied Physics, 18.* https://doi.org/10.1016/j.cap.2018.05.008

Wang, B., Li, X., Luo, B., Yang, J., Wang, X., Song, Q., Chen, S., & Zhi, L. (2013). Pyrolyzed bacterial cellulose: A versatile support for lithium ion battery anode materials. *Small (Weinheim an Der Bergstrasse, Germany), 9.* https://doi.org/10.1002/smll.201300692

Wang, C., Pan, Z., & Zeng, J. (2014). Structure, morphology and properties of benzyl starch nanocrystals. *Arabian Journal for Science and Engineering, 39*(9), 6703–6710. https:// doi.org/10.1007/s13369-014-1201-9

Wang, H., & Zhang, K.-Q. (2013). Photonic crystal structures with tunable structure color as colorimetric sensors. *Sensors, 13*(4). https://doi.org/10.3390/s130404192

Wang, Y., Chen, Z., Tang, J., & Lin, N. (2019). Tunable optical materials based on self-assembly of polysaccharide nanocrystals. In N. Lin, J. Tang, A. Dufresne, & M. K. C. Tam (Eds.), *Advanced Functional Materials from Nanopolysaccharides* (pp. 87–136). Springer Singapore. https://doi.org/10.1007/978-981-15-0913-1_3

Wang, Z., Carlsson, D., Tammela, P., Hua, K., Zhang, P., Nyholm, L., & Strømme, M. (2015). Surface modified nanocellulose fibers yield conducting polymer-based flexible supercapacitors with enhanced capacitances. *ACS Nano, 9*. https://doi.org/10.1021/acsnano.5b02846

Wen, K., Zhang, Z., Jiang, X., He, J., & Yang, J. (2020). Image representation of structure color based on edge detection algorithm. *Results in Physics, 19*, 103441. https://doi.org/10.1016/j.rinp.2020.103441

Wetzel, B., Haupert, F., & Qiu Zhang, M. (2003). Epoxy nanocomposites with high mechanical and tribological performance. *Composites Science and Technology, 63*(14), 2055–2067. https://doi.org/10.1016/S0266-3538(03)00115-5

Wu, C., Sun, J., Chen, M., Ge, Y., Ma, J., Hu, Y., Pang, J., & Yan, Z. (2019). Effect of oxidized chitin nanocrystals and curcumin into chitosan films for seafood freshness monitoring. *Food Hydrocolloids, 95*, 308–317. https://doi.org/10.1016/j.foodhyd.2019.04.047

Wu, J., Zhang, K., Girouard, N., & Meredith, J. C. (2014). Facile route to produce chitin nanofibers as precursors for flexible and transparent gas barrier materials. *Biomacromolecules, 15*(12), 4614–4620. https://doi.org/10.1021/bm501416q

Wu, X., Cao, S., Ghim, D., Jiang, Q., Singamaneni, S., & Jun, Y.-S. (2021). A thermally engineered polydopamine and bacterial nanocellulose bilayer membrane for photothermal membrane distillation with bactericidal capability. *Nano Energy, 79*, 105353. https://doi.org/10.1016/j.nanoen.2020.105353

Wu, Z.-Y., Li, C., Liang, H.-W., Chen, J.-F., & Yu, S.-H. (2013). Ultralight, flexible, and fire-resistant carbon nanofiber aerogels from bacterial cellulose. *Angewandte Chemie (International Ed. in English), 52*, 2925–2929. https://doi.org/10.1002/anie.201209676

Xiao-peng, L., Li, Y., Li, X., Song, D., Min, P., Hu, C., Zhang, H.-B., Koratkar, N., & Yu, Z.-Z. (2019). Highly sensitive, reliable and flexible piezoresistive pressure sensors featuring polyurethane sponge coated with MXene sheets. *Journal of Colloid and Interface Science, 542*. https://doi.org/10.1016/j.jcis.2019.01.123

Xu, C., Chen, C., & Wu, D. (2018). The starch nanocrystal filled biodegradable poly(ε-caprolactone) composite membrane with highly improved properties. *Carbohydrate Polymers, 182*, 115–122. https://doi.org/10.1016/j.carbpol.2017.11.001

Xu, K., Liu, C., Kang, K., Zheng, Z., Wang, S., Tang, Z., & Yang, W. (2018). Isolation of nanocrystalline cellulose from rice straw and preparation of its biocomposites with chitosan: Physicochemical characterization and evaluation of interfacial compatibility. *Composites Science and Technology, 154*, 8–17. https://doi.org/10.1016/j.compscitech.2017.10.022

Xu, X., Liu, F., Jiang, L., Zhu, J. Y., Haagenson, D., & Wiesenborn, D. (2013). Cellulose nanocrystals vs. cellulose nanofibrils: A comparative study on their microstructures and effects as polymer reinforcing agents. *ACS Applied Materials & Interfaces, 6*. https://doi.org/10.1021/am302624t

Yan, M., Li, S., Zhang, M., Li, C., Dong, F., & Li, W. (2013). Characterization of surface acetylated nanocrystalline cellulose by single-step method. *BioResources, 8*(4), 6330–6341.

Yang, M., Zhang, X., Guan, S., Dou, Y., & Gao, X. (2020). Preparation of lignin containing cellulose nanofibers and its application in PVA nanocomposite films. *International Journal of Biological Macromolecules, 158*, 1259–1267. https://doi.org/10.1016/j.ijbiomac.2020.05.044

Yao, J., Huang, H., Mao, L., Li, Z., Zhu, H., & Liu, Y. (2017). Structural and optical properties of cellulose nanocrystals isolated from the fruit shell of Camellia oleifera Abel. *Fibers and Polymers, 18*(11), 2118–2124. https://doi.org/10.1007/s12221-017-7489-9

Yataka, Y., Suzuki, A., Iijima, K., & Hashizume, M. (2020). Enhancement of the mechanical properties of polysaccharide composite films utilizing cellulose nanofibers. *Polymer Journal, 52*(6), 645–653. https://doi.org/10.1038/s41428-020-0311-3

Yin, Z., Zeng, J., Wang, C., & Pan, Z. (2015). Preparation and properties of cross-linked starch nanocrystals/polylactic acid nanocomposites. *International Journal of Polymer Science*, *2015*, e454708. https://doi.org/10.1155/2015/454708

Yingkamhaeng, N., Intapan, I., & Sukyai, P. (2018). Fabrication and characterisation of functionalised superparamagnetic bacterial nanocellulose using ultrasonic-assisted in situ synthesis. *Fibers and Polymers*, *19*(3), 489–497. https://doi.org/10.1007/s12221-018-7738-6

Yu, J., Ai, F., Dufresne, A., Gao, S., Huang, J., & Chang, P. R. (2008). Structure and mechanical properties of poly(lactic acid) filled with (starch nanocrystal)-graft-poly(ε-caprolactone). *Macromolecular Materials and Engineering*, *293*(9), 763–770. https://doi.org/10.1002/mame.200800134

Zhai, W., Xia, Q., Zhou, K., Yue, X., Ren, M., Zheng, G., Dai, K., Liu, C., & Shen, C. (2019). Multifunctional flexible carbon black/polydimethylsiloxane piezoresistive sensor with ultrahigh linear range, excellent durability and oil/water separation capability. *Chemical Engineering Journal*, *372*. https://doi.org/10.1016/j.cej.2019.04.142

Zhang, S., Liu, H., Yang, S., Shi, X., Zhang, D., Shan, C., Mi, L., Liu, C., Shen, C., & Guo, Z. (2019). Ultrasensitive and highly compressible piezoresistive sensor based on polyurethane sponge coated with cracked cellulose nanofibril/silver nanowire layer. *ACS Applied Materials & Interfaces*, *11*. https://doi.org/10.1021/acsami.9b00900

Zhang, S., Sun, K., Liu, H., Chen, X., Zheng, Y., Shi, X., Zhang, D., Mi, L., Liu, C., & Shen, C. (2020). Enhanced piezoresistive performance of conductive WPU/CNT composite foam through incorporating brittle cellulose nanocrystal. *Chemical Engineering Journal*, *387*, 124045. https://doi.org/10.1016/j.cej.2020.124045

Zhao, X., Xu, Z., Xu, H., Lin, N., & Ma, J. (2021). Surface-charged starch nanocrystals from glutinous rice: Preparation, crystalline properties and cytotoxicity. *International Journal of Biological Macromolecules*, *192*, 557–563. https://doi.org/10.1016/j.ijbiomac.2021.10.024

Zheng, G., Cui, Y., Karabulut, E., Wågberg, L., Zhu, H., & Hu, L. (2013). Nanostructured paper for flexible energy and electronic devices. *MRS Bulletin/Materials Research Society*, *38*, 320–325. https://doi.org/10.1557/mrs.2013.59

Zhong, T., Wolcott, M. P., Liu, H., & Wang, J. (2019). Developing chitin nanocrystals for flexible packaging coatings. *Carbohydrate Polymers*, *226*, 115276. https://doi.org/10.1016/j.carbpol.2019.115276

Zhu, H., Hu, L., Cumings, J., Huang, J., Chen, Y., Preston, C., & Rohrbach, K. (2013). Highly transparent and flexible nanopaper transistor. *ACS Nano*, *7*. https://doi.org/10.1021/nn304407r

4 Standard Reinforcement Methods, Mechanism, Compatibility, and Surface Modification of Nanofillers in Polymers

Kriti Sharma and G. L. Devnani
Harcourt Butler Technical University

Sanya Verma
Indian Institute of Technology, New Delhi

Mohit Nigam
RBSETC

CONTENTS

4.1 Introduction 116
4.2 Types of Nanofillers... 117
4.3 Organic Nanofillers 120
    4.3.1 Nanofibers.. 120
        4.3.1.1 Nanocellulose.. 120
        4.3.1.2 Nanolignin 120
        4.3.1.3 Chitin/Chitosan... 120
        4.3.1.4 Organic–Inorganic Hybrid Nanofillers............................. 121
4.4 Inorganic Nanofillers 121
    4.4.1 Carbon-Based Nanofillers 121
        4.4.1.1 Carbon Nanotubes 121
        4.4.1.2 Graphene.. 121
    4.4.2 Nanoclays.. 122
        4.4.2.1 Layered Nanoclays.. 122
        4.4.2.2 Nonlayered Silicates 122
    4.4.3 Metal Oxide Nanofillers 122
    4.4.4 Metal Nanoparticles.. 122
    4.4.5 Other Nanoparticles.. 122

DOI: 10.1201/9781003279372-4

4.5   Surface Modification Techniques of Nanofillers 125
4.6   Types of Surface Modification Techniques 126
       4.6.1   Physical 126
4.7   Heat Treatment 126
       4.7.1   Plasma Treatment 127
       4.7.2   Low Atmospheric Pressure Discharge 127
       4.7.3   High Atmospheric Pressure Discharge.. 127
       4.7.4 γ Irradiation... 127
4.8   Chemical Treatment.. 128
       4.8.1   Coupling Agent Treatment.. 128
       4.8.2   Graft Polymerization 128
4.9   Mechanisms of Reinforcement of Nanofillers in PNCs 129
       4.9.1   For Particulate Fillers (Carbon Black, Silica, ZnO) 129
       4.9.2   For Tubular Fillers (Carbon Nanotubes [CNTs], Nanofibers)........... 130
       4.9.3   For Layered Filler (Nanoclays) 130
               4.9.3.1   Mechanisms 131
4.10   Compatibilization of Nanofillers 132
4.11   Processing of Polymer Nanocomposites.. 132
       4.11.1 In Situ Polymerization 133
       4.11.2 Blending/Direct Mixing 133
               4.11.2.1 Solution Mixing 133
               4.11.2.2 Melt Blending.. 134
       4.11.3 Bottom-Up 135
       4.11.4 Sol-Gel... 135
4.12   Characterization/Evaluation Techniques.. 135
       4.12.1 Structural and Morphological Characterization............................. 135
       4.12.2 Wide-Angle X-Ray Diffraction 136
       4.12.3 Small-Angle X-Ray Diffraction/Small-Angle X-Ray Scattering 136
       4.12.4 Scanning Electron Microscopy 137
       4.12.5 Raman Spectroscopy 137
       4.12.6 Atomic Force Microscopy (AFM).. 137
4.13   Thermal, Mechanical, Rheological and Other Techniques of
       Characterization.. 137
4.14   Benefits and Drawbacks.. 138
4.15   Potential Applications of Polymer Nanocomposites 139
4.16   Conclusion 140
References... 141

4.1   INTRODUCTION

The need and usage of special materials or combination materials have been ever increasing. Engineering materials at nano, micro, atomic or molecular scales give us the freedom and control to mix certain materials to get the desired properties. Such materials are called composites. Polymer composites are those composites where one of the materials is a polymer matrix. These polymer composites when reinforced using nanoscale particles are called polymer nanocomposites (PNCs). These consist of a polymer or copolymer having nanoparticles or nanofillers dispersed in the polymer matrix where the nanosized particles or nanofillers are classified into various types according to their shape and material.

In the last decades, it has been observed that the addition of low content of these nanofillers into the polymer improves their properties (mechanical, thermal, barrier and flammability) without affecting their processability. Nanocomposites can also show unique design possibilities, which offer excellent advantages in creating functional materials with desired properties for specific applications. The performance of nanocomposites may depend on a number of nanoparticle features such as size, specific surface area, aspect ratio, volume fraction used, compatibility with the matrix and dispersion.

The nanoparticle dispersion and distribution state remain the key challenge in order to obtain the full potential of enhancement of properties because of the hydrophilicity of the nanofillers and hydrophobicity of the polymer matrix. But, due to the recent development in the field of nanotechnology and growing interest in polymer matrix composites, nanosized fillers are distributed homogeneously (known as filler–polymer nanocomposites). This is mainly achieved through surface modification of nanofillers via physical, chemical or plasma treatments and further addition of compatibilizers to increase the dispersion of nanofillers into the polymer matrix and adhesion of the nanofiller–polymer surface (Bitinis et al., 2011; Müller et al., 2017).

Uniform dispersion can be characterized by different states at nano-, micro- and microscopic scales. To obtain nanocomposites with different polymers, various types of nanoparticles have been used such as clay (Oliveira et al., 2012; Zaïri et al., 2011), graphene (de Melo et al., 2018), nanotube (Maron et al., 2018) and nanocellulose (Sonia & Priya Dasan, 2013). Techniques like wide-angle X-ray diffraction (WAXD) and small-angle X-ray scattering (SAXS), electron microscopy, Raman spectroscopy and atomic force microscopy (AFM) and thermal, mechanical and rheological techniques like Fourier-transformed infrared (FTIR), rheometry, differential scanning calorimeter (DSC) and thermogravimetric (TGA), thermomechanical (TMA) and dynamic modulus analysis (DMA) have been used to evaluate the degree of dispersion and distribution of the nanoparticles in the polymeric matrices (Beatrice et al., 2010).

Depending on the degree of separation of the nanoparticles, three types of nanocomposite morphologies are possible (Alexandre & Dubois, 2000) (Figure 4.1).

Due to the large surface area of contact between the matrix and nanoparticles, exfoliated nanocomposites have maximum reinforcement. This is the main difference between nanocomposites and conventional composites.

The aim of this chapter is to review the understanding of the types of nanofillers, mechanism, compatibilization, and surface modification of nanofillers in polymers (Bhattacharya et al., 2008).

4.2 TYPES OF NANOFILLERS

Nanofillers are nanoscopic (i.e., size 1–100 nm) particles used as additives or fillers in reinforced polymer composites. Such reinforced composites where the

Conventional composites (or microcomposites)

- Composites of separate phases are obtained with properties similar to traditional composites when the polymer is unable to intercalate between the silicate layers.

Intercalated nanocomposites

- Well-ordered intercalated layers of polymer and clay are obatined when one or more extended polymer chain is intercalated between the layers of the silicate.

Exfoliated nanocomposites

- Exfoliated PNCs are obatined when the silicate layers are completely and uniformly dispersed in a continuous polymer matrix.

FIGURE 4.1   Represent three types of nanocomposites.

nanofillers are dispersed into the matrix of polymer composites are called polymer nanocomposites or PNCs. The surface modification of nanofillers in addition to their small size and large surface-to-volume ratio results in a highly enhanced interaction of nanofillers with the polymer matrix, which in turn enhances some of the mechanical (like stiffness), thermal (kinetic motion), electrical (like electrical resistivity), barrier (like gas storage), chemical (like catalytic reactivity) and other resulting properties (like electromagnetic forces) of PNCs which increase with an increase in the interfacial interactions, adhesion, motion of particles, etc (Guo, 2016).

On the basis of their dimension, nanofillers can be classified into three types (Rangaraj Vengatesan & Mittal, 2015):

   i. one nanoscale dimension (nanoplatelet)
   ii. two nanoscale dimension (nanofiber)
   iii. three nanoscale dimension (nanoparticulate) (Figure 4.2)

On the basis of the particulate material, nanofillers can be broadly classified into three types (Jamróz et al., 2019; Rallini & Kenny, 2017):

   i. inorganic nanofillers (metals and metal oxides)
   ii. organic nanofillers (natural biopolymers)
   iii. organic–inorganic hybrid (hybrid of organic and inorganic) (Figure 4.3)

Depending upon the requirements and availability, the desired material and shape of nanofillers can be decided. Using appropriate chemical and mechanical methods, the nanofillers can be manufactured into their desired form, shape or dimension (Dantas de Oliveira & Augusto Gonçalves Beatrice, 2019; Fu et al., 2019; Jamróz et al., 2019b; Lim et al., 2021; Rallini & Kenny, 2017b; Shankar & Rhim, 2018).

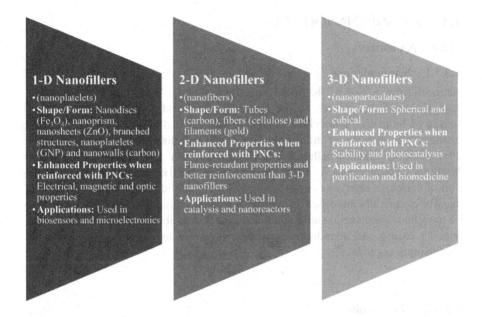

FIGURE 4.2   Represents the classification and properties of nanofillers on the basis of their dimensions.

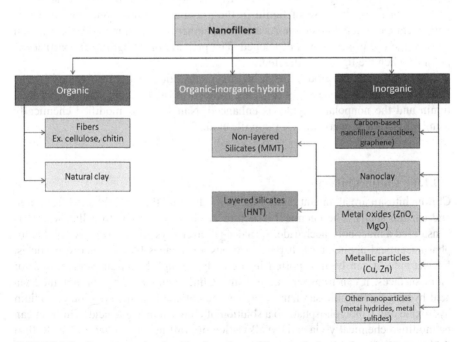

FIGURE 4.3   Represents the classification of nanofillers on the basis of particulate material.

4.3   ORGANIC NANOFILLERS

4.3.1   NANOFIBERS

4.3.1.1   Nanocellulose

Nanocellulose is one of the widely used nanofibers. As a source of renewable energy, there is an interest in the use of biomass and cellulose is a promising source of biomass. Nanocellulose fibrils are produced via mechanical shearing (using a homogenizer or microfluidizer) of the cellulose fiber slurry (Dufresne, 2017). Nanocrystals of cellulose can be obtained via acid hydrolysis (*Cellulose Intercr y s Talline*, n.d.). In 1983, the first report on the mechanical destructuration of cellulose fiber was published in two companion papers. Nanocellulose is usually chemically modified using esterification, amidation, etc., and can also be modified using surfactants. Since nanofibers are biodegradable, environment-friendly to use and abundantly available, their use as nanofillers is becoming increasingly popular.

4.3.1.2   Nanolignin

After cellulose, the abundantly available natural polymer is lignin. It is considered an amorphous polymer that is highly branched and also has high purity. Lignin is obtained by the most important industrial method known as kraft method in terms of quantity production. In a solution of sodium hydroxide and Na_2S (black liquor), lignin is solubilized by the use of the kraft method (Gilca et al., 2014). Due to the presence of sulfur, the use of lignin in the polymer is limited and may interfere with chemical functionalization. As a result, another industrial method of extraction without the use of sulfur was developed. The preparation of lignin nanoparticles is possible when the lignin is extracted.

Nanolignin can be produced via hydroxymethylation of lignin obtained by the alkaline extraction process. By reducing the hydroxyls, the compatibility between lignin and the nonpolar polymer is enhanced. Nanolignin is modified chemically using alkylation or acetylation (Frangville et al., 2012).

4.3.1.3   Chitin/Chitosan

Chitin/chitosan are abundant materials. As food waste, they are obtained from crab and shrimp shells. The chemical structure of chitin is similar to cellulose and is considered a mucopolysaccharide. As a drug delivery system, chitosan is suitable for pharmaceutical applications. It poses various advantages like nonmetric particles, for example, it can be transported in the body through blood circulation, and for more surfaces, it can transport more drugs. Chitin nanofibers can be obtained via acid hydrolysis and chitosan nanofibers can be obtained via the addition of sodium hydroxide and tripolyphosphate to a solution of chitosan in acetic acid. Chitosan can be modified chemically via carboxyalkylation or grafting copolymerization (Rallini & Kenny, 2017).

4.3.1.4　Organic–Inorganic Hybrid Nanofillers

Organic–inorganic hybrid nanofillers have the properties like tenacity derived from the organic material while properties like thermal stability and rigidity are derived from the inorganic material. An example of organic–inorganic hybrid nanofiller is POSS or polyhedral oligomeric silsesquioxanes. These are popular as they are compatible with a lot of different types of polymers and can be directly grafted to use with the polymer. Due to their high compatibility with the polymer, the mechanical and fire retardancy properties are highly enhanced in the POSS-reinforced PNCs.

4.4　INORGANIC NANOFILLERS

4.4.1　Carbon-Based Nanofillers

4.4.1.1　Carbon Nanotubes

Carbon nanotubes (CNTs) are tubular structures having their cylindrical diameter in nanometers and a size length in micrometers. They are classified into multiwalled carbon nanotubes (MWCNTs) or single-walled carbon nanotubes (SWCNTs) based on the method of their preparation (Machado et al., 2014; Torres, 2011). MWCNTs are cylindrical shells of concentric shells of graphene sheets while SWCNTs are solid cylinders of rolled-up graphene sheets. PNCs based on CNTs have highly enhanced electrical, thermal and mechanical properties. The main problem with CNTs is that they tend to agglomerate which results in a decrease in their dispersion into the polymer matrix. The surface of the CNT is chemically modified to introduce functional groups on the side wall to increase polymer–filler interactions. For instance, CNT treated with HNO_3/H_2SO_4 mix (nitric acid–sulfuric acid mixture) gives rise to the formation of carboxylic acid groups on the surface of the CNT leading to better interaction with the polymer matrix (Oliveira et al., 2016; Torres, 2011).

4.4.1.2　Graphene

Andre. K. Geim and Konstantin S. Novoselov discovered graphene in 2004. Graphene nanoparticles are in the form of a sheet structure where the carbon atoms are bonded via sp^2 hybridization. Graphene nanofillers are manufactured using graphites through thermal expansion, micromechanical exfoliation, CVD (chemical vapor deposition) or chemical reduction of graphene oxide. Some properties of graphene are that 1 TPa is Young's modulus of graphene, 5000 W/m.fK is thermal conductivity and 125 GPa is fracture strength. Due to high surface area and high gas impermeability value of graphene, the graphene-reinforced PNCs have enhanced mechanical, thermal, electrical and gas barrier properties. On the exfoliation of bulk graphite into individual sheets, the use of graphene was dependent. To produce individual exfoliated graphene sheets, various chemical–mechanical routes were developed such as chemical exfoliation and CVD. Depending on the presence of defects and the purity, each method has its own advantages and disadvantages (Mittal & Chaudhry, 2015; Wan & Chen, 2012).

4.4.2 NANOCLAYS

4.4.2.1 Layered Nanoclays

These include layered silicates (montmorillonite or MMT), clay minerals (hydrous aluminum phyllosilicates) and traces of metal oxides and organic matter. Nanoclay-reinforced PNCs show an enhanced increase in stiffness, barrier properties and flame retardancy at a very low amount of nanoclay volume fraction due to their lamellar structure and high surface area. Nanoclays are usually organically modified or silanized before polymer composite preparation (Kotal & Bhowmick, 2015).

4.4.2.2 Nonlayered Silicates

These are usually tubular in shape and hollow and have a high aspect ratio and high surface area. Nonlayered silicate-reinforced PNCs show enhanced mechanical, thermal and flammability properties.

4.4.3 METAL OXIDE NANOFILLERS

These include silica (SiO_2), alumina (Al_2O_3), etc. They have a high specific surface area. Metal oxide nanofillers are obtained via the process of high-temperature hydrolysis (to obtain nonporous nanoparticles) or sol-gel process or precipitation process (to obtain porous nanoparticles). Metal oxide nanoparticles tend to agglomerate and hence are modified chemically or using heat treatment (Dantas de Oliveira & Augusto Gonçalves Beatrice, 2019b).

4.4.4 METAL NANOPARTICLES

These include metals like Cu (copper), Zn (zinc) and Ag (silver). Silver nanoparticles can be obtained via chemical reduction or laser irradiation. Nanosilver particles are usually physically modified using surfactants like diacid (Dantas de Oliveira & Augusto Gonçalves Beatrice, 2019b; Rallini & Kenny, 2017).

4.4.5 OTHER NANOPARTICLES

PNCs can also be made with metal hydride alloys. These are produced as free metal powder particles from repeating hydriding–dehydriding cycle and consequent pulverization (Dantas de Oliveira & Augusto Gonçalves Beatrice, 2019b) (Table 4.1).

TABLE 4.1
Summarizes Few Common Nanofillers Used

Name	Types	Manufacturing Process	Surface Modification	Advantages
Carbon-based nanofillers	Carbon nanotubes (CNTs)	Laser furnace, chemical vapor deposition (CVD)	Treatment with (HNO_3/H_2SO_4)	High chemical reactivity and electrical conductivity
	Graphene sheet	CVD, chemical/mechanical exfoliation	Treatment with potassium, sonication	
	Carbon nanofibers (CNFs)	Electrospinning of polymers followed by CVD	Oxidation with strong acids	
	Carbon black	Thermal decomposition by oils (to obtain furnace black)	Oxidation, halogenation, grafting with alkyl chains	
Nanoclays	Layered silicates	Chemical exfoliation	Silanization, use of alkylphosphonium salts	
	Nonlayered silicates			
Natural fibers	Cellulose nanofibrils (CNFs)	Mechanically induced destructuration strategy via mechanical shearing		Biodegradable and easily and abundantly available naturally
	Cellulose nanocrystals (CNCs)	Strong acid hydrolysis treatment	Treatment with sulfuric acid, amidation, esterification, and etherification (during acid hydrolysis)	
	Lignin	Hydroxymethylation or precipitation using hydrochloric acid of extracted lignin	Alkylation, acetylation	

(Continued)

TABLE 4.1 (*Continued*)
Summarizes Few Common Nanofillers Used

Name	Types	Manufacturing Process	Surface Modification	Advantages
	Chitin/chitosan	Dissolve chitosan in acetic acid, adjust pH with sodium hydroxide and add an aqueous solution of tripolyphosphate	Carboxyalkylation, graft copolymerization	
Metallic particles	Metallic nanoparticles	Utilization of metals in solid state storage of the gas		
	Nano-oxides (like silica (SiO_2), alumina (Al_2O_3), and titania (TiO_2))	High-temperature hydrolysis to obtain fumed oxides	Heat treatment/chemical modification to remove the hydrophilic group	
	Nanocarbides [Silicon carbide (SiC) and boron carbide (B4C), zirconium carbide (ZrC)]	Thermal plasma-assisted methods, vaporization–condensation process		
Others	Other nanoparticles (POSS, silver nanoparticles)	Chemical reduction of metal salts and laser irradiation	Graft polymerization for POSS, modification using diacid for nanosilver	Compatible or miscible with many solvents and polymer

Source: Suter et al., 2015; Bitinis et al., 2011; Zaïri et al., 2011.

4.5 SURFACE MODIFICATION TECHNIQUES OF NANOFILLERS

In general, nanofillers and the polymer matrix are incompatible due to the following reasons:

i. Agglomeration of nanofillers

Due to their small size, nanofillers have high surface energy which leads to their agglomeration.

ii. Nonuniform dispersion

Due to the agglomeration of nanofillers, the dispersion of nanofillers into the polymer matrix becomes nonhomogeneous or nonuniform.

iii. Poor interfacial interaction

Due to the surface of nanofillers being hydrophilic and the surface of polymers being hydrophobic, the interfacial interaction and consecutive adhesion/bonding of nanofillers and the polymer matrix become weak during the fabrication step. Due to the incompatibility of nanofillers and the polymer matrix, the properties of PNCs are compromised. The described problems can be overcome by modifying the surface of the nanofillers. Surface modification techniques modify the surface of the nanofillers in such a way that the hydrophilicity of the surface of the nanomaterial is increased and nanoparticles do not agglomerate which in turn leads to better dispersion and interfacial interaction and consequently more enhanced properties of PNCs. Nanomaterials whose surface have been modified are commonly referred to as surface modification nanomaterials or SMNs (Dantas de Oliveira & Augusto Gonçalves Beatrice, 2019; Fu et al., 2019; Kolosov et al., 2020; Nesic & Seslija, 2017; Rallini & Kenny, 2017b; Rangaraj Vengatesan & Mittal, 2015b; Rong et al., 2006; Saman et al., 2021; Tanahashi, 2010; Tanaka et al., 2004) (Figure 4.4).

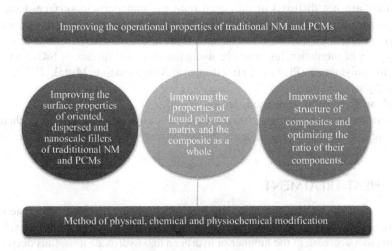

FIGURE 4.4 The considered approach to improving the technological properties of composites (NM: nanomaterial; PCM: polycomposite material).

4.6 TYPES OF SURFACE MODIFICATION TECHNIQUES

4.6.1 PHYSICAL

This involves physical modification of the nanofillers using surfactants or macromolecules or polymeric chains of macromolecules which contain two parts:

    i. one or more polar group(s)
    ii. an aliphatic chain

Due to the presence of a polar group(s) in the surfactant, an electrostatic interaction occurs between the nanofiller surface of high energy and the polar component of the surfactant, which results in selective adsorption of the polar groups onto the surface of the nanoparticle. In some cases, an ionic bond is formed as well. This in turn decreases the extent of interparticle interaction between the nanoparticles resulting in a decrease in the agglomeration of nanoparticles.

For inorganic nanofillers, encapsulation can be carried out:

    i. using preformed polymers
    ii. using in situ polymers
    iii. using the emulsion polymerization process

Encapsulation of inorganic particles with preformed polymers uses hyperdispersants while encapsulation of inorganic nanoparticles with in situ polymers uses a polymeric dispersant that consists of a functional group (like –OH and –COOH) and a macromolecular chain (like polyester and polyether). Hyperdispersants and polymeric dispersants can anchor more strongly to the surface of the nanoparticles and reduce the tendency of re-agglomeration of the nanoparticles due to the interference from their long polymeric chains, as compared to the traditional surfactants. However, hyperdispersants only encapsulate the agglomerated nanoparticles and they are not diffused or adsorbed onto the nanoparticle surface. Emulsion polymerization involves surface adsorption of monomer and consecutive polymerization occurring at the adsorbed surface layer of the nanoparticles due to surface adsorption of monomer, for example, using the modified surface of SiO_2 with oleic acid for polylactide (PLA) matrix (Rangaraj Vengatesan & Mittal, 2015b; Rong et al., 2006).

Advantage: Physical methods are easier and faster to implement.

Drawback: The modifiers (surfactants) are easily desorbed with no chemical bonding to hold them in place, hence leading to insignificant surface modification.

4.7 HEAT TREATMENT

This is used to enhance the electrical properties of nanomaterials. The use of an appropriate amount of heat dries the moisture on the surface of the nanofillers, hence leading to a decrease in the number of hydrogen and hydroxide ions, thus decreasing the overall conductivity of the nanomaterial resulting in better insulation.

Drawback: If the amount of heat exposure to the nanomaterial exceeds the minimum amount of heat that the nanofiller can withstand, then the surface structure and morphology of the nanofiller can go through irreparable damage.

4.7.1 Plasma Treatment

Plasma treatment involves coating the surface of the nanofiller with plasma. It uses either a glow plasma discharge (uniform discharge) or filamentary plasma discharge (nonuniform discharge). Uniform discharge leads to more enhanced dispersibility of nanoparticles in the polymer matrix. Plasma is the fourth state of matter, where an ionized substance starts exhibiting a very high electrical conductivity. Plasma must be chosen from its different types (based on electron density and electron temperature) and effectiveness. Cold plasma treatment can be done under two conditions (Saman et al., 2021).

4.7.2 Low Atmospheric Pressure Discharge

It involves the use of low-pressure plasma (made under pressure lower than atmospheric pressure) and low-pressure plasma reactors (like microwaves at 2.53 GHz) inside a closed vacuum system. Uniform discharge is easily achieved under low gas pressure.

Low-pressure plasma treatment is a complicated and expensive method. Moreover, low-pressure conditions are difficult to achieve, maintain and control as part of a continuous process (Laroussi, 2009; Mandolfino, 2019).

4.7.3 High Atmospheric Pressure Discharge

It overcomes the problems with low-pressure plasma discharge treatment as it does not use a closed vacuum system. Under atmospheric conditions, nanoparticles are exposed to the plasma plume forms. It can be carried out using two regimes:

   i. microdischarge or filamentary plasma (formed using corona discharge, dielectric barrier discharge and plasma jet configuration)
   ii. atmospheric pressure glow discharge

Advantage: This is a very effective method of enhancing the properties, mainly the insulation properties.

Drawbacks: Selection of the most suitable plasma is difficult. Plasma discharge treatment also involves the generation of harmful by-products (Brandenburg, 2018; Cristofolini et al., 2020; Zhang et al., 2017).

4.7.4 γ Irradiation

The nanoparticles are exposed to γ rays, and hence, both the outside and inside surfaces of the nanoparticle agglomerates are equally activated due to which monomers are able to interact with the nanoparticles to a higher extent. Silica, iron, nickel and cobalt nanoparticles are used for γ irradiation grafting.

Advantages: Chemical modification ensures covalent bond formation between the modifier and the nanofiller; hence, desorption of the modifier from the nanoparticles is eliminated.

Drawbacks: The use of chemicals in this method is an unsustainable practice owing to the nondecomposability and toxicity of chemicals. Moreover, the by-products formed in the process are sometimes unknown and might be even more harmful to the environment.

4.8   CHEMICAL TREATMENT

This involves the use of chemical agents and solvents to alter the surface of the nanoparticles.

4.8.1   COUPLING AGENT TREATMENT

Coupling agents like silane, zirconate, etc., are used to increase the extent of adhesion between the nanofillers and the polymer matrices.

Drawback: It is hard to figure out the uniform surface coverage of coupling agents on nanofillers with the available methods.

4.8.2   GRAFT POLYMERIZATION

It involves the interaction of graft monomer particles with the polymer matrix under the desired graft conditions to form a nanocomposite microstructure (radical, anionic and cationic polymerization) consisting of an agglomeration of strong nanoparticles and the grafted and the ungrafted polymers. Graft polymerization can be done using two routes (Vengatesan & Mittal, 2015; Rong et al., 2006):

   i. **Grafting from:** It involves the use of active compounds as an initiator which are attached to the inorganic particle surface covalently. The monomers are polymerized from these active compounds from scratch. This is of two types:

      i.   Using polymerizable groups (e.g. silane grafting)
      ii. Using initiating groups
            While using polymerizable groups is easier, polymerization through initiating groups leads to a higher grafting percentage and efficiency.

   ii. **Grafting to:** Polymers are not synthesized from scratch using monomers. Ready-made polymers containing reactive end groups are available which interact with the functional groups on the nanoparticles' surface. Grafting considerably improves the dispersibility of nanoparticles in organic solvents and their compatibility with polymers.

         'Grafting from' is more widely used due to the easy penetration of small monomers into the nanoparticle agglomerates rather than larger polymer penetration in 'grafting to'. Moreover, 'grafting to' leads to a polymer coating on nanoparticles resulting in further attachment of the polymer during the composite preparation although 'grafting in' allows to control the weight of the grafting polymer.

4.9 MECHANISMS OF REINFORCEMENT OF NANOFILLERS IN PNCs

Reinforcement mechanisms in nanocomposites are complex and are still a topic of research. There are many different mechanisms proposed for nanocomposites, but a conclusion has not been reached yet. According to the different types and shapes of the nanofillers being used, the reinforcing mechanism of rubber nanocomposites differs.

Reinforcement of particulate nanofillers can be explained based on the following elements:

   i. **Polymer network (strain independent):** Describes the increase in modulus due to the polymer cross-link network of the unvulcanized as well as the vulcanized rubber.
  ii. **Hydrodynamic effects (strain independent):** Describe the increase in the modulus of the rubber as a function of the volume fraction of the filler when nanoparticles are added.
 iii. **Filler–composite interactions (strain independent):** Describe the formation of bound rubber or immobilized rubber due to the trapping of the polymer rubber chains inside the void of the filler–filler network. Subsequently, the immobilized rubber starts acting as filler particles contributing to the reinforcement mechanism due to the prevention of the separation of the nanofiller and rubber giving rise to a 'constrained compatibility' between the nanofiller and the rubber.
  iv. **Filler–filler interactions (dependent on strain):** Describe a decrease in the modulus at high strain due to breaking of the filler–filler bonds.

According to the shape of the nanofillers being used, the reinforcing mechanism of rubber nanocomposites differs.

4.9.1 FOR PARTICULATE FILLERS (CARBON BLACK, SILICA, ZNO)

Deformation at low strains of the particulate nanofiller agglomerates is very common in nanoparticle-reinforced rubber. Payne effect which involves the reformation of nanofiller agglomerates or network structure at high filler concentrations is used to describe the filler interactions with the rubber and the filler itself. Reinforcement of nanoparticles in rubber is mainly attributed to the increase in the modulus of the rubber composite due to the constrained volume of the rubber around the nanoparticles which is mainly described by filler–polymer interaction in the case of nanoparticles.

The reinforcement efficiency for particulate fillers mainly depends on the:

i. Size of the nanoparticle
ii. Structure of the nanoparticle
iii. Surface area of the nanoparticle
iv. Filler–rubber interaction
v. Filler–filler interaction
vi. Dispersion Level

4.9.2 FOR TUBULAR FILLERS (CARBON NANOTUBES [CNTs], NANOFIBERS)

The bonding between tubular fillers and polymer is weak due to the tendency of nanotubes to form agglomerates resulting in a nonuniform dispersion of nanotubes into the polymer matrix. Tubular fillers like CNTs are reinforced to the polymer through the formation of covalent bonds due to open-end oxidation or functionalization of the walls/surface of the modified tubular-walled surface of CNTs. Additionally, modification of polymer surface too leads to better dispersion and adhesion-like reinforcement of modified CNTs with epoxidation of natural rubber gives a much better adhesion than reinforcement of CNTs with natural rubber.

Nanomechanical interlocking of CNTs is a method involving a change in the configuration or surface morphology of the nanotubes to increase the filler–rubber interactions through noncovalent bond formation between the nanotubes and the rubber. Reinforcement of tubular fillers in rubber is mainly attributed to the effect of the formation of cross-links or networks and physical entanglements with rubber.

The reinforcement efficiency for tubular fillers mainly depends on the:

i. Aspect ratio
ii. Dispersion Level
iii. Alignment
iv. Extent of interfacial stress transfer

4.9.3 FOR LAYERED FILLER (NANOCLAYS)

Layered fillers like nanoclays consisting of layered silicates have two layers and the gap in between these two layers is referred to as 'gallery'. This gap is filled by the addition of cations to counterbalance the negative charges of the silicate sheet. Then, these cations are organically modified by replacing the cations with long organic chains (like alkyl ammonium or alkyl phosphonium ions) consisting of organic cations. This allows the dispersion of the layered silicates into the polymer matrix as intercalated (for shorter alkyl chains lesser than 8 carbon atoms) or exfoliated structures (for longer alkyl chains). The intercalation increases the interlayer distance giving rise to better dispersion of the modified layered silicates. Low surface energy polymers are adsorbed to the high surface energy layered silicates by entering in between the gaps or the galleries and subsequently into the interlayer spacing. Immobilization of the rubber around the nanofiller surface also contributes to the reinforcement of the nanofiller into the polymer. However, at low strains, the hydrodynamic effect mainly describes the reinforcement mechanism and the increase in the modulus of the rubber–nanoclay composite.

Reinforcement for nanofillers having a high aspect ratio is described by physical entanglements rather than the filler–polymer interactions. For example, CNTs being tubular have a high aspect ratio and have physical entanglements influencing the reinforcement more than the filler–rubber interactions in low aspect ratio carbon black or silica.

The reinforcement efficiency for nanoclay fillers mainly depends on the:

   i. Extent of exfoliation or intercalation
  ii. Degree of immobilization of polymer or volume of polymer constrained
 iii. Alignment

4.9.3.1 Mechanisms

Now, many mechanisms have been proposed for rubber reinforcement by nanofillers based on the shape of the nanofillers, but the common features found in all types of nanofiller–rubber composite which contribute to the reinforcement mechanism are:

   i. Payne effect describing the bonding–unbonding or agglomeration–deagglomeration of the filler network
  ii. Formation of percolated nanofiller networks at higher concentrations of the nanofiller (percolation phenomenon)
 iii. Immobilized or bound rubber contributing to increased filler–rubber interactions

Reinforced rubber composite mainly depends on the stiffness of the reinforced composite, i.e., the more the stiffness, the better the reinforcement. Hence, the Payne effect, formation of percolated structure, formation of immobilized rubber, etc., all lead to an increase in the stiffness, and subsequently, the reinforcement takes place.

At low strains, linear viscoelastic behavior can be observed due to a very high degree of reinforcement because of the physical entanglements around the surface of the nanofillers or the restriction of the conformational freedom of the polymer chains on the nanofiller surface. Percolated structure and nanofiller network contribute to the reinforcement at low strains. Hence, at low strains, a very high modulus is observed. At high strains, the percolated structures and nanofiller network deform and the modulus is decreased suddenly due to the release of polymer chains, and this release of strain and subsequent re-adsorption of the released polymer chains restates the modulus back. The increase in modulus is attributed to the Payne effect which involves the reformation of nanofiller agglomerates or network structures at high filler concentrations. The nonlinear behavior in the mechanism of polymer matrix-filler bonding and unbonding is associated with the network structure of filler going through an agglomeration–deagglomeration process.

In general, the reinforcement efficiency depends on the following factors:

   i. Particle size of nanofiller
  ii. Structure of nanofiller

   iii. Surface area of nanofiller

   iv. Filler–composite interactions: Higher the filler–composite interactions, i.e., the higher the bonding between the filler and the composite, the higher the reinforcement.

   v. Filler–filler interactions: Lower filler–filler interactions (or larger interlayer distance between the layers of nanofillers), i.e., interparticle interactions of the nanofiller, give rise to higher filler–composite interaction, hence a higher degree of reinforcement.

   vi. Dispersion level of nanofiller in the polymer matrix: The higher the uniformity and the homogeneity of dispersion of the nanofiller into the polymer matrix, the higher the reinforcement.

   vii. Volume of the polymer matrix in the proximity of nanofillers

   viii. Extent of network or cross-linking

   ix. Interfacial adhesion: The higher the interfacial adhesion between the nanofiller surface and the polymer surface, the higher the reinforcement (Donnet, 1998; Maiti et al., 2008; Rezende et al., 2010; Robertson et al., 2011; Sahakaro, 2017).

4.10 COMPATIBILIZATION OF NANOFILLERS

The differences in the chemical nature of polymers and nanofillers give rise to incompatibility between the polymer and nanofiller which in turn lead to low enhancement of properties of the final polymer nanocomposite. Nanofillers are subject to thermomechanical degradation while polymers may undergo cross-linking (Mistretta et al., 2014). Degradation of substances during surface modification involving decomposition of the modifier and the resulting interaction between the degraded products and polymers leads to less interfacial interaction between the nanofiller and polymer. This degradation must be minimized to enhance the properties of PNCs. This is done by either increasing the number of processing cycles or carrying out the processing in severe conditions for a longer time. The compatibility of the nanofiller is further enhanced during the processing step of PNCs (Dantas de Oliveira & Augusto Gonçalves Beatrice, 2019c; Mistretta et al., 2014). Research has shown that polymer blends of polyesters and polyamides using clay nanofillers can be compatibilized with compatibilizers like maleic anhydride, oxazoline, etc (Dantas de Oliveira & Augusto Gonçalves Beatrice, 2019; Taguet et al., 2014; Vrsaljko et al., 2015).

4.11 PROCESSING OF POLYMER NANOCOMPOSITES

Polymer nanocomposites can be produced by two methods:

   i. in situ polymerization

   ii. blending

For the preparation of good quality polymer nanocomposite samples, using a proper processing method is critical to achieve high performance of PNCs. The selection of a proper processing method is based on the type of polymeric matrix and nano-fillers and desired properties for the final product (Dantas de Oliveira & Augusto Gonçalves Beatrice, 2019; Donatella et al., 2016; Fu et al., 2019; Gou et al., 2012; Nayak et al., 2020; Rallini & Kenny, 2017b; Tanahashi, 2010b).

4.11.1 IN SITU POLYMERIZATION

In situ polymerization or in situ intercalative polymerization involves the dis-persion of the nanofiller particles into a monomer or monomeric solution (of the selected polymer) followed by its polymerization to obtain the final PNC product. Polymerization is carried out via standard methods of heat/high-tem-perature treatment/radiation or via an initiator (organic/inorganic or a catalyst). After polymerization, the polymer and nanofiller are covalently bonded or poly-mer molecules can also wrap themselves around the nanofiller particles. Clay-reinforced PNCs, carbon-based reinforced PNCs, metal-polymer PNCs, etc., can be prepared via in situ polymerization. In situ intercalative polymerization is especially used to manufacture clay-based PNCs where the nanoclay having an organic filler intercalated between the interlayer space/galleries is dispersed into the polymer matrix.

Advantages: In situ polymerization is safe to use for thermally unstable or insolu-ble polymers being used in PNC preparation. Polymers with high loading of nanofill-ers can also be prepared via the in situ method because the polymers can be grafted onto the surface of the nanofiller. Other benefits also include cost effectiveness, high transparency, good interfacial adhesion of the nanofiller and controllable particle morphology.

Drawbacks: The main problem of this method is the increase in the viscosity of the polymer–filler mixture as the polymerization progresses. Removal of solvent also becomes a challenge if in situ polymerization is carried out in the presence of a solvent. Moreover, nanofiller particles tend to agglomerate easily; hence, surface modification of nanofillers is essential before processing via in situ.

4.11.2 BLENDING/DIRECT MIXING

Direct mixing involves the breaking down of the nanofiller agglomerates due to the mixing process of nanofillers and polymers. The mixing can be carried out using two methods:

    i. Mixing in the presence of a solvent or solution mixing
    ii. Mixing in the absence of a solvent or melt compounding

4.11.2.1 Solution Mixing

Solution mixing involves the dispersion of the nanofiller into a polymer solution fol-lowed by evaporation or coagulation or precipitation of the solvent to get the final PNC product. Sonication, ultrasonic irradiation, magnetic stirring, shear mixing or other forms of agitation are used to disperse the nanoparticles in the polymer solution for deagglomeration. PNCs manufactured through solution mixing are generally obtained as thin films or sheets. Solution mixing is used to produce layered silicate-reinforced PNCs, CNT-reinforced PNCs, etc.

Advantages: This method is easy to use and a relatively good dispersion of nano-filler into the polymer matrix is obtained. Being a solution-based method, the overall viscosity of the nanofiller-polymer-solvent mixture is reduced which facilitates dispersion and mixing.

Drawbacks: Due to the use of the solvent, the selection of polymer is subject to its solubility in the solvent and hence the selection is limited to only a few polymers. The use of chemical solvents and their subsequent disposal pose a threat to the environment.

4.11.2.2 Melt Blending

Melt blending is a comparatively new and well-established method carried out in the presence of an inert gas like argon, neon, etc. In this, nanofillers are dispersed into the polymer matrix above the glass transition temperature of the polymer or into the molten polymer. It makes use of high-force shear compounding devices such that shear is applied on the nanoparticles by the polymer during the mixing in the melt form by the dragging viscous force. As a result of this shear force, the breakdown of the agglomerates of the nanofillers occurs, thus achieving a uniform and homogeneous dispersion. The breakdown of the agglomerates can be explained through rupture models. One rupture model states that since the area of contact of the particles with their neighbors is very small at the cross section of the nanofiller agglomerate, rupture first occurs there and rises to split the agglomerate into two equal halves. Another rupture model called the 'onion' peeling' model states that the pressure or stresses generated at any point on the surface of the agglomerates are large enough to eliminate the smaller particles or aggregate of smaller particles off of the larger agglomerates. Removal of the smaller nanofiller agglomerates slowly forms a cloud of agglomerates near the initial agglomerate, preventing a further reduction in the size of the initial nanofiller agglomerate. A further reduction in size occurs when the aggregates are swept off from the cloud and are replaced by fresh aggregates from the agglomerate. Melt blending is the preferred method for the preparation of PNCs of thermoplastics and elastomeric polymeric matrix.

Advantages: The main advantage of this method is that since it does not require the use of a chemical solvent, it is environment-friendly and is also simple, easy and convenient to use. Well-dispersed PNCs are obtained only at lower nanofiller loads. Moreover, melt blending gives end users the liberty to choose specifications for the final product like the choice of polymer grade, level of reinforcement, etc., because it allows the direct formulation of PNCs using the shearer/mixer which shifts the process downstream. The specificity of interactions of filler–polymer and intercalation is also increased as the polymer–solvent interactions are eliminated due to the absence of the solvent.

Disadvantage: The limitation of this technique is that it uses a high temperature, which can damage the modified surface of the nanofillers.

4.11.3 Bottom-Up

The bottom-up approach involves the creation of a second phase directly in the polymer matrix. Typically, a precursor (a salt, an alkoxide) dispersed in the polymer matrix undergoes a series of chemical reactions to obtain the second phase.

Advantages: The bottom-up approach allows direct control of the interphase of the particles. For instance, to increase filler–polymer interfacial interactions, suitable covalent bonds can be created.

4.11.4 Sol-Gel

Sol-gel or in situ particle processing is mainly used to produce organic–inorganic hybrid PNCs. Sol-gel either proceeds through the mixing of a precursor with a copolymer of the polymer matrix or the mixing of a precursor with the polymer matrix itself; a combination of both can also be used. Basically, it involves in situ precipitation of the nanofiller in the polymer matrix through hydrolysis and condensation reactions of the precursors. In the sol-gel method, nanoparticles in solid form are spread over a monomer forming a colloidal solution called sol, and interconnecting regions, called gel, between the phases are created during the polymerization reaction spreading a 3D network of filler–polymer throughout the liquid phase. Crystals are created in the polymer matrix due to nucleation, and finally, PNCs are created from these crystals. Silica-reinforced PNCs, titania-reinforced PNCs, metal-polymer PNCs, etc., can be prepared using the sol-gel processing technique. The sol-gel method is also fit for amorphous polymers like rubbers.

Advantage: It can be used for insoluble polymers and for the preparation of organic–inorganic hybrid PNCs.

Disadvantages: The process control is difficult and needs to be continuously monitored and maintained due to the complex chemical reactions taking place during the sol-gel PNC preparation process. Moreover, chemical wastes are nonbiodegradable and harmful to the environment, and hence, it should be ensured that they are properly disposed.

4.12 CHARACTERIZATION/EVALUATION TECHNIQUES

Characterization/evaluation techniques are used to understand the basic physical and chemical properties of PNCs. Figure 4.5 shows various characterization/evaluation techniques.

4.12.1 Structural and Morphological Characterization

Good dispersion and distribution are the key factors for improving the properties of nanocomposites. During the preparation of nanocomposites, different structures and morphologies are generated, and they depend on various parameters. The most important is the method of preparation, particle concentration and the nature of

FIGURE 4.5   Various characterization/evaluation techniques.

particle functionalization (Marini & Bretas, 2017; Rallini & Kenny, 2017b; Zhao et al., 2005). Some commonly used techniques are:

   i. wide-angle X-ray diffraction (WAXD)
  ii. small-angle X-ray diffraction (SAXD)
 iii. scanning electron microscopy (SEM)
  iv. Raman spectroscopy
   v. atomic force microscopy (AFM)

4.12.2   WIDE-ANGLE X-RAY DIFFRACTION

WAXD involves analysis of the diffraction pattern. It identifies the crystalline phases by measuring the space between the crystalline layers using Bragg's law. This method provides information about the crystalline structure, unit cell size, chemical composition as well as chemical stoichiometry. This method is the most widely used and is simple and easily available. The industrial application of WAXD includes its use in the evaluation of intercalation/exfoliation of clays in thermoplastics, elastomeric matrices and thermosets.

4.12.3   SMALL-ANGLE X-RAY DIFFRACTION/SMALL-ANGLE X-RAY SCATTERING

SAXS is an analytical characterization technique. In SAXS, the intensities of X-rays scattered by a sample are measured as a function of the scattering angle. These measurements are done at very small angles. AFM and SEM are also required to characterize the dispersion and distribution of nanoparticles.

4.12.4 Scanning Electron Microscopy

SEM is a type of electron microscope. In this, images of the sample surface are obtained by scanning the surface with a focused beam of electrons.

4.12.5 Raman Spectroscopy

Raman spectroscopy has proven to be a useful investigation of carbon-based material properties. This technique is based on the vibrational spectra. It uses a laser light that interacts with molecular vibration/phonons. The energy of these laser phonons is shifted up or down. It evaluates energy shifts to provide information on vibrational modes in conventional Raman spectroscopy systems. It is suitable for the analysis of nanomaterials. Their vibrational spectra allow the identification of phases and phase transitions, or shape determination of nanostructures, in nanoparticles. This technique is particularly suitable for the analysis of graphene nanoplatelets.

4.12.6 Atomic Force Microscopy (AFM)

Images of any type of surface including polymers, ceramics, composites, glasses and also biological samples can be obtained. Measurement and localization of adhesion strength, magnetic forces and mechanical properties can be done using AFM. This technique is performed using a sharp tip about 10–20 nm in diameter attached to a cantilever. AFM consists of a cantilever in silicon or silicon nitride provided with a probe capable of interacting with the sample surface. It may work in two modes: contact mode and trapping mode. The AFM tip is continuously in contact with the surface in the process of the contact mode while in tapping mode the AFM cantilever is vibrated above the sample surface so the tip is in contact with the surface only intermittently.

4.13 THERMAL, MECHANICAL, RHEOLOGICAL AND OTHER TECHNIQUES OF CHARACTERIZATION

To enhance the properties of polymeric materials like conductivity, strength, etc., dispersed nanoparticles are used. A large surface-to-volume ratio facilitates increased interaction between the matrix and fillers, thus enhancing the overall material properties. In this section, we will discuss the effect of various fillers on mechanical, electrical, thermal, rheological and other techniques. So, as we go for further characterization of PNCs, some techniques are commonly used, i.e., DSC, FTIR, rheometry, TGA, TMA and DMA. Viscoelastic measurements are more sensitive to the nano- and the mesoscale structure of polymers; when combined with techniques like WAXD, TEM, DSC, TGA and DMA, it will provide the understanding of the state and mechanism of dispersion of the nanoparticles. Variations in polymer composition and structure highly affect the rheological properties of PNCs. In dynamic mechanical spectrum, nanocomposites exhibit a change of pattern, as a function of the degree of exfoliation/dispersion. No. of particles per unit volume, use for determining the characteristics response of nanocomposites. These nanocomposites show solid-like behavior and slower response because of the presence of nanofillers. This type of behavior is explained by the development of grafting percolated nanoparticle network structure. Its formation is a consequence of physical interactions between dispersed nanoparticles, polymeric chains and surfactants, which promote a considerable resistance to flow.

4.14   BENEFITS AND DRAWBACKS

PNCs are composite materials, and hence, they give the users the freedom to manufacture PNCs of desired properties according to their requirements.

Several benefits are associated with PNCs:

    i. low cost
   ii. superior mechanical, flammability and thermal properties
  iii. increased biodegradability of biodegradable polymers due to nanofiller reinforcement
   iv. high temperature capability.

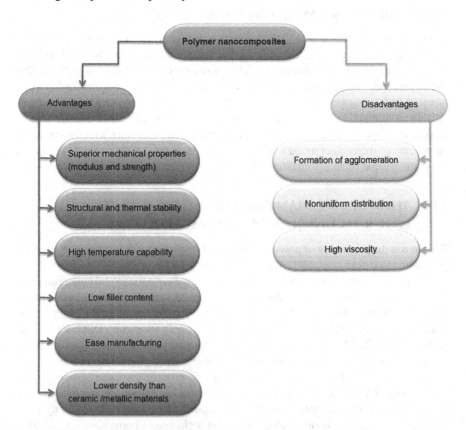

FIGURE 4.6   The benefits and drawbacks of PNCs.

But, the development of PNCs also poses several challenges. The biggest limitation of PNCs is nanoparticle agglomeration. But the majority of the time, nanofillers cannot be directly used with the polymer matrix due to their incompatibility with the polymer. Hence, before the processing step, surface modification of nanofillers becomes necessary for better reinforcement, adding an extra step of costing time and money. Figure 4.6 represents the benefits and drawbacks of PNCs (Al-Haik et al., 2010; Njuguna et al., 2014; Tyagi & Tyagi, 2014).

4.15 POTENTIAL APPLICATIONS OF POLYMER NANOCOMPOSITES

PNCs have various applications in an increasing number of commercial products, for example, structural applications, industrial applications such as automotive parts, food packaging and constructions (Ray, 2014), aerospace, etc., the development of high-performance composites, biomedical applications (Suter et al., 2015), coating applications, environment protection, etc (Bitinis et al., 2011; Zaïri et al., 2011) (Table 4.2).

TABLE 4.2
Shows Various Modes of Applications

Various Modes of Applications	Uses
Structural applications	Polymeric materials are mostly used in structural applications because of vast applications from domestic to aerospace. PNCs have exceptional mechanical properties and also good electrical conductivities. So, it is good for structural applications
Develop high-performance composites	To reduce weight and increase strength considerably
Industrial applications	Automotive: It is used in gas tanks and interior and exterior panels. Aerospace: Use of high-performance and flame-retardant panels Food packaging: Containers and wrapping films in food packaging. The use of nanocomposite packaging increases the shelf life of various types of foods. Construction: It is used in building sectional and structural panels
Environment protection	Water-laden atmosphere is the most damaging environment polymer materials can encounter. The major advantage is the ability of absorption of less water. Use environment-friendly nanocomposites that have strong anticorrosion properties .It plays a significant role in the development of a sustainable environment
Biomedical applications	Biodegradable polymers fulfill the need for biomedical applications. PLA is a biodegradable polymer and is used in food packaging applications and biomedical products

Source: Suter et al., 2015; Bitinis et al., 2011; Zaïri et al., 2011.

4.16   CONCLUSION

In this chapter, the commonly used nanofillers, their types, surface modification techniques, reinforcement mechanisms and compatibilization have been reviewed. The processing methods and characterization techniques of nanofiller-reinforced composites have also been explained. PNCs are more in demand than conventional polymer composites due to the small size of nanofillers contributing to the light-weight of PNCs, the enhanced stiffness and other mechanical and barrier properties. Most nanofillers in popular use are inorganic and nonbiodegradable. With the focus on sustainable technology, nanofibers like nanocellulose and nanolignin should be used more often. But, even nonbiodegradable nanofillers increase the biodegradability of the biodegradable polymer composites due to a very less amount of nanofillers required for the reinforcement, another class of nanofillers called hybrid nanofiller which is a combination of both organic and inorganic nanofillers is also becoming increasingly popular due to their direct grafting, high compatibility with many polymers and more enhanced properties of the hybrid nanofiller-reinforced PNCs.

In general, nanofillers cannot be directly used with the polymer due to their incompatibility with the polymer because of the tendency of agglomeration of nanofillers which results in a nonuniform dispersion of the nanofiller particles in the polymer matrix and low adhesion and interfacial interaction between the nanofiller and polymer surface. Hence, the nanofillers' surface needs to be modified either by physically using surfactants or by giving heat treatment, chemically using chemical coupling agents or graft polymerization or by giving plasma treatment. Surface modification of nanofillers usually functionalizes the surface of nanofillers that introduces functional groups on the surface of the nanofiller which increases covalent or noncovalent or van der Waals or hydrogen bonding depending upon the type of the functionalization group introduced. For better adhesion and interaction, surface modification of polymers can also be done. Compatibilizers are also added to the nanofiller and polymer matrix to prevent the degradation of substances during surface modification. The addition of a modifier during surface modification undergoes thermal decomposition which results in an interaction between the degraded products and the polymers, thus decreasing the nanofiller and polymer interactions leading to lower adhesion.

The mechanism of reinforcement of nanofillers has been sufficiently researched, but yet, no conclusion has been reached to confirm a general mechanism. This chapter reviews nanofiller reinforcement rubber which mainly depends on filler–filler and filler–polymer interactions. More research is recommended to identify reinforcement mechanisms at nano, atomic and molecular levels for different nanofillers reinforced in different polymers.

The processing of PNCs basically involves the dispersion of nanofillers into the polymer matrix. The selection of the final processing technique should be done based on the availability of the nanofiller and the polymer matrix and most importantly in order to achieve a good dispersion of the nanofiller into the polymer matrix for producing a good quality PNC. Therefore, a single PNC processing method cannot be used to produce all types of PNCs. PNCs can be obtained according to our desired shape like thin films or sheets, using methods like in situ polymerization, direct mixing, sol-gel, etc. PNCs of good quality can only be processed for only a limited amount of nanofillers, but to achieve the full potential of the properties of PNCs, the problem of low dispersion at high nanofiller loading should be solved. Moreover, the processing of PNCs leads to the degradation of the modified nanofiller surface. Research needs to be done to eliminate such degradation problems. Most of the conventional processing techniques require surface modification of nanofillers to eliminate the problem of agglomeration; sol-gel, however, can be used without a modified nanofiller as the aggregates are broken using high-force compounding devices. Finally, we use characterization techniques for PNCs to understand the mechanical, thermal, rheological, electrical and flammability properties by measuring or detecting the crystallographic structures of nanofillers, chemical composition, particle concentration, dispersion and distribution pattern of nanofillers, etc., using various techniques like WAXD, SAXD, Raman spectroscopy, etc. After identifying and comprehending the properties of PNCs, one can utilize them in suitable applications and can also research to further improve the properties.

Although PNCs being composite materials have an edge over the use of conventional materials given the flexibility to choose the polymer and the nanofiller, the surface modification technique and the processing step according to our requirements, yet, they also pose a limitation on the available choices due to the incompatibility of polymer and nanofiller, hence a further restriction to use the most suitable processing method for maximum processing efficiency. Therefore, more research is needed to find the best combinations of polymer, nanofiller and processing methods in order to manufacture the best quality PNCs.

REFERENCES

Al-Haik, M., Luhrs, C. C., Reda Taha, M. M., Roy, A. K., Dai, L., Phillips, J., & Doorn, S. (2010). Hybrid carbon fibers/carbon nanotubes structures for next generation polymeric composites. *Journal of Nanotechnology.* https://doi.org/10.1155/2010/860178

Alexandre, M., & Dubois, P. (2000). Polymer-layered silicate nanocomposites: preparation, properties and uses of a new class of materials. *Materials Science and Engineering: R: Reports, 28,* 1–63.

Beatrice, C. A. G., Branciforti, M. C., Alves, R. M. V., & Bretas, R. E. S. (2010). Rheological, mechanical, optical, and transport properties of blown films of polyamide 6/residual monomer/montmorillonite nanocomposites. *Journal of Applied Polymer Science, 116*(6), 3581–3592. https://doi.org/10.1002/app.31898

Bhattacharya, S. N., Kamal, M. R., Musa, R., & Gupta, R. K. (2008). *Polymeric Nanocomposites: Theory and Practice.* Carl Hanser Publishers.

Bitinis, N., Hernandez, M., Verdejo, R., Kenny, J. M., & Lopez-Manchado, M. A. (2011). Recent advances in clay/polymer nanocomposites. *Advanced Materials, 23*(44), 5229–5236. https://doi.org/10.1002/adma.201101948

Brandenburg, R. (2018). Corrigendum: Dielectric barrier discharges: progress on plasma sources and on the understanding of regimes and single filaments (Plasma Sources Science and Technology (2017) 26 (053001) DOI: 10.1088/1361-6595/aa6426). In *Plasma Sources Science and Technology* (Vol. 27, Issue 7). Institute of Physics Publishing. https://doi.org/10.1088/1361-6595/aaced9

Cellulose Intercr y s Talline. (n.d.).

Cristofolini, A., Popoli, A., & Neretti, G. (2020). A multi-stage model for dielectric barrier discharge in atmospheric pressure air. *International Journal of Applied Electromagnetics and Mechanics, 63*(S1), S21–S29. https://doi.org/10.3233/JAE-209120

Dantas de Oliveira, A., & Augusto Gonçalves Beatrice, C. (2019). Polymer nanocomposites with different types of nanofiller. In *Nanocomposites - Recent Evolutions.* IntechOpen. https://doi.org/10.5772/intechopen.81329

de Melo, C. C. N., Beatrice, C. A. G., Pessan, L. A., de Oliveira, A. D., & Machado, F. M. (2018). Analysis of nonisothermal crystallization kinetics of graphene oxide - reinforced polyamide 6 nanocomposites. *Thermochimica Acta*, *667*, 111–121. https://doi. org/10.1016/j.tca.2018.07.014

Donatella, D., Silvestre, C., Cimmino, S., Marra, A., & Pezzuto, M. (2016). Processing, structure, and morphology in polymer nanocomposites. *Polymer Morphology: Principles, Characterization, and Processing*, 374–396.

Donnet, J. B. (1998). Black and white fillers and tire compound. *Rubber chemistry and technology*, 71(3), 323–341.

Dufresne, A. (2017). Cellulose nanomaterial reinforced polymer nanocomposites. In *Current Opinion in Colloid and Interface Science* (Vol. 29, pp. 1–8). Elsevier Ltd. https://doi. org/10.1016/j.cocis.2017.01.004

Frangville, C., Rutkevičius, M., Richter, A. P., Velev, O. D., Stoyanov, S. D., & Paunov, V. N. (2012). Fabrication of environmentally biodegradable lignin nanoparticles. *ChemPhysChem*, *13*(18), 4235–4243. https://doi.org/10.1002/cphc.201200537

Fu, S., Sun, Z., Huang, P., Li, Y., & Hu, N. (2019). Some basic aspects of polymer nanocomposites: A critical review. *Nano Materials Science*, *1*(1), 2–30. https://doi.org/10.1016/j. nanoms.2019.02.006

Gilca, I. A., Ghitescu, R. E., Puitel, A. C., & Popa, V. I. (2014). Preparation of lignin nanoparticles by chemical modification. *Iranian Polymer Journal (English Edition)*, *23*(5), 355–363. https://doi.org/10.1007/s13726-014-0232-0

Gou, J., Zhuge, J., & Liang, F. (2012). Processing of polymer nanocomposites. In *Manufacturing Techniques for Polymer Matrix Composites (PMCs)* (pp. 95–119). Elsevier. https://doi. org/10.1533/9780857096258.1.95

Guo, Q. (2016). *Polymer Morphology: Principles, Characterization, and Processing*. John Wiley & Sons.

Jamróz, E., Kulawik, P., & Kopel, P. (2019). The effect of nanofillers on the functional properties of biopolymer-based films: A review. In *Polymers* (Vol. 11, Issue 4). MDPI AG. https://doi.org/10.3390/polym11040675

Kolosov, A. E., Sivetskii, V. I., Kolosova, E. P., Vanin, V. v., Gondlyakh, A. v., Sidorov, D. E., Ivitskiy, I. I., & Symoniuk, V. P. (2020). Use of physicochemical modification methods for producing traditional and nanomodified polymeric composites with improved operational properties. *International Journal of Polymer Science*, *2019*. https://doi. org/10.1155/2019/1258727

Kotal, M., & Bhowmick, A. K. (2015). Polymer nanocomposites from modified clays: Recent advances and challenges. In *Progress in Polymer Science* (Vol. 51, pp. 127–187). Elsevier Ltd. https://doi.org/10.1016/j.progpolymsci.2015.10.001

Laroussi, M. (2009). Low-temperature plasmas for medicine? In *IEEE Transactions on Plasma Science* (Vol. 37, Issue 6 PART 1, pp. 714–725). https://doi.org/10.1109/TPS.2009.2017267

Lim, J. V., Bee, S. T., Sin, L. T., Ratnam, C. T., & Hamid, Z. A. A. (2021). A review on the synthesis, properties, and utilities of functionalized carbon nanoparticles for polymer nanocomposites. In Pavel Urbáne (Ed.), *Polymers* (Vol. 13, Issue 20, 3547). MDPI. https:// doi.org/10.3390/polym13203547

Machado, I. R. L., Mendes, H. M. F., Alves, G. E. S., & Faleiros, R. R. (2014). Carbon nanotubes: Potential use in veterinary medicine. *Ciencia Rural*, *44*(10), 1823–1829. https:// doi.org/10.1590/0103-8478cr20140003

Maiti, M., Bhattacharya, M., & Bhowmick, A. K. (2008). Elastomer nanocomposites. *Rubber Chemistry and Technology*, *81*(3), 384–469.

Mandolfino, C. (2019). Polypropylene surface modification by low pressure plasma to increase adhesive bonding: Effect of process parameters. *Surface and Coatings Technology*, *366*, 331–337. https://doi.org/10.1016/j.surfcoat.2019.03.047

Marini, J., & Bretas, R. E. S. (2017). Optical properties of blown films of PA6/ MMT nanocomposites. *Materials Research*, *20*, 53–60. https://doi.org/10.1590/ 1980-5373-MR-2017-0280

Maron, G. K., Noremberg, B. S., Alano, J. H., Pereira, F. R., Deon, V. G., Santos, R. C. R., Freire, V. N., Valentini, A., & Carreno, N. L. V. (2018). Carbon fiber/ epoxy composites: Effect of zinc sulphide coated carbon nanotube on thermal and mechanical properties. *Polymer Bulletin*, *75*(4), 1619–1633. https://doi.org/10.1007/ s00289-017-2115-y

Mistretta, M. C., Morreale, M., & la Mantia, F. P. (2014). Thermomechanical degradation of polyethylene/polyamide 6 blend-clay nanocomposites. *Polymer Degradation and Stability*, *99*(1), 61–67. https://doi.org/10.1016/j.polymdegradstab.2013.12. 009

Mittal, V., & Chaudhry, A. U. (2015). Polymer - graphene nanocomposites: Effect of polymer matrix and filler amount on properties. *Macromolecular Materials and Engineering*, *300*(5), 510–521. https://doi.org/10.1002/mame.201400392

Müller, K., Bugnicourt, E., Latorre, M., Jorda, M., Sanz, Y. E., Lagaron, J. M., Miesbauer, O., Bianchin, A., Hankin, S., Bölz, U., Pérez, G., Jesdinszki, M., Lindner, M., Scheuerer, Z., Castelló, S., & Schmid, M. (2017). Review on the processing and properties of polymer nanocomposites and nanocoatings and their applications in the packaging, automotive and solar energy fields. In *Nanomaterials* (Vol. 7, Issue 4). MDPI AG. https://doi. org/10.3390/nano7040074

Nayak, R. K., Mahato, K. K., & Ray, B. C. (2019). Processing of Polymer-Based Nanocomposites. *Reinforced Polymer Composites: Processing, Characterization and Post Life Cycle Assessment*, 55–75.

Nesic, A. R., & Seslija, S. I. (2017). The influence of nanofillers on physical–chemical properties of polysaccharide-based film intended for food packaging. In *Food Packaging* (pp. 637–697). Elsevier. https://doi.org/10.1016/b978-0-12-804302-8.00019-4

Njuguna, J., Ansari, F., Sachse, S., Zhu, H., & Rodriguez, V. M. (2014). Nanomaterials, nanofillers, and nanocomposites: Types and properties. In *Health and Environmental Safety of Nanomaterials: Polymer Nancomposites and Other Materials Containing Nanoparticles* (pp. 3–27). Elsevier Ltd. https://doi.org/10.1533/9780857096678.1.3

Oliveira, A. D., Beatrice, C. A. G., Passador, F. R., & Pessan, L. A. (2016). Polyetherimide-based nanocomposites materials for hydrogen storage. *AIP Conference Proceedings*, *1779*. https://doi.org/10.1063/1.4965497

Oliveira, A. D., Larocca, N. M., Paul, D. R., & Pessan, L. A. (2012). Effects of mixing protocol on the performance of nanocomposites based on polyamide 6/acrylonitrile-butadiene-styrene blends. *Polymer Engineering and Science*, *52*(9), 1909–1919. https://doi. org/10.1002/pen.23152

Rallini, M., & Kenny, J. M. (2017). Nanofillers in polymers. In *Modification of Polymer Properties* (pp. 47–86). Elsevier Inc. https://doi.org/10.1016/B978-0-323-44353-1.00003-8

Ray, S. S. (2014). Recent trends and future outlooks in the field of clay-containing polymer nanocomposites. *Macromolecular Chemistry and Physics*, *215*(12), 1162–1179. https:// doi.org/10.1002/macp.201400069

Rezende, C. A., Bragança, F. C., Doi, T. R., Lee, L. T., Galembeck, F., & Boué, F. (2010). Natural rubber-clay nanocomposites: Mechanical and structural properties. *Polymer*, *51*(16), 3644–3652. https://doi.org/10.1016/j.polymer.2010.06.026

Robertson, C. G., Lin, C. J., Bogoslovov, R. B., Rackaitis, M., Sadhukhan, P., Quinn, J. D., & Roland, C. M. (2011). Flocculation, reinforcement, and glass transition effects in silica-filled styrene-butadiene rubber. *Rubber Chemistry and Technology*, *84*(4), 507–519. https://doi.org/10.5254/1.3601885

Rong, M. Z., Zhang, M. Q., & Ruan, W. H. (2006). Surface modification of nanoscale fillers for improving properties of polymer nanocomposites: A review. *Materials Science and Technology*, *22*(7), 787–796. https://doi.org/10.1179/174328406X101247

Sahakaro, K. (2017). Mechanism of reinforcement using nanofillers in rubber nanocomposites. In *Progress in Rubber Nanocomposites* (pp. 81–113). Elsevier Inc. https://doi. org/10.1016/B978-0-08-100409-8.00003-6

Saman, N. M., Ahmad, M. H., & Buntat, Z. (2021). Application of cold plasma in nanofillers surface modification for enhancement of insulation characteristics of polymer nanocomposites: A review. *IEEE Access, 9*, 80906–80930. https://doi.org/10.1109/ACCESS.2021.3085204

Shankar, S., & Rhim, J.-W. (2018). Bionanocomposite films for food packaging applications. In *Reference Module in Food Science*. Elsevier. https://doi.org/10.1016/B978-0-08-100596-5.21875-1

Sonia, A., & Priya Dasan, K. (2013). Celluloses microfibers (CMF)/poly (ethylene-co-vinyl acetate) (EVA) composites for food packaging applications: A study based on barrier and biodegradation behavior. *Journal of Food Engineering, 118*(1), 78–89. https://doi.org/10.1016/j.jfoodeng.2013.03.020

Suter, J. L., Groen, D., & Coveney, P. V. (2015). Chemically specifi C multiscale modeling of clay-polymer nanocomposites reveals intercalation dynamics, tactoid self-assembly and emergent materials properties. *Advanced Materials, 27*(6), 966–984. https://doi.org/10.1002/adma.201403361

Taguet, A., Cassagnau, P., & Lopez-Cuesta, J. M. (2014). Structuration, selective dispersion and compatibilizing effect of (nano)fillers in polymer blends. In *Progress in Polymer Science* (Vol. 39, Issue 8, pp. 1526–1563). Elsevier Ltd. https://doi.org/10.1016/j.progpolymsci.2014.04.002

Tanahashi, M. (2010). Development of fabrication methods of filler/polymer nanocomposites: With focus on simple melt-compounding-based approach without surface modification of nanofillers. *Materials, 3*(3), 1593–1619. https://doi.org/10.3390/ma3031593

Tanaka, T., Montanari, G. C., & Mulhaupt, R. (2004). Polymer nanocomposites as dielectrics and electrical insulation-perspectives for processing technologies, material characterization and future applications. In *IEEE Transactions on Dielectrics and Electrical Insulation* (Vol. 11, Issue 5). IEEE.

Torres, T. (2011). Carbon nanotubes and related structures. Synthesis, characterization, functionalization, and applications. Edited by Dirk M. Guldi and Nazario Martín. *Angewandte Chemie International Edition, 50*(7), 1473–1474. https://doi.org/10.1002/anie.201006930

Tyagi, M., & Tyagi, D. (2014). Polymer nanocomposites and their applications in electronics industry. In *International Journal of Electronic and Electrical Engineering* (Vol. 7, Issue 6). IRPH. http://www.irphouse.com

Vengatesan, M. R., & Mittal, V. (2015). Surface modification of nanomaterials for application in polymer nanocomposites: an overview. In Vikas Mittal (Ed.), *Surface Modification of Nanoparticle and Natural Fiber Fillers* (pp. 1–28). Wiley-VCH Verlag GmbH & Co.

Vrsaljko, D., Macut, D., & Kovačevic, V. (2015). Potential role of nanofillers as compatibilizers in immiscible PLA/LDPE Blends. *Journal of Applied Polymer Science, 132*(6). https://doi.org/10.1002/app.41414

Wan, C., & Chen, B. (2012). Reinforcement and interphase of polymer/graphene oxide nanocomposites. *Journal of Materials Chemistry, 22*(8), 3637–3646. https://doi.org/10.1039/c2jm15062j

Zaïri, F., Gloaguen, J. M., Naït-Abdelaziz, M., Mesbah, A., & Lefebvre, J. M. (2011). Study of the effect of size and clay structural parameters on the yield and post-yield response of polymer/clay nanocomposites via a multiscale micromechanical modelling. *Acta Materialia, 59*(10), 3851–3863. https://doi.org/10.1016/j.actamat.2011.03.009

Zhang, Y., Wang, H. Y., Zhang, Y. R., & Bogaerts, A. (2017). Formation of microdischarges inside a mesoporous catalyst in dielectric barrier discharge plasmas. *Plasma Sources Science and Technology, 26*(5). https://doi.org/10.1088/1361-6595/aa66be

Zhao, J., Morgan, A. B., & Harris, J. D. (2005). Rheological characterization of polystyrene-clay nanocomposites to compare the degree of exfoliation and dispersion. *Polymer, 46*(20), 8641–8660. https://doi.org/10.1016/j.polymer.2005.04.038

5 Nanocellulose-Based Green Fillers for Elastomers—Influence of Geometry and Chemical Modification on Properties of Nanocomposites

Milanta Tom and Sabu Thomas
Mahatma Gandhi University

Bastien Seantier and Yves Grohens
Université Bretagne Sud

P. K. Mohamed, S. Ramakrishnan, and Job Kuriakose
Apollo Tyres Ltd.

CONTENTS

5.1   Introduction 146
5.2   Nanocellulose 147
5.3   Cellulose Nanocrystal (CNC) 149
5.4   Cellulose Nanofiber (CNF) 149
5.5   Modifications of Nanocellulose 150
5.6   Esterification/Acetylation 152
5.7   Silylation 152
5.8   Carboxylation/Oxidation... 153
5.9   Sulfonation 153
5.10 Carbamation (Urethanization) 153
5.11 Phosphorylation 154

DOI: 10.1201/9781003279372-5

5.12 Amidation... 154
5.13 Grafting of Molecules or Macromolecular Groups 154
5.14 Non-Covalent Surface Modification 155
5.15 Structure–Property Relationship.. 156
5.16 Morphological Analysis... 156
5.17 X-Ray Diffraction (XRD) 157
5.18 FTIR... 159
5.19 Rheology Studies 160
5.20 Mechanical Properties... 160
5.21 Dynamic Mechanical Analysis (DMA) 161
5.22 Fracture Mechanism.. 163
5.23 Thermal Properties... 164
5.24 Contact Angle... 165
5.25 Swelling Properties 165
5.26 Transparency 165
5.27 Conclusion 165
References.. 166

5.1 INTRODUCTION

The development of bio-based composites has drawn attraction to lower the consumption of petroleum-based materials with green fillers such as natural fibers, cellulose, chitosan, lignin, biochar, etc. These fillers stand as counter measures to tackle the environmental concern. Cellulose is an important sustainable candidate for designing bio-based composites with its myriad of inherent properties, especially the nanoderivative, where its properties are comparable to those of synthetic fillers like Kevlar.

Nanocellulose is an extract from various cellulosic resources such as plants (Omran et al., 2021), animals, bacteria (Noor et al., 2020), and agricultural waste (Mateo et al., 2021). The nanoscale-structured cellulosic materials possess improved mechanical properties (e.g., high strength, stiffness, and modulus), physical properties (e.g., lightweight, high surface-area-to-volume ratio), and chemical properties (e.g., biocompatibility) (Nair et al., 2018). With potentially reactive hydroxyl functional groups on its surface, nanocellulose can readily be modified with the desired functional groups (Habibi, 2014). Considering these properties, nanocellulose has attracted considerable attention from various industries such as nanomedicine, biocomposites, and automotive.

The role of nanocellulose as a filler is determined by its morphology and physiochemical properties (Camargos & Rezende, 2021). The geometrical dimensions of the prepared nanocellulose vary depending on the source and extraction conditions (George & Sabapathi, 2015). Two important entities under the label of nanocellulose are cellulose nanocrystal (CNC) and cellulose nanofiber (CNF). Through proper control over the preparation conditions, modulation of morphological features can be implemented to both CNCs and CNFs, which have major effects on the mechanical and rheological property. The selection of either of the nanocellulose depends on the application. Furthermore, tailoring the ionizable groups and stability of nanocellulose give good inference about the shelf life of nanocellulose formulations as well as its use as an additive in nanocomposites.

Elastomers with their characteristic flexibility, extensibility, and other prominent properties play ubiquitous role in areas like automobile, construction, consumer products, wire and cable, and industrial and medical products. Fillers are quintessential part of elastomers to achieve notable mechanical reinforcement to meet the performance requirements. Fillers with dimensions in the range of nanoscale have marked reinforcing effect owing to their large surface area which leads to a large interphase or boundary area between the matrix and nanofiller (Jamróz et al., 2019). The abundant availability of plant biomass and the superior mechanical properties of nanocellulose have made it a desirable reinforcing material for elastomer nanocomposites (Camargos & Rezende, 2021).

The most widely studied elastomer, natural rubber (NR), obtained from renewable biomass (Hevea Brasiliensis) endowed with high mechanical strength, flexibility, elasticity, resilience, abrasion resistance, and biodegradability (Feldman, 2012; Mariano et al., 2016; Trovatti et al., 2013) is one of the important elastomers in the development of nanocomposites. Nanocomposites of synthetic rubber like butadiene rubber, styrene–butadiene rubber, neoprene, the polysulfide rubbers (thiokols), butyl rubber, silicones, etc., and thermoplastic elastomers like styrenic copolymers, polyurethane elastomers, polyester elastomers, etc., are of great importance because of their extensive applications in diverse high-technology arena.

The lower compatibility of nanocellulose in hydrophobic matrix can be improved by suitable surface treatments and modifications. The treatments on nanocellulose surface would render them with enhanced properties and broaden the applications (Omran et al., 2021; Tavakolian et al., 2020). The main objective of this chapter is to present a comprehensive overview of the current state-of-the-art on the development of green materials applications. In addition, this chapter shows clear distinctions between short rigid CNCs and long relatively tough CNFs in terms of their reinforcing effects and mechanisms, thus providing reference guidance for the development of polymer nanocomposites containing nanocellulose.

5.2 NANOCELLULOSE

Nanocellulose, nanostructured form of cellulose, is one of the most prominent green materials of modern times. The biodegradability and regenerative specifications along with large surface area, supreme strength, high transparency, light density, immense biocompatibility, and chemical improvement make it a good choice for functional composite materials. The abundant hydroxyl functional groups allow a wide range of functionalizations, to develop materials with tunable features (Trache et al., 2020). Cellulose is a linear chain polysaccharide consisting of repeated β-(1→4)-D-glucopyranose units. Cellulose was first extracted by Anselme Payen in 1838 and coined the term 'cellulose' and later Herman Staudinger established its chemical structure (Santmartí & Lee, 2018). The cellobiose units are linked together to form elementary fibrils, a crystalline structure of cellulose. These fibrils are bundled together to produce microfibrils, which in turn form macrofibrils or cellulosic fibers. The properties like hydrophilicity, chirality, ease of chemical functionalization,

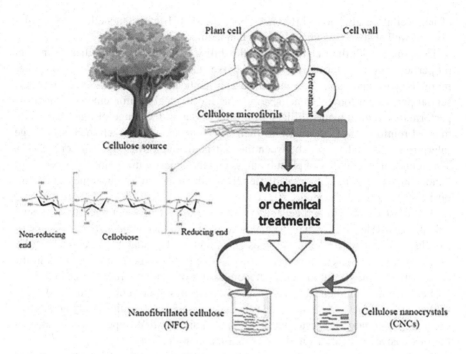

FIGURE 5.1 Schematic representation of extraction of nanocellulose from wood source (Haron et al., 2021), reprinted with permission.

insolubility in most aqueous solvents, and infusibility of cellulosic materials are set forth by intra- and intermolecular chemical groups. Nanocellulose can be extracted from either woody or non-woody cellulosic sources like herbaceous plants, grass, agricultural crops and their by-products, animal, algae and bacterial sources, waste paper, among others (Noor et al., 2020; Trache & Thakur, 2020). A schematic representation of cellulose from woody source to the fundamental molecule is displayed in Figure 5.1. The properties of nanocellulose are found to dependent on its natural source, pretreatment, extraction methodologies, and reaction parameters (Blanco et al., 2018; Dufresne, 2013; Trache et al., 2017). Generally, extraction of nanocellulose from lignocellulosic sources requires pretreatments like delignification and bleaching processes to eliminate non-cellulosic components (Fernandes et al., 2013). Pretreatments promote the accessibility toward cellulose-rich fraction and increase the porosity, the inner surface, and reactivity, but decrease the degree of polymerization (Kargarzadeh et al., 2017). Lesser number of pretreatments are required in the case of nanocellulose extracted from animal source (Trache et al., 2017), whereas bacterial cellulose does not contain extractives, hemicellulose, and lignin with respect to vegetable cellulose, and thus does not necessitate specific pretreatments. Nevertheless, its production in industrial scale remains relatively expensive (Oun et al., 2020).

The high aspect ratio of nanocellulose allows the superior transfer of stress from the matrix to the filler, and the huge specific surface area leads to a low concentration of the filler to gain a reinforcing result (Wang et al., 2019). As a result of strongly reactive hydroxyl groups, there is a strong tendency in the cellulose material to connect together forming wide network through both intermolecular and intramolecular hydrogen bonds (Rana et al., 2021). Furthermore, the plenteous availability of hydroxyl groups enables functionalizing of nanocellulose via convenient molecules for effectual reinforcement.

5.3 CELLULOSE NANOCRYSTAL (CNC)

Cellulose nanocrystals (CNCs) were first isolated in 1947 by Nickerson and Habrie by sulfuric acid hydrolysis (Nickerson and Habrie, 1947). These are highly crystalline nanoscale cellulose with a characteristic rod or whisker shape of 3–5 nm width and 50–500 nm length (Chin et al., 2018; Kaur et al., 2021; Vanderfleet & Cranston, 2021). The dimensions of CNCs depend on various parameters like cellulosic source, extraction conditions, etc. The most common isolation technique of these nanocrystals is the chemical process involving acid hydrolysis. Though different types of acids and mixed acids can be utilized for nanocrystal extraction, sulfuric acid hydrolysis is the most common. This is due to uniformity, reproducibility, high yield, improved stability of nanocellulose dispersion by anionic sulfate group, and predictable properties of CNCs (Vanderfleet & Cranston, 2021). CNCs are characterized by very high mechanical and thermal properties due to their high crystalline content. The colloidal stability, size, and crystallinity index of CNCs can influence their performance. Higher colloidal stability facilitates uniform dispersion of CNCs with predictable and consistent performance, which also influences material processing, viscosity, and shelf life (Lin et al., 2019). The high surface-area-to-volume ratios of these nanoscale materials provide greater number of active sites for chemical reactions, thus improving adhesion between CNCs and other material/chemical of interest (Barnat-Hunek et al., 2019). Relatively high aspect ratio of these nanofillers is advantageous in reinforcing polymeric materials. In addition, high aspect ratios allow self-assembly, adopt mesh structures to stabilize interfaces, and form percolated networks. Broadly, crystalline domains are more resistant to chemical, mechanical, and enzymatic treatments compared to the amorphous ones (Thakur et al., 2021). Crystallinity of CNCs is primarily linked to mechanical properties but can also affect chemical reactivity and thermal stability (Agarwal et al., 2016). The existence of large number of hydroxyl groups facilitates easy surface tuning of nanocellulose to achieve desired properties (Calvino et al., 2020). The schematic representation of CNCs and CNFs is represented in Figure 5.2.

5.4 CELLULOSE NANOFIBER (CNF)

CNFs are long entangled cellulosic chains that contain both their crystalline and amorphous regions (Shaghaleh et al., 2018). Fibrillation of biomass into CNF has been achieved through enzymatic, chemical, and mechanical treatments, or a combination thereof, and the morphology of the nanomaterial produced is specific to the production process (Pennells et al., 2020). Mechanical fibrillation is the most common extraction method which includes high-pressure homogenization, Microfluidization,

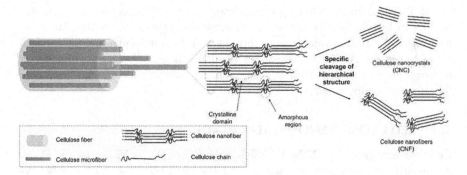

FIGURE 5.2 Schematic representation of cellulose nanocrystals (CNCs) and cellulose nanofibers (CNFs) (Fiorati et al., 2020), reprinted with permission.

ball milling, cryocrushing, ultrasonication, grinding, etc. (K. Zhang et al., 2019). According to TAPPI, CNF should have diameter of around 5–30 nm, but with an aspect ratio greater than 50 (TAPPI, 2011). CNF suspensions are found to have gel properties even at low cellulose concentrations, and they usually form porous entangled network structure (Carter et al., 2021).

The morphologies and dimensions of CNFs can vary substantially, depending on the degrees of fibrillation and any pretreatment involved. CNFs contain amorphous cellulose and are not as highly crystalline as CNCs. The fibrillation process enables the exposure of inner surface and leads to gradual breakdown of inter- and intramolecular hydrogen bonds. This in turn will enhance the reactivity and accessibility of the cellulose structure. By selecting the appropriate fibrillation method and optimum operating conditions, the CNFs can be obtained in different morphologies ranging from the micro- to nanofibrillar levels. Adoption of pretreatments like acid hydrolysis, enzymatic or oxidation pretreatment, etc., before mechanical treatments can decrease the energy consumption (K. Zhang et al., 2019). CNFs are often characterized by their exceptional features like tensile strength of 0.3–22 GPa, Young's modulus of 58–180 GPa, very large aspect ratio, and flexibility, which would impart greater reinforcing efficacy to these nanofibers (Ghasemlou et al., 2021; Mateo et al., 2021).

5.5 MODIFICATIONS OF NANOCELLULOSE

The emergence of intense researches on these bionanomaterials for their utilization in high-value applications found that surface modification or functionalization can impart excellent compatibility with the matrix material, thus playing a pivotal role in controlling the properties of the nanocomposites. There are a number of surface modification techniques to improve the interface without impairment of the intrinsic properties of nanocellulose. The two important routes of different modifications on nanocellulose are represented in Figure 5.3, and common modifications of nanocellulose are given in Figure 5.4.

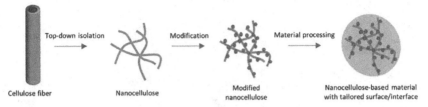

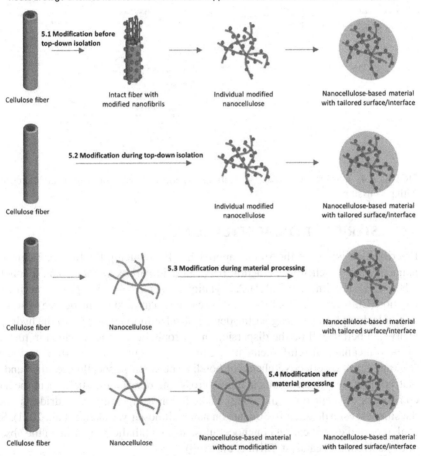

FIGURE 5.3 Schematic representation of routes of modifications on nanocellulose (Yang et al., 2021), reprinted with permission.

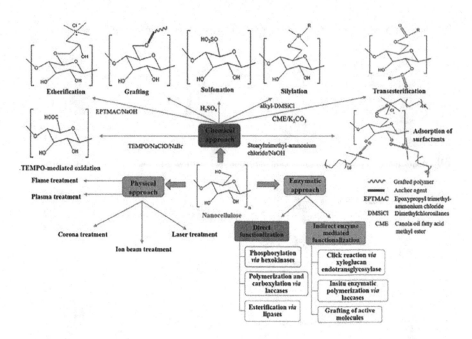

FIGURE 5.4   Different types of modifications on nanocellulose (Trache et al., 2020), reprinted with permission.

5.6   ESTERIFICATION/ACETYLATION

Esterification is one of the simple approaches for reducing the hydroxyl density of nanocellulose by chemical reaction between acid carboxylic acid) and alcohol (or other –OH) (Hakimi et al., 2021). Acetylation is the most widely studied esterification reaction (Chin et al., 2018). Esterification can be quantified by the degree of substitution of hydroxyl groups. Higher the degree of substitution, higher will the hydrophobicity and better will be the dispersion in hydrophobic matrices, which in turn lead to more mechanical reinforcement in composites (Yin et al., 2020). Esterification reaction develops effective chemical bonding between nanocellulose filler and NR matrix interface, which results in higher cross-link density and stiffness in the nanocomposites (Noremylia et al., 2022). Acetylation using acetic anhydride is one of the simplest esterification treatments on nanocellulose (Marakana et al., 2021). Silva et al. reported esterification of nanocellulose using phthalic anhydride in the absence of any solvent and catalyst (Silva et al., 2018).

5.7   SILYLATION

In the process of silylation, silyl groups are introduced onto the surface of nanocellulose through substitution reactions. Silanes are effective coupling agents as they have strong affinity for hydroxyl groups. Permanent grafting of silane moieties can be achieved through condensation of Si-OH with C-OH during temperature curing process (Rui et al., 2015). Organosilanes such as 3-aminopropyltriethoxysilane

(APTES), bis-(3-triethoxysilylpropyl) tetrasulfide (TESPT), and 3-isocyanateprop-yltriethoxysilane (IPTS) have been fruitfully used to reduce the hydrophilic nature of hydroxyl-rich nanocellulose (Roy et al., 2021). Silyl-modified CNC showed better dispersion and interfacial adhesion to polymer matrix. CNFs produced through pre-treatment involving alkalis possess more reactive sites, enabling increased silylation (Sharma et al., 2019).

5.8 CARBOXYLATION/OXIDATION

Oxidation is one of the standing out methods to impart hydrophobicity to nanocel-lulose by selectively converting the hydroxyl groups present on the C-6 of the glucose unit of cellulose into negatively charged carboxylic as well as aldehyde groups (Habibi, 2014). The most widely applied method for oxidation is TEMPO (2,2,6,6-tetrameth-ylpiperidine–1–oxyl) (Neves et al., 2021). TEMPO-mediated oxidation of cellulosic fibers is considered as a pretreatment for the preparation of nanocellulose (Haunreiter et al., 2022). During TEMPO-mediated oxidation, the primary C6 hydroxyl get com-pletely or selectively converted to sodium C6-carboylate groups. Therefore, TEMPO-oxidized nanocellulose consists of glucosyl and sodium glucuronosyl units. The carboxylate contents increase from ≈ 0.1 mmol/g for the starting wood cellulose to ≈ 1.7 mmol/g after the TEMPO-mediated oxidation, depending on the oxidation con-ditions. TEMPO oxidation is often used to selectively activate the primary hydroxyl groups of cellulose nanoparticles by their conversion to carboxylic acids (Dufresne et al., 2021). The high dimensional stability of TEMPO-oxidized nanocellulose makes them suitable for applications in electronic devices (Isogai, 2021).

5.9 SULFONATION

In this process, sulfuric acid catalyzes the hydrolysis to replace hydroxyl groups with sulfate half-ester moieties. The negative charge of sulfate half-ester groups has a positive impact on the dispersion of nanocellulose and suppressing hydrogen bond formation by exerting electrostatic repulsion (Dhali et al., 2021). The sulfonation of nanocellulose can achieve via three routes. First one is direct sulfonation during sulfuric acid hydrolysis of cellulose, and second one is the post-sulfonation of nano-cellulose. The mechanism of dispersion of nanocellulose after sulfonation is similar to that of nanocellulose produced by sulfuric acid hydrolysis (Raza & Abu-Jdayil, 2022). The third strategy involves a two-step process of periodate oxidation and sul-fonation (Zhu & Lin, 2019). In contrast to sulfonation, sulfoethylation introduces sulfate half-ester groups to nanocellulose surface, which could be used to enhance the mechanical properties due to the maintenance of the fibril-like morphology (Z. Li et al., 2021).

5.10 CARBAMATION (URETHANIZATION)

Urethanization involves the reaction between isocyanate and surface –OH groups of nanocellulose to form covalent bond (urethane linkage) (Raza & Abu-Jdayil, 2022). The higher concentration of hydroxyl groups on the nanocellulose surface provides the active reaction sites for isocyanates to form carbamates. Toluene-2,4-diisocyanate (TDI) is a widely used isocyanate for nanocellulose modification. The reaction between isocyanates and the hydroxyl groups of nanocellulose and the resultant formation of polyurethane bonds can be facilitated with the help of amines such as triethylamine as excellent catalysts although they could self-polymerize as a side reaction (Abushammala & Mao, 2019).

5.11   PHOSPHORYLATION

Phosphorylation is the surface treatment of nanocellulose with phosphorylating agent such as phosphorus oxychloride, phosphorus pentoxide, phosphoric acid, diammonium hydrogen phosphate, and organophosphates. The degree of substitution depends on several factors such as molar ratio of phosphorylating agent and anhydroglucose units of cellulose, time, and temperature (Ram & Shanmuganathan, 2021). The modification results in negatively charged nanocellulose surface, exhibiting better dispersion properties. Another special feature of phosphorylated nanocellulose is the improvement in heat resistance and flame retardancy. It has better prospects than other surface modification techniques since it can increase nanocellulose dispersion along with the improvement in physical properties like fire retardancy, mechanical strength, recovery, and reusability of non-toxic reagent (Raza & Abu-Jdayil, 2022).

5.12   AMIDATION

Amidation modification is usually performed as second-step reaction on pre-oxidized nanocellulose, targeting carboxylic functionalities. This amine-modified nanocellulose can be infused with a hydrophobic alkyl chain of TEMPO-oxidized nanocellulose to enhance the hydrophobicity (Jaffar et al., 2022). The amidation reaction can take place both in organic solvents (e.g., dimethylformamide, abbreviated as DMF) and aqueous solutions. In the aqueous reaction system, the carboxylated nanocellulose is reacted with N-hydroxysuccinimide (NHS) to improve the reactivity of carboxyl groups on nanocellulose surface with the amide groups. The final step of amidation modification is to mix the NHS-activated carboxylated nanocellulose with the amide-containing molecules. The amine groups of the molecules can directly react with NHS-activated carboxylated nanocellulose, and conjugate the molecules onto the nanocellulose surface with the linkage of amide bonds. If the reaction route chooses organic solvents instead of aqueous medium, the amine-containing molecules can be directly conjugated onto the carboxylated surface of nanocellulose (Huang et al., 2019).

5.13   GRAFTING OF MOLECULES OR MACROMOLECULAR GROUPS

Grafting of molecules or macromolecules onto the surface of nanocellulose is common to bring on non-polar properties (Peng et al., 2021). Modification by grafting mechanism can be done through three routes, namely 'grafting from', 'grafting to',

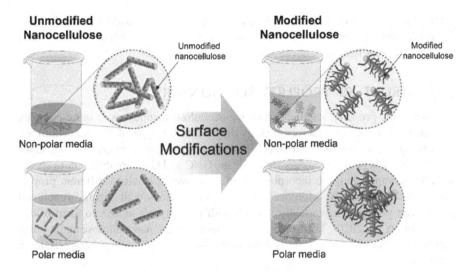

Unmodified Nanocellulose

Unmodified nanocellulose

Non-polar media

Polar media

Modified Nanocellulose

Modified nanocellulose

Non-polar media

Polar media

Surface Modifications

FIGURE 5.5 Illustration of interaction of nanocellulose before and after surface modification (Ghasemlou et al., 2021), reprinted with permission.

and 'grafting through'. In 'grafting to' approach, presynthesized polymer chain coupled onto the surface of nanocellulose, enabling good control over the properties of grafted molecule while compromising graft density due to steric hindrance. The 'grafting from' route involves direct polymerization on the surface of nanocellulose. Even though 'grafting from' approach can achieve polymer grafts with higher density, the moieties obtained through 'grafting from' approach can disperse in organic solvents due to lower steric hindrance exerted by small monomers and less shielding of the reactive sites on the surface and the lower viscosity of the medium (Zhu & Lin, 2019). In the 'grafting-through' route, the functionalization of cellulosic material occurs through polymerizable species, which is then mixed with a co-monomer and then the polymerization reaction is initiated (Thakur et al., 2021). The illustration of interactions of nanocellulose before and after modification is represented in Figure 5.5.

5.14 NON-COVALENT SURFACE MODIFICATION

This method involves tuning the surface by physical interactions or adsorption of molecules or macromolecules. The physical bonding can occur through charge–charge interactions, hydrogen bonds, Van der Waals forces, and solvency and association forces (Tortorella et al., 2020). Cationic and anionic surfactants are used commonly for non-covalent surface modifications. The hydrophilic end of the surfactant molecule may adsorb on the surface of nanocellulose, whereas the hydrophobic end may extend out providing a non-polar surface and lowering the surface tension of the nanoparticle (Putro et al., 2019). The adsorption of surfactants will be better if the nanocellulose has a charged surface either positive or negative, and the most commonly used surfactant is cetyltrimethylammonium bromide CTAB (Dufresne et al.,

2021). The adsorption of macromolecules like polyethylene oxide and block copolymers of polyethylene glycol has improved the dispersibility, thermal stability, and mechanical properties of nanocellulose (Lunardi et al., 2021).

5.15   STRUCTURE–PROPERTY RELATIONSHIP

The structure–property relationship of nanocellulose composites is quite complex as it depends on several factors such as dispersion, the type of the nanocellulose, morphology, and interfacial interactions between the polymeric matrix and the nanocellulose (Talebi et al., 2022). As we know the CNC and CNF are different in shape, size, and composition, the incorporation of these would result in different properties in elastomer matrix. Moreover, surface modifications with specific functional groups can physically/chemically interact with the polymer matrix to form bridge (Mahendra et al., 2020). The influence of geometry of the nanocellulose and modifications on elastomer composites is discussed through the following characterization techniques.

5.16   MORPHOLOGICAL ANALYSIS

a. SEM

The short rod-like structures of CNC and long entangled chain like structures of CNF can easily be identified in morphological analysis like SEM (He et al., 2020). The morphological analysis indicated CNF samples with much higher number of bridging compared to CNC. This is because of the CNFs' interfiber entanglements and larger aspect ratio (Isogai & Zhou, 2019). It is essential to analyze the orientation and arrangement of nanocellulose as they can influence the properties of nanocomposites (X. Zhang et al., 2020). Figure 5.6 represents SEM images of CNC and CNF.

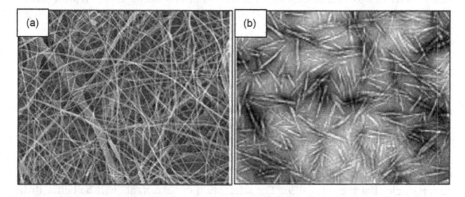

FIGURE 5.6   SEM image of CNF (Sathishkumar et al., 2014) and CNC (Habibi et al., 2010), reprinted with permission.

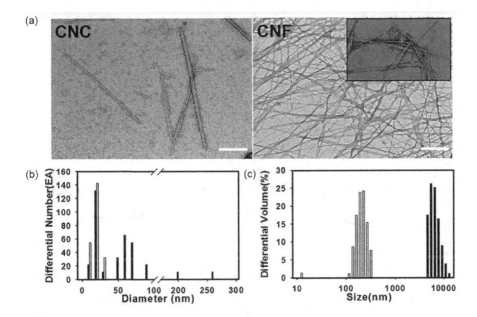

FIGURE 5.7 (a) TEM images of CNC and CNF, (b) diameter distribution of CNC (gray), and CNF (black) (c) size distribution of CNC (gray) and CNF (black) (Yu et al., 2016), reprinted with permission.

b. TEM

The morphological difference between different types of nanocellulose can be clearly observed in TEM analysis. CNFs exhibit a complex, highly entangled structure. Twisted/untwisted, curled/straight, and entangled/separate nanofibrils with diameters ranging from 6 to 100 nm in diameter can be identified. The highly entangled structure of CNFs significantly increases resistance to flow and results in gel-like behavior. CNC suspension shows much lower viscosity than the CNF sample even at higher fiber concentration because of the former's low aspect ratio and lack of entanglement. Due to CNFs' complex network morphology, it is difficult to measure the length and diameter of individual CNF with high accuracy (Barkane et al., 2021; Camargos & Rezende, 2021; Flauzino Neto et al., 2016; Mahendra et al., 2020). TEM images and size distribution of CNC and CNF are represented in Figure 5.7.

5.17 X-RAY DIFFRACTION (XRD)

The crystalline nature of molecules is analyzed using XRD. The mode of extraction of nanocellulose can affect the crystallinity (Trifol et al., 2016). Nanocellulose, both CNC and CNF, shows higher peak intensity than cellulose due to the major amorphous content during isolation process (Xu et al., 2013). CNC shows higher degree

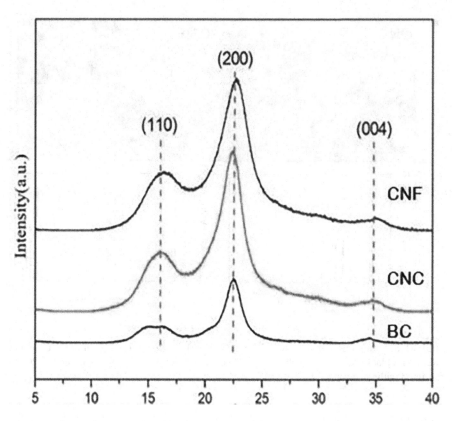

FIGURE 5.8 XRD patterns of bagasse cellulose (BC), CNC, and CNF (B. Zhang et al., 2019), reprinted with permission.

of crystallinity as most of the amorphous regions are cleaved during the hydroly-sis treatment releasing individual crystallites (Kaur et al., 2021). CNCs generally show diffraction peaks at 15.1°, 17.5°, 22.7°, and 34.3°, representing cellulose I crystal planes (101), (10$\bar{1}$), (002), and (040), respectively, and diffractions from cellulose II are present at angles of 12.5°, 20.1°, 22.7°, and 34.3° (Cheng et al., 2019). Treatments on cellulosic fibers transform cellulose I to cellulose II. The XRD patterns of CNFs are broad peaks which are shifted to lower angles. This may because of superposition of the crystalline diffraction peaks upon increasingly strong amorphous diffractions (Gopi et al., 2017). Figure 5.8 represents XRD patterns of bagasse cellulose, CNC, and CNF. It is observed that the high-pressure mechanical grinding used in CNF manufacturing could deform cellulose crystals leading to broadened and shifted dif-fraction peaks (Yu et al., 2016). Zhang et al. observed that high-intensity shearing treatment could produce more regular arrangement of fiber in the crystallization area (Zhang et al., 2019).

5.18 FTIR

The functional composition of CNC and CNF is same, but varies as a result of extraction method adopted (Ahmad et al., 2017). The –OH stretching vibration of the nanocellulose was revealed at the wide band between 3700 and 3200 cm^{-1} (Cheng et al., 2019). The spectra of all samples showed the characteristic C–H stretching vibration around 2900 cm^{-1}. The higher absorption of CNF may be because of possibilities of higher hydrogen bonding within the hydroxyl groups of large entangled structure (He et al., 2020). In the case of nanocellulose-incorporated nanocomposites, the higher absorption indicates better interaction between filler and matrix (Barkane et al., 2021). The absorption bands attributed to nanocellulose at 1316 and 1430 cm^{-1} representing CH$_2$ wagging and C–H stretching vibrations. The separate peak at 898 cm^{-1} is related to C–O–C asymmetric stretching (Blanco et al., 2018). The vibration peak detected at 1366 cm^{-1} in CNC and CNF samples is related to the bending vibration of the C–H and C–O bonds in the polysaccharide aromatic rings (Han et al., 2013). Figure 5.9 represents the FTIR spectra of CNC, CNF, and cellulose.

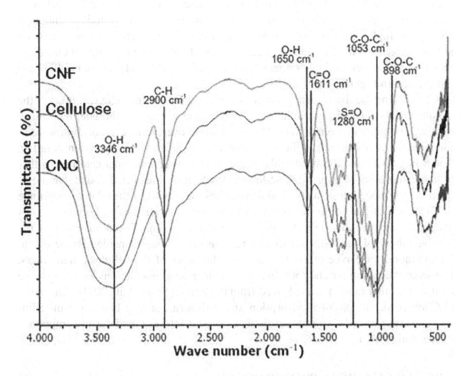

FIGURE 5.9 FTIR spectra of cellulose, CNC, and CNF (Teixeira et al., 2021), reprinted with permission.

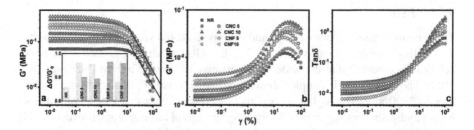

FIGURE 5.10 Storage modulus G′, loss modulus G″, and tan δ as a function of strain amplitude, and solid symbols and hollow symbols represent without and with ionic liquid (Yasin et al., 2021), reprinted with permission.

5.19 RHEOLOGY STUDIES

Assuming the surface functionalities to be same for CNC and CNF, the differences observed are likely due to the distinct specific surface area, degree of entanglement, aspect ratio, which can contribute to the difference in rheological properties. It was observed that CNFs present high steady shear viscosity and gel-like behavior (Camargos & Rezende, 2021). Mariano et al. observed the magnitude of modulus values was higher for CNF suspension as nanofibers could create stronger network at low concentrations (Mariano et al., 2018). The rheological properties of elastomer nanocomposites containing CNC and CNF nanofillers with and without modifier are represented in Figure 5.10.

Considering the rheological properties imparted nanocellulose in composites, the cellulosic nanofillers do not influence critical value of strain amplitude, but reduce relative amplitude of Payne effect. The weak strain overshoot of nanocellulose-reinforced compounds indicated the intermolecular association via hydrogen bonding. The filling effect and the structural complexation between polymer chain and either type of nanocellulose improve modulus and lower tan δ. Yasin et al. found that modification on nanocellulose, its type, and cross-linking can affect the relative amplitude of Payne effect due to the restriction on segmental movements of polymer chains and influence on rubber–filler interaction under straining. The destruction of the hydrogen-bonded networks is partially responsible for the intensified Payne effect. Furthermore, the Payne effect is related to the types of the cellulosic nanofillers. However, the nanofiller and the filler–rubber interfacial interactions exert influence on the molecular motion as evidenced from the figure (Yasin et al., 2021). According to Chen et al., the physical adsorption and hydrogen bonding between nanocellulose and cross-linked elastomer resulted in increased storage modulus of compounds (Chen et al., 2015).

5.20 MECHANICAL PROPERTIES

Nanocellulose in all forms has found to be efficient reinforcing filler that can induce a significant improvement in mechanical properties and is comparative to conventional fillers. Two important factors that have to be considered to achieve optimum

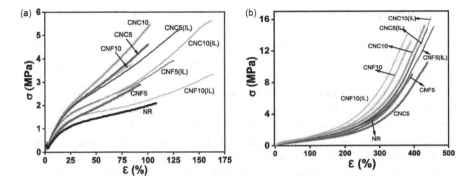

FIGURE 5.11 Stress–strain curves of CNC- and CNF-reinforced elastomer composites with and without ionic liquid modification (Yasin et al., 2021), reprinted with permission.

mechanical properties are that (i) good interfacial interaction between filler and matrix and (ii) effective transfer of involved stress (Han et al., 2019; Jarnthong et al., 2015; Jiang et al., 2020; Ogunsona et al., 2020; Singh et al., 2020). Considering the influence of geometry of nanocellulose, the mechanical properties of nanocomposites containing CNF were greater than those containing CNC. Conversely, nanocomposites containing CNC exhibited higher Young's modulus at lower concentrations of filler loading. This could be the result of higher crystallinity and degree of polymerization of CNCs (Talebi et al., 2022). The stress–strain curves of CNC- and CNF-loaded elastomer composites are given in Figure 5.11.

The difference in the elongation properties of both types of nanocellulose arises from the structural difference crystals and fibers. Higher elongation properties are due to efficient stress transfer throughout the polymer nanocomposite, even distribution of applied stress, and minimization of the areas of stress concentrations (Hakimi et al., 2021). The CNF-reinforced composites have outstanding tensile strength and elongation when compared to CNC-reinforced composites due to the higher entanglement of the CNF molecules, formation of percolation networks, and much larger aspect ratio (Chang et al., 2021). The short rod-like structures of CNC stand behind CNF in stress transfer (Mahendra et al., 2020). Furthermore, long fibers reach percolation at low filler contents and the fiber–fiber interactions due to percolation contribute to further improvement in the mechanical properties of composites. Therefore, nanofillers with high aspect ratio outweigh fiber strength in the case of mechanical properties (Zhao et al., 2022; Xu et al., 2013).

5.21 DYNAMIC MECHANICAL ANALYSIS (DMA)

The DMA is susceptible to molecular motions, and it can be utilized to analyze the effect of CNC and CNF in nanocomposites. The incorporation of nanocellulose in either morphology will result in an increase in storage modulus due to the restriction of free molecular motion of the matrix chains by the nanocellulose molecules (Lavoratti et al., 2015). The glass transition temperature and tan δ intensity were higher for CNF-incorporated nanocomposites because of larger number of hydrogen bonding and higher penetration on the texture than CNC (Xu et al., 2013). According to Dufresne, enhancement of the storage modulus confirms good dispersion of nanocellulose in the polymer matrix. Good interaction of nanocellulose filler with the matrix is evidenced by the movement of tan δ peak to a higher temperature (Dufresne, 2000).

Barkane et al. reported a contradictory view on the viscoelastic properties of CNC- and CNF-reinforced nanocomposites. They observed that at lower concentrations, CNF would act like separate reinforcing agents, but at higher loading they form entangled mesh-like structures, providing optimum reinforcement. The short rod-like nanoparticles of CNC, with its high crystalline nature, provide sufficient rigidity to the nanocomposites. As represented in figure, CNC exhibits an improved storage modulus when compared to CNF, which attributes to the efficient dispersion of CNC in polymer matrix. The higher loadings of nanocellulose restrict polymer segmental motions which require higher energy for phase transitions, which in turn results in higher energy dissipation. This has been reflected in the loss modulus curves of nanocomposites. The lower tan δ peaks of nanocellulose-loaded composites signify its influence on polymer chain mobility (Barkane et al., 2021). Figure 5.12 represents storage modulus, loss modulus, and loss factor tan δ of nanocomposites with different wt% of CNC and CNF.

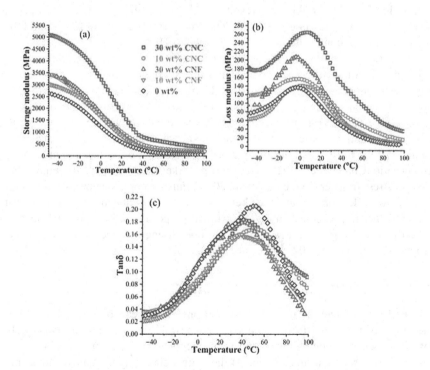

FIGURE 5.12 (a) Storage modulus, (b) loss modulus, and (c) loss factor tan δ of nanocomposites with different wt% of CNC and CNF (Barkane et al., 2021), reprinted with permission.

5.22   FRACTURE MECHANISM

The influence of geometry of nanocellulose is evident in fracture mechanism of nanocomposites. The fibrils of CNF can bridge the crazes throughout the samples, which contributes to the increased tensile strength and strain-at-failure. This is possibly ascribed to CNFs' interfiber entanglements and much larger aspect ratio when compared to CNC. Figure illustrates the fracture mechanisms of nanocomposite reinforced by CNCs and CNFs. The larger lengths and higher flexibility of CNF fibril make it possible to link a craze at multiple locations. The higher aspect ratio, formation of network structures, and entanglements facilitate the formation of a large number of visible pullout and fibril interlocking on the fracture in the case of CNF-reinforced nanocomposite (Xu et al., 2013). The illustration of fracture mechanism is given in Figure 5.13. Dominic et al. observed plastic deformation and fracture stress transfer in nanocellulose-reinforced NR composites, which is an indication of effective stress transfer from the matrix to the filler (Dominic et al., 2020).

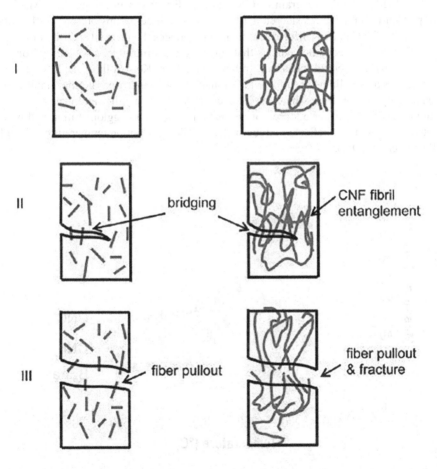

FIGURE 5.13   Illustration of fracture mechanism of CNC- and CNF-incorporated polymer nanocomposites (Xu et al., 2013), reprinted with permission.

5.23 THERMAL PROPERTIES

Thermal stability is very important to evaluate the feasibility of using nanocellulose as effective filler for elastomers (Gan et al., 2020). The study of degradation behavior allows to optimize the designing and processing conditions of composites and to develop polymer nanocomposites with enhanced thermal stability (W. Li et al., 2014). The decomposition temperature of nanocellulose is approximately 200°C–300°C. Differences in the source of nanocellulose, geometry, composition, types of matrices, processing techniques, and the drying process could affect the thermal stability of the nanocomposites (Blanco et al., 2018).

Hu et al. observed that CNC has better thermal stability than CNF. Because of high crystallinity of CNC, the chemical groups within crystal area need more energy to overcome the interaction between adjacent molecules (Hu et al., 2021). The degradation temperature of CNF was observed in a range from 191°C to 364°C, whereas decomposition process of CNC occurred between 368°C and 482°C (Mahendra et al., 2020). The thermogram of CNC and CNF is given in Figure 5.14. Another important point to be considered is that remnant sulfate groups present on the surface of the CNC due to sulfuric acid treatment reduces the thermal stability of CNC. Another contradictory behavior is that sulfate groups introduced during hydrolysis with sulfuric acid could act as flame retardant (Rahimi Kord Sofla et al., 2016). It was found that nanocellulose-reinforced elastomer showed improved thermal stability to certain loading of filler, where the interaction between the filler and matrix is optimum. Later on, further addition of nanocellulose leads to agglomeration and poor interaction, which is reflected on lower thermal decomposition temperatures (Abdul Rashid et al., 2018).

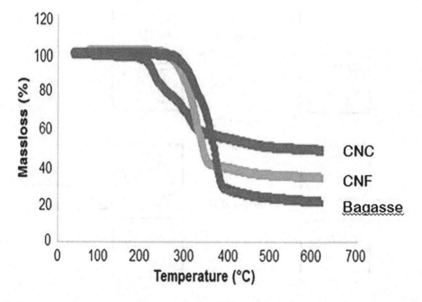

FIGURE 5.14 Thermogram of CNC and CNF along with its macroscale source bagasse (Rahimi Kord Sofla et al., 2016), reprinted with permission.

5.24 CONTACT ANGLE

The degree of water wettability is a measure of hydrophilicity of nanocomposites, which is determined by the contact angle measurements. The contact angle of nanocomposites is also affected by the morphology of the filler content. The nanocomposites containing CNC were found to have higher contact angle than CNF-incorporated nanocomposites. This is because of the higher rigidity of CNC than CNF (Talebi et al., 2022). The hydrophobic modification of nanocellulose can reduce the surface energy and enhance the roughness that can even lead to superhydrophobic surfaces. Hydrophobic or superhydrophobic surfaces are preferred to enhance the interaction between the nanocellulose and the elastomer (Panchal et al., 2019).

5.25 SWELLING PROPERTIES

The incorporation of nanocellulose has resulted in decrease in absorption of solvents due to the restriction of movement of solvent molecules due to the presence of nanocellulose (Abraham et al., 2013). It is also reported that the nanocellulose has the ability to restrict the polymer chain mobility and thereby decrease diffusion and permeation through the nanocomposite (Mathew, 2017). Another interesting observation is that incorporation of CNC has lower swelling value when compared to CNF at equal contents. This may be attributed to the higher crystallinity of CNC when compared to CNF (Talebi et al., 2022). According to Celestini et al., the incorporation of 1 wt% of cellulose nanoparticles in NR reduced swelling and prevented the appearance of breakage of the elastomeric matrix even after long duration of immersion in solvents. This observation is independent of the filler source or the amount used (Celestini et al., 2021). Dominic et al. observed that there is an increase in swelling at higher concentrations of CNFs in NR composites due to the agglomeration of nanocellulose and poor interaction between the filler and matrix (M. C. D. Dominic et al., 2020).

5.26 TRANSPARENCY

The optical transparency is a matter of concern in the case of elastomer sensors. It has been observed that CNCs exhibit higher transparency than do CNFs. This behavior is most likely due to smaller sizes and lack of entanglement. This property is advantageous to optical applications where several common nanocomposites such as carbon and clay nanocomposites have to be avoided because of their lack of transparency (Camargos & Rezende, 2021). Zhang et al. verified the transparency by determining the transmittance of nanocomposites. They found that transmittance decreased with the increase in concentration of nanocellulose. Higher concentrations will lead to agglomeration of nanocellulose, hence reducing the transmittance (K. Zhang et al., 2019).

5.27 CONCLUSION

Nanocellulose has brought a new facet in reinforcing elastomers with its extraordinary properties. The extraction of nanocellulose from various cellulosic materials has been optimized through years, and large-scale production at lower cost is becoming common. The valuable inherent traits of nanocellulose have a great dependence on its source and method of extraction. Nanocellulose-based elastomer composites exhibit a higher tensile strength than, for example, carbon fibers or aramid fibers such as Kevlar. Nanocellulose, in its different forms, is efficient reinforcing filler for elastomers which could replace the conventional petroleum-based fillers. The strong dependence of the properties of nanocellulose on the geometry has to be taken into account for its consideration in various applications. There is a bright future for nanocellulose as well as nanocellulose-reinforced elastomers as it benefits both industry and society, being environmentally friendly, low cost, and high mechanical performance.

REFERENCES

Abdul Rashid, E. S., Muhd Julkapli, N., & Yehye, W. A. (2018). Nanocellulose reinforced as green agent in polymer matrix composites applications. *Polymers for Advanced Technologies, 29*(6), 1531–1546. https://doi.org/10.1002/pat.4264

Abraham, E., Thomas, M. S., John, C., Pothen, L. A., Shoseyov, O., & Thomas, S. (2013). Green nanocomposites of natural rubber/nanocellulose: Membrane transport, rheological and thermal degradation characterisations. *Industrial Crops and Products, 51*, 415–424. https://doi.org/10.1016/j.indcrop.2013.09.022

Abushammala, H., & Mao, J. (2019). A review of the surface modification of cellulose and nanocellulose using aliphatic and aromatic mono- and di-isocyanates. *Molecules, 24*(2782), 1–18.

Agarwal, U. P., Ralph, S. A., Reiner, R. S., & Baez, C. (2016). Probing crystallinity of never-dried wood cellulose with Raman spectroscopy. *Cellulose, 23*(1), 125–144. https://doi.org/10.1007/s10570-015-0788-7

Ahmad, I., Thomas, S., Wiley, A. D. J., Ioelovich, M., & Kinds, V. (2017). *Handbook of Nanocellulose and Cellulose Nanocomposites*. Wiley.

Barkane, A., Kampe, E., Platnieks, O., & Gaidukovs, S. (2021). Cellulose nanocrystals vs. Cellulose nanofibers: A comparative study of reinforcing effects in uv-cured vegetable oil nanocomposites. *Nanomaterials, 11*(7). https://doi.org/10.3390/nano11071791

Barnat-Hunek, D., Grzegorczyk-Frańczak, M., Szymańska-Chargot, M., & Łagód, G. (2019). Effect of eco-friendly cellulose nanocrystals on physical properties of cement mortars. *Polymers, 11*(12). https://doi.org/10.3390/polym11122088

Blanco, A., Monte, M. C., Campano, C., Balea, A., Merayo, N., & Negro, C. (2018). Nanocellulose for industrial use: Cellulose nanofibers (CNF), cellulose nanocrystals (CNC), and bacterial cellulose (BC). In: Chaudhery Mustansar Hussain (ed) *Handbook of Nanomaterials for Industrial Applications,* 74–126. Elsevier Inc. https://doi.org/10.1016/B978-0-12-813351-4.00005-5

Calvino, C., Macke, N., Kato, R., & Rowan, S. J. (2020). Development, processing and applications of bio-sourced cellulose nanocrystal composites. *Progress in Polymer Science, 103*. https://doi.org/10.1016/j.progpolymsci.2020.101221

Camargos, C. H. M., & Rezende, C. A. (2021). Structure–property relationships of cellulose nanocrystals and nanofibrils: Implications for the design and performance of nanocomposites and all-nanocellulose systems. *ACS Applied Nano Materials, 4*(10), 10505–10518. https://doi.org/10.1021/acsanm.1c02008

Carter, N., Grant, I., Dewey, M., Bourque, M., & Neivandt, D. J. (2021). Production and characterization of cellulose nanofiber slurries and sheets for biomedical applications. *Frontiers in Nanotechnology, 3*(December), 1–11. https://doi.org/10.3389/fnano.2021.729743

Celestini, V., Ribeiro, W. B., Damo, T., Lavoratti, A., Zattera, A. J., & Brandalise, R. N. (2021). Changes in the rheometric, morphological and mechanical properties of nitrile rubber composites by the use of different concentrations of cellulose nanofibers. *Journal of Elastomers and Plastics*, 1–18. https://doi.org/10.1177/00952443211017183

Chang, B. P., Gupta, A., Muthuraj, R., & Mekonnen, T. H. (2021). Bioresourced fillers for rubber composite sustainability: Current development and future opportunities. *Green Chemistry*, *23*(15), 5337–5378. https://doi.org/10.1039/d1gc01115d

Chen, Y., Xu, C., & Cao, X. (2015). Dynamic rheology studies of carboxylated butadiene – styrene rubber/cellulose nanocrystals nanocomposites: Vulcanization process and network structures. *Polymer Composites*, *36*(4), 623–629. https://doi.org/10.1002/pc

Cheng, G., Zhou, M., Wei, Y. J., Cheng, F., & Zhu, P. X. (2019). Comparison of mechanical reinforcement effects of cellulose nanocrystal, cellulose nanofiber, and microfibrillated cellulose in starch composites. *Polymer Composites*, *40*, E365–E372. https://doi.org/10.1002/pc.24685

Chin, K. M., Sung Ting, S., Ong, H. L., Omar, M., Ting, S. S., Ong, H. L., Omar, M., Sung Ting, S., Ong, H. L., & Omar, M. (2018). Surface functionalized nanocellulose as a veritable inclusionary material in contemporary bioinspired applications: A review. *Journal of Applied Polymer Science*, *135*(13). https://doi.org/10.1002/app.46065

Dhali, K., Ghasemlou, M., Daver, F., Cass, P., & Adhikari, B. (2021). A review of nanocellulose as a new material towards environmental sustainability. *Science of the Total Environment*, *775*, 145871. https://doi.org/10.1016/j.scitotenv.2021.145871

Dominic, M., Joseph, R., Sabura Begum, P. M., Kanoth, B. P., Chandra, J., & Thomas, S. (2020). Green tire technology: Effect of rice husk derived nanocellulose (RHNC) in replacing carbon black (CB) in natural rubber (NR) compounding. *Carbohydrate Polymers*, *230*. https://doi.org/10.1016/j.carbpol.2019.115620

Dominic, M. C. D., Joseph, R., Begum, P. M. S., Joseph, M., Padmanabhan, D., Morris, L. A., Kumar, A. S., & Formela, K. (2020). Cellulose nanofibers isolated from the Cuscuta Reflexa plant as a green reinforcement of natural rubber. *Polymers*, *12*(4). https://doi.org/10.3390/POLYM12040814

Dufresne, A. (2000). Dynamic mechanical analysis of the interphase in bacterial polyester/cellulose whiskers natural composites. *Composite Interfaces*, *7*(1), 53–67. https://doi.org/10.1163/156855400300183588

Dufresne, A. (2013). Nanocellulose: A new ageless bionanomaterial. *Materials Today*, *16*(6), 220–227. https://doi.org/10.1016/j.mattod.2013.06.004

Dufresne, A., de Souza, A. G., de Lima, G. F., Colombo, R., Rosa, D. S., Xie, J., Liu, S., Tortorella, S., Buratti, V. V., Maturi, M., Sambri, L., Franchini, M. C., Locatelli, E., Bhagia, S., Bornani, K., Agarwal, R., Satlewal, A., Ďurkovič, J., Lagaňa, R., ... Adhikari, B. (2021). 5. Chemical modification of nanocellulose. *Nanocellulose*, *15*(June), 35–42. https://doi.org/10.1016/j.progpolymsci.2021.101418

Feldman, D. (2012). Elastomer nanocomposite; Properties. *Journal of Macromolecular Science, Part A: Pure and Applied Chemistry*, *49*(9), 784–793. https://doi.org/10.1080/10601325.2012.703537

Fernandes, E. M., Pires, R. A., Mano, J. F., & Reis, R. L. (2013). Bionanocomposites from lignocellulosic resources: Properties, applications and future trends for their use in the biomedical field. *Progress in Polymer Science*, *38*(10–11), 1415–1441. https://doi.org/10.1016/j.progpolymsci.2013.05.013

Fiorati, A., Bellingeri, A., Carlo, P., Corsi, I., & Venditti, I. (2020). Silver nanoparticles for water pollution monitoring and treatments: Ecosafety challenge and cellulose-based hybrids solution. *Polymers*, *12*(1635). https://doi.org/10.3390/POLYM12081635

Flauzino Neto, W. P., Mariano, M., da Silva, I. S. V., Silvério, H. A., Putaux, J. L., Otaguro, H., Pasquini, D., & Dufresne, A. (2016). Mechanical properties of natural rubber nanocomposites reinforced with high aspect ratio cellulose nanocrystals isolated from soy hulls. *Carbohydrate Polymers*, *153*, 143–152. https://doi.org/10.1016/j.carbpol.2016.07.073

Gan, P. G., Sam, S. T., Abdullah, M. F., & Omar, M. F. (2020). Thermal properties of nanocellulose-reinforced composites: A review. *Journal of Applied Polymer Science*, *137*(11). https://doi.org/10.1002/app.48544

George, J., & Sabapathi, S. N. (2015). Cellulose nanocrystals: Synthesis, functional properties, and applications. *Nanotechnology, Science and Applications*, *8*, 45–54. https://doi. org/10.2147/NSA.S64386

Ghasemlou, M., Daver, F., Ivanova, E. P., Habibi, Y., & Adhikari, B. (2021). Surface modifications of nanocellulose: From synthesis to high-performance nanocomposites. *Progress in Polymer Science*, *119*, 101418. https://doi.org/10.1016/j.progpolymsci.2021.101418

Gopi, S., Balakrishnan, P., Divya, C., Valic, S., Govorcin Bajsic, E., Pius, A., & Thomas, S. (2017). Facile synthesis of chitin nanocrystals decorated on 3D cellulose aerogels as a new multi-functional material for waste water treatment with enhanced anti-bacterial and anti-oxidant properties. *New Journal of Chemistry*, *41*(21). https://doi.org/10.1039/ c7nj02392h

Habibi, Y. (2014). Key advances in the chemical modification of nanocelluloses. *Chemical Society Reviews*, *43*(5), 1519–1542. https://doi.org/10.1039/c3cs60204d

Habibi, Y., Lucia, L. A., & Rojas, O. J. (2010). Cellulose nanocrystals: Chemistry, self-assembly, and applications. *Chemical Reviews*, *110*(6), 3479–3500. https://doi.org/10.1021/ cr900339w

Hakimi, N. M. F., Hua, L. S., Chen, L. W., Mohamad, S. F., Osman Al Edrus, S. S., Park, B. D., & Azmi, A. (2021). Surface modified nanocellulose and its reinforcement in natural rubber matrix nanocomposites: A review. *Polymers*, *13*(19), 1–24. https://doi.org/10.3390/ polym13193241

Han, J., Lu, K., Yue, Y., Mei, C., Huang, C., Wu, Q., & Xu, X. (2019). Nanocellulose-templated assembly of polyaniline in natural rubber-based hybrid elastomers toward flexible electronic conductors. *Industrial Crops and Products*, *128*(November 2018), 94–107. https:// doi.org/10.1016/j.indcrop.2018.11.004

Han, J., Zhou, C., Wu, Y., Liu, F., & Wu, Q. (2013). Self-assembling behavior of cellulose nanoparticles during freeze-drying: Effect of suspension concentration, particle size, crystal structure, and surface charge. *Biomacromolecules*, *14*(5), 1529–1540. https://doi. org/10.1021/bm4001734

Haron, G. A. S., Mahmood, H., Noh, M. H., Alam, M. Z., & Moniruzzaman, M. (2021). Ionic liquids as a sustainable platform for nanocellulose processing from bioresources: Overview and current status. *ACS Sustainable Chemistry and Engineering*, *9*(3), 1008– 1034. https://doi.org/10.1021/acssuschemeng.0c06409

Haunreiter, K. J., Dichiara, A. B., & Gustafson, R. (2022). Nanocellulose by ammonium persulfate oxidation: An alternative to TEMPO-mediated oxidation. *ACS Sustainable Chemistry & Engineering*, *10*(12), 3882–3891. https://doi.org/10.1021/acssuschemeng.1c07814

He, Y., Boluk, Y., Pan, J., Ahniyaz, A., Deltin, T., & Claesson, P. M. (2020). Comparative study of CNC and CNF as additives in waterborne acrylate-based anti-corrosion coatings. *Journal of Dispersion Science and Technology*, *41*(13), 2037–2047. https://doi.org /10.1080/01932691.2019.1647229

Hu, C., Zhou, Y., Zhang, T., Jiang, T., Meng, C., & Zeng, G. (2021). Morphological, thermal, mechanical, and optical properties of hybrid nanocellulose film containing cellulose nanofiber and cellulose nanocrystals. *Fibers and Polymers*, *22*(8), 2187–2193. https:// doi.org/10.1007/s12221-021-0903-3

Huang, J., Ma, X., Yang, G., & Alain, D. (2019). Introduction to nanocellulose. *Nanocellulose: From Fundamentals to Advanced Materials*, *Mcc*, 1–20. https://doi. org/10.1002/9783527807437.ch1

Isogai, A. (2021). Emerging nanocellulose technologies: Recent developments. *Advanced Materials*, *33*(28), 1–10. https://doi.org/10.1002/adma.202000630

Isogai, A., & Zhou, Y. (2019). Diverse nanocelluloses prepared from TEMPO-oxidized wood cellulose fibers: Nanonetworks, nanofibers, and nanocrystals. *Current Opinion in Solid State and Materials Science*, *23*(2), 101–106. https://doi.org/10.1016/j.cossms. 2019.01.001

Jaffar, S. S., Saallah, S., Misson, M., Siddiquee, S., Roslan, J., Saalah, S., & Lenggoro, W. (2022). Recent development and environmental applications of nanocellulose-based membranes. *Membranes*, *12*(3). https://doi.org/10.3390/membranes12030287

Jamróz, E., Kulawik, P., & Kopel, P. (2019). The effect of nanofillers on the functional properties of biopolymer-based films: A review. *Polymers*, *11*(4), 1–43. https://doi.org/10.3390/polym11040675

Jarnthong, M., Wang, F., Wei, X. Y., Wang, R., & Li, J. H. (2015). Preparation and properties of biocomposite based on natural rubber and bagasse nanocellulose. *MATEC Web of Conferences*, *26*, 1–4. https://doi.org/10.1051/matecconf/20152601005

Jiang, W., Shen, P., Yi, J., Li, L., Wu, C., & Gu, J. (2020). Surface modification of nanocrystalline cellulose and its application in natural rubber composites. *Journal of Applied Polymer Science*, *137*(39). https://doi.org/10.1002/app.49163

Kargarzadeh, H., Mariano, M., Huang, J., Lin, N., Ahmad, I., Dufresne, A., & Thomas, S. (2017). Recent developments on nanocellulose reinforced polymer nanocomposites: A review. *Polymer*, *132*, 368–393. https://doi.org/10.1016/j.polymer.2017.09.043

Kaur, P., Sharma, N., Munagala, M., & Rajkhowa, R. (2021). Nanocellulose: Resources, physiochemical properties, current uses and future applications. *Frontiers in Nanotechnology*, *3*(November), 1–17. https://doi.org/10.3389/fnano.2021.747329

Lavoratti, A., Scienza, L. C., & Zattera, A. J. (2015). Dynamic mechanical analysis of cellulose nanofiber/polyester resin composites. *ICCM International Conferences on Composite Materials*, *2015-July* (April 2016).

Li, W., Wu, Q., Zhao, X., Huang, Z., Cao, J., Li, J., & Liu, S. (2014). Enhanced thermal and mechanical properties of PVA composites formed with filamentous nanocellulose fibrils. *Carbohydrate Polymers*, *113*, 403–410. https://doi.org/10.1016/j.carbpol.2014.07.031

Li, Z., Zhang, Y., Anankanbil, S., & Guo, Z. (2021). Applications of nanocellulosic products in food: Manufacturing processes, structural features and multifaceted functionalities. *Trends in Food Science and Technology*, *113*(December 2020), 277–300. https://doi.org/10.1016/j.tifs.2021.03.027

Lin, K. H., Hu, D., Sugimoto, T., Chang, F. C., Kobayashi, M., & Enomae, T. (2019). An analysis on the electrophoretic mobility of cellulose nanocrystals as thin cylinders: Relaxation and end effect. *RSC Advances*, *9*(58), 34032–34038. https://doi.org/10.1039/c9ra05156b

Lunardi, V. B., Soetaredjo, F. E., Putro, J. N., Santoso, S. P., Yuliana, M., Sunarso, J., Ju, Y., & Ismadji, S. (2021). Nanocellulose: Sources, pretreatment, isolations, modification, and its application as the drug carriers. *Polymers*, *13*, 2052.

Mahendra, I. P., Wirjosentono, B., Tamrin, T., Ismail, H., Mendez, J. A., & Causin, V. (2020). The effect of nanocrystalline cellulose and TEMPO-oxidized nanocellulose on the compatibility of polypropylene/cyclic natural rubber blends. *Journal of Thermoplastic Composite Materials*. https://doi.org/10.1177/0892705720959129

Marakana, P. G., Dey, A., & Saini, B. (2021). Isolation of nanocellulose from lignocellulosic biomass: Synthesis, characterization, modification, and potential applications. *Journal of Environmental Chemical Engineering*, *9*(6), 106606. https://doi.org/10.1016/j.jece.2021.106606

Mariano, M., El Kissi, N., & Dufresne, A. (2016). Cellulose nanocrystal reinforced oxidized natural rubber nanocomposites. *Carbohydrate Polymers*, *137*, 174–183. https://doi.org/10.1016/j.carbpol.2015.10.027

Mariano, M., El Kissi, N., & Dufresne, A. (2018). Cellulose nanomaterials: Size and surface influence on the thermal and rheological behavior. *Polímeros*, *28*(2), 93–102. https://doi.org/10.1590/0104-1428.2413

Mateo, S., Peinado, S., Morillas-Gutiérrez, F., La Rubia, M. D., & Moya, A. J. (2021). Nanocellulose from agricultural wastes: Products and applications—A review. *Processes*, *9*(9). https://doi.org/10.3390/pr9091594

Mathew, A. P. (2017). Elastomeric Nanocomposites Reinforced with Nanocellulose and Nanochitin. In: H. Kargarzadeh, I. Ahmad, S. Thomas, & A. Dufresne (eds) *Handbook of Nanocellulose and Cellulose Nanocomposites*, 217–234. Wiley. https://doi.org/10.1002/9783527689972.ch6

Nair, S. S., Chen, H., Peng, Y., Huang, Y., & Yan, N. (2018). Polylactic acid biocomposites reinforced with nanocellulose fibrils with high lignin content for improved mechanical, thermal, and barrier properties. *ACS Sustainable Chemistry and Engineering*, 6(8), 10058–10068. https://doi.org/10.1021/acssuschemeng.8b01405

Neves, R. M., Ornaghi, H. L., Zattera, A. J., & Amico, S. C. (2021). Recent studies on modified cellulose/nanocellulose epoxy composites: A systematic review. *Carbohydrate Polymers*, 255(November 2020). https://doi.org/10.1016/j.carbpol.2020.117366

Nickerson, R. F., & Habrie, J. A. (1947). Cellulose intercrystalline structure. *Industrial and Engineering Chemistry*, 39(11), 1507–1512.

Noor, S. M., Anuar, A. N., Tamunaidu, P., Goto, M., Shameli, K., & Halim, M. H. A. (2020). Nanocellulose production from natural and recyclable sources: A review. *IOP Conference Series: Earth and Environmental Science*, 479(1). https://doi.org/10.1088/1755-1315/479/1/012027

Noremylia, M. B., Hassan, M. Z., & Ismail, Z. (2022). Recent advancement in isolation, processing, characterization and applications of emerging nanocellulose: A review. *International Journal of Biological Macromolecules*, 206(January), 954–976. https://doi.org/10.1016/j.ijbiomac.2022.03.064

Ogunsona, E., Hojabr, S., Berry, R., & Mekonnen, T. H. (2020). Nanocellulose-triggered structural and property changes of acrylonitrile-butadiene rubber films. *International Journal of Biological Macromolecules*, 164, 2038–2050. https://doi.org/10.1016/j.ijbiomac.2020.07.202

Omran, A. A. B., Mohammed, A. A. B. A., Sapuan, S. M., Ilyas, R. A., Asyraf, M. R. M., Saeid, S., Koloor, R., & Petr, M. (2021). Micro- and nanocellulose in polymer composite materials: A review. *Polymers*, 13(2), 1–35.

Oun, A. A., Shankar, S., & Rhim, J. W. (2020). Multifunctional nanocellulose/metal and metal oxide nanoparticle hybrid nanomaterials. *Critical Reviews in Food Science and Nutrition*, 60(3), 435–460.

Panchal, P., Ogunsona, E., & Mekonnen, T. (2019). Trends in advanced functional material applications of nanocellulose. *Processes*, 7(1). https://doi.org/10.3390/pr7010010

Peng, S., Luo, Q., Zhou, G., & Xu, X. (2021). Recent advances on cellulose nanocrystals and their derivatives. *Polymers*, 13(19), 3247. https://doi.org/10.3390/polym13193247

Pennells, J., Godwin, I. D., Amiralian, N., & Martin, D. J. (2020). Trends in the production of cellulose nanofibers from non-wood sources. *Cellulose*, 27(2), 575–593. https://doi.org/10.1007/s10570-019-02828-9

Putro, J. N., Ismadji, S., Gunarto, C., Yuliana, M., Santoso, S. P., Soetaredjo, F. E., & Ju, Y. H. (2019). The effect of surfactants modification on nanocrystalline cellulose for paclitaxel loading and release study. *Journal of Molecular Liquids*, 282, 407–414. https://doi.org/10.1016/j.molliq.2019.03.037

Rahimi Kord Sofla, M., Brown, R. J., Tsuzuki, T., & Rainey, T. J. (2016). A comparison of cellulose nanocrystals and cellulose nanofibres extracted from bagasse using acid and ball milling methods. *Advances in Natural Sciences: Nanoscience and Nanotechnology*, 7(3). https://doi.org/10.1088/2043-6262/7/3/035004

Ram, F., & Shanmuganathan, K. (2021). Nanocellulose-based materials and composites for fuel cells. In: Sabu Thomas & Yasir Beeran Pottathara (eds) *Nanocellulose Based Composites for Electronics,* 259–293. Elsevier Inc. https://doi.org/10.1016/b978-0-12-822350-5.00011-4

Rana, A. K., Frollini, E., Thakur, V. K., Pasquini, D., de Teixeira, E. M., da Curvelo, A. A. S., Belgacem, M. N., & Dufresne, A. (2021). Cellulose nanocrystals: Pretreatments, preparation strategies, and surface functionalization. *International Journal of Biological Macromolecules*, 182(3), 1554–1581. https://doi.org/10.1016/j.ijbiomac.2021.05.119

Raza, M., & Abu-Jdayil, B. (2022). Cellulose nanocrystals from lignocellulosic feedstock: A review of production technology and surface chemistry modification. *Cellulose*, *29*(2), 685–722. https://doi.org/10.1007/s10570-021-04371-y

Roy, K., Pongwisuthiruchte, A., Chandra Debnath, S., Potiyaraj, P., Syuhada, D. N., Azura, A. R., Mokhothu, T. H., & John, M. J. (2021). Bio-based fillers for environmentally friendly composites. *Polymers*, *1–8*(20), 243–270. https://doi.org/10.3390/polym13203600

Rui, H. Y., Chen, C. G., Liu, L., Yao, X. Y. J., Yu, H. Y., Chen, R., Chen, G. Y., Liu, L., Yang, X. G., & Yao, J. M. (2015). Silylation of cellulose nanocrystals and their reinforcement of commercial silicone rubber. *Journal of Nanoparticle Research*, *17*(9), 1–13. https://doi.org/10.1007/s11051-015-3165-4

Santmartí, A., & Lee, K.-Y. (2018). Crystallinity and thermal stability of nanocellulose. *Nanocellulose and Sustainability*, *January*, 67–86. https://doi.org/10.1201/9781351262927-5

Sathishkumar, P., Kamala-kannan, S., Cho, M., Su, J., Hadibarata, T., Razman, M., & Oh, B. (2014). Laccase immobilization on cellulose nanofiber: The catalytic efficiency and recyclic application for simulated dye effluent treatment. *Journal of Molecular Catalysis. B, Enzymatic*, *100*, 111–120. https://doi.org/10.1016/j.molcatb.2013.12.008

Shaghaleh, H., Xu, X., & Wang, S. (2018). Current progress in production of biopolymeric materials based on cellulose, cellulose nanofibers, and cellulose derivatives. *RSC Advances*, *8*(2), 825–842. https://doi.org/10.1039/c7ra11157f

Sharma, A., Thakur, M., Bhattacharya, M., Mandal, T., & Goswami, S. (2019). Commercial application of cellulose nano-composites – A review. *Biotechnology Reports*, *21*(2018), e00316. https://doi.org/10.1016/j.btre.2019.e00316

Silva, L. S., Ferreira, F. J. L., Silva, M. S., Citó, A. M. G. L., Meneguin, A. B., Sábio, R. M., Barud, H. S., Bezerra, R. D. S., Osajima, J. A., & Silva Filho, E. C. (2018). Potential of amino-functionalized cellulose as an alternative sorbent intended to remove anionic dyes from aqueous solutions. *International Journal of Biological Macromolecules*, *116*, 1282–1295. https://doi.org/10.1016/j.ijbiomac.2018.05.034

Singh, S., Dhakar, G. L., Kapgate, B. P., Maji, P. K., Verma, C., Chhajed, M., Rajkumar, K., & Das, C. (2020). Synthesis and chemical modification of crystalline nanocellulose to reinforce natural rubber composites. *Polymers for Advanced Technologies*, *31*(12), 3059–3069. https://doi.org/10.1002/pat.5030

Talebi, H., Ghasemi, F. A., & Ashori, A. (2022). The effect of nanocellulose on mechanical and physical properties of chitosan-based biocomposites. *Journal of Elastomers and Plastics*, *54*(1), 22–41. https://doi.org/10.1177/00952443211017169

TAPPI. (2011). ISO Technical Specification ISO/TS 27687, Nanotechnologies – Terminology and definitions for nano - objects – Nanoparticle, nanofiber and nanoplate. 1–6.

Tavakolian, M., Jafari, S. M., & Van De Ven, T. G. M. (2020). A review on surface - Functionalized cellulosic nanostructures as biocompatible antibacterial materials. *Nano-Micro Letters*, *12*(1), 1–23. https://doi.org/10.1007/s40820-020-0408-4

Teixeira, L. T., Braz, W. F., Correia de Siqueira, R. N., Pandoli, O. G., & Geraldes, M. C. (2021). Sulfated and carboxylated nanocellulose for Co+2 adsorption. *Journal of Materials Research and Technology*, *15*, 434–447. https://doi.org/10.1016/j.jmrt.2021.07.123

Thakur, V., Guleria, A., Kumar, S., Sharma, S., & Singh, K. (2021). Recent advances in nano-cellulose processing, functionalization and applications: A review. *Materials Advances*, *2*(6), 1872–1895. https://doi.org/10.1039/d1ma00049g

Tortorella, S., Buratti, V. V., Maturi, M., Sambri, L., Franchini, M. C., & Locatelli, E. (2020). Surface-modified nanocellulose for application in biomedical engineering and nano-medicine: A review. *International Journal of Nanomedicine*, *15*, 9909–9937. https://doi.org/10.2147/IJN.S266103

Trache, D., Hussin, M. H., Haafiz, M. K. M., & Thakur, V. K. (2017). Recent progress in cellulose nanocrystals: Sources and production. *Nanoscale*, *9*(5), 1763–1786. https://doi.org/10.1039/c6nr09494e

Trache, D., Tarchoun, A. F., Derradji, M., Hamidon, T. S., Masruchin, N., Brosse, N., & Hussin, M. H. (2020). Nanocellulose: From fundamentals to advanced applications. *Frontiers in Chemistry*, 8(May). https://doi.org/10.3389/fchem.2020.00392

Trache, D., & Thakur, V. K. (2020). Nanocellulose and nanocarbons based hybrid materials: Synthesis, characterization and applications. *Nanomaterials*, 10(9), 1–5. https://doi.org/10.3390/nano10091800

Trifol, J., Plackett, D., Sillard, C., Hassager, O., Daugaard, A. E., Bras, J., & Szabo, P. (2016). A comparison of partially acetylated nanocellulose, nanocrystalline cellulose, and nanoclay as fillers for high-performance polylactide nanocomposites. *Journal of Applied Polymer Science*, 133(14). https://doi.org/10.1002/app.43257

Trovatti, E., Carvalho, A. J. F., Ribeiro, S. J. L., & Gandini, A. (2013). Simple green approach to reinforce natural rubber with bacterial cellulose nanofibers. *Biomacromolecules*, 14(8), 2667–2674. https://doi.org/10.1021/bm400523h

Vanderfleet, O. M., & Cranston, E. D. (2021). Production routes to tailor the performance of cellulose nanocrystals. *Nature Reviews Materials*, 6(2), 124–144. https://doi.org/10.1038/s41578-020-00239-y

Wang, J., Liu, X., Jin, T., He, H., Liu, L., Wang, J., Liu, X., Jin, T., He, H., & Liu, L. (2019). Preparation of nanocellulose and its potential in reinforced composites: A review preparation of nanocellulose and its potential in. *Journal of Biomaterials Science, Polymer Edition*, 30(11), 919–946. https://doi.org/10.1080/09205063.2019.1612726

Xu, X., Liu, F., Jiang, L., Zhu, J. Y., Haagenson, D., & Wiesenborn, D. P. (2013). Cellulose nanocrystals vs. cellulose nanofibrils: A comparative study on their microstructures and effects as polymer reinforcing agents. *ACS Applied Materials and Interfaces*, 5(8), 2999–3009. https://doi.org/10.1021/am302624t

Yang, X., Biswas, S. K., Han, J., Tanpichai, S., Li, M. C., Chen, C., Zhu, S., Das, A. K., & Yano, H. (2021). Surface and interface engineering for nanocellulosic advanced materials. *Advanced Materials*, 33(28). https://doi.org/10.1002/adma.202002264

Yasin, S., Hussain, M., Zheng, Q., & Song, Y. (2021). Effects of ionic liquid on cellulosic nanofiller filled natural rubber bionanocomposites. *Journal of Colloid And Interface Science*, 591, 409–417. https://doi.org/10.1016/j.jcis.2021.02.029

Yin, Y., Lucia, L. A., Pal, L., Jiang, X., & Hubbe, M. A. (2020). Lipase-catalyzed laurate esterification of cellulose nanocrystals and their use as reinforcement in PLA composites. *Cellulose*, 27(11), 6263–6273. https://doi.org/10.1007/s10570-020-03225-3

Yu, S. Il, Min, S. K., & Shin, H. S. (2016). Nanocellulose size regulates microalgal flocculation and lipid metabolism. *Scientific Reports*, 6(October), 1–9. https://doi.org/10.1038/srep35684

Zhang, B., Huang, C., Zhao, H., Wang, J., Yin, C., Zhang, L., & Zhao, Y. (2019). Effects of cellulose nanocrystals and cellulose nanofibers on the structure and properties of polyhydroxybutyrate nanocomposites. *Polymers*, 11(12). https://doi.org/10.3390/polym11122063

Zhang, K., Barhoum, A., Xiaoqing, C., Li, H., & Samyn, P. (2019). Cellulose nanofibers: Fabrication and surface functionalization techniques. In: A. Barhoum, M. Bechelany, & A. Makhlouf (eds) *Handbook of Nanofibers*, 409–449. Springer. https://doi.org/10.1007/978-3-319-53655-2_58

Zhang, X., Xiong, R., Kang, S., Yang, Y., & Tsukruk, V. V. (2020). Alternating stacking of nanocrystals and nanofibers into ultrastrong chiral biocomposite laminates. *ACS Nano*, 14(11), 14675–14685. https://doi.org/10.1021/acsnano.0c06192

Zhao, K., Tian, X., Xing, J., Huang, N., Zhang, H., & Huanying Zhao, W. W. (2022). Tunable mechanical behavior of collagen-based films: A comparison of celluloses in different geometries. *Bioresource Technology Reports*, 100310. https://doi.org/10.1016/j.ijbiomac.2022.05.191

Zhu, G., & Lin, N. (2019). Surface chemistry of nanocellulose. In: J. Huang, A. Dufresne, & N. Lin (eds) *Nanocellulose: From Fundamentals to Advanced Materials*, 115–150. Wiley. https://doi.org/10.1002/9783527807437.ch5

6 Recent Advances in Organic Nanofillers-Derived Polymer Hydrogels

M. Iqbal Ishra, B. S. S. Nishadani,
M. G. R. Tamasha Jayawickrama, and
M. M. M. G. P. G. Manthilaka
University of Peradeniya
Institute of Materials Engineering and Technopreneurships

CONTENTS

6.1 Introduction 174
6.2 Types of Polymer Hydrogels and Nanofillers 175
    6.2.1 Classification of Hydrogels 175
    6.2.2 Types of Nanofillers 177
    6.2.3 Types of Organic Nanofillers 179
6.3 Effect of Organic Nanofillers on the Functional Properties of Polymer
    Hydrogel Materials 181
6.4 Functional Applications of Polymer Hydrogel-Based Materials with
    Organic Nanofillers 183
    6.4.1 Drug Delivery Applications 183
    6.4.2 Gene Delivery Applications 185
    6.4.3 Tissue Engineering Applications 186
    6.4.4 Energy Application 187
    6.4.5 Food and Environment Applications 188
6.5 Other Applications 189
6.6 The Future of Nanomaterials 190
References 191

DOI: 10.1201/9781003279372-6

6.1 INTRODUCTION

Nanotechnology could bring another revolution to the world of material science (L. Zhang et al., 2019). Nanotechnology involves the manipulations of matter on an atomic scale. The first-ever concept was presented in 1959 by the famous professor of physics Dr. Richard p. Feynman (Ranjan et al., 2018). The term nanotechnology had been coined by Norio Taniguchi in 1974. The field of nanotechnology is one of the most popular areas for current research and development in various technical disciplines (Polarz & Smarsly, 2002). This obviously includes polymer science and technology, and even in this field, the investigations cover a broad range of topics. This would include nanoelectronics as the critical dimension scale for modern devices is now below 100 nm. The general class of organic nanofiller-derived polymer hydrogel materials is a fast-growing research area. Significant effort is focused on the ability to obtained control of the nanoscale structure via innovative synthetic approaches (Park, 2007). The properties of nanofiller-incorporated materials depend not only on the properties of their individual parents but also on their morphology and interfacial characteristic (Xing & Tang, 2021). Therefore, organic nanofillers promise new applications in many fields.

A hydrogel is a 3D network of hydrophilic polymers that can swell in a fluid environment and hold a large amount of water while maintaining the structures due to the chemical or physical cross-linking of individual polymer chains (C. Zhao et al., 2020). Organic nanofiller-based hydrogel is a new generation of materials with a combination of bio-based polymer and fillers that have at least 1nm scale dimension (Abu Elella et al., 2021). In polymer nanocomposites films, the biopolymers act as a matrix, while organic nanofillers are dispersed there to improve the functional properties. These show unique properties combining the advantages of fillers and polymer hydrogels (Bellet et al., 2021). The organic nanofillers have a set of improved properties, such as mechanical, antimicrobial, and antioxidant properties and physical properties (Shin et al., 2018). These properties do not occur naturally in the hydrogels themselves; therefore, they are gained due to the addition of nanocomposites. The components of polymer hydrogels are characterized by high availability and good biodegradability (Pacelli et al., 2018).

The limitations of polymer-based hydrogels can be reduced by mixing two or more components. Currently, a novel method to improve the properties of hydrogel is the use of nanofillers, which not only can fulfill the reinforcing function but also could act as an active agent for hydrogel matrix that can extend the shelf life of products through exhibiting antimicrobial and antioxidative properties (Aranaz et al., 2021). The aim of this chapter is to summarize recent advances and achievements regarding the addition of organic nanofillers into polymer hydrogels with a focus on their functional properties and possible uses in different research areas, which include drug delivery, enzyme delivery medical and pharmaceutical application, tissue engineering, wound dressing, food packing and design, environment applications, energy application, and material synthesis for various research fields (A. Kumar et al., 2019).

6.2 TYPES OF POLYMER HYDROGELS AND NANOFILLERS

6.2.1 Classification of Hydrogels

Hydrogels are a type of three-dimensional network of hydrophilic cross-linked polymer that can absorb and store large amounts of water (Feldman, 2019). The hydrophilic groups present in the polymer network can be hydrated under neutral conditions to produce a gel structure (Heifler et al., 2021). The cross-linked network structure does not dissolve but can highly swell in water and respond to the fluctuation of environmental stimuli (Ng et al., 2020). The hydrophilic groups (like –OH, –CONH, –CONH$_2$, and –SO$_3$H) present in the polymeric backbone enable the absorption of a large quantity of water, while the cross-linked network structure suppresses their dissolution (Pacelli et al., 2018). Hydrogels show both solid-like and liquid properties (C. Zhang et al., 2020). In general, a low-concentration aqueous hydrophilic polymer solution with no entangled structures exhibits Newtonian mechanical behavior (Abu Elella et al., 2021). Cross-linked hydrogels possess rheological properties with extremely high viscosity and high elasticity similar to solids. Due to their water insolubility, hydrogels are the research hotspots of swollen polymer networks, as well as their application studies (Vashist et al., 2018). Hydrogels have been applied to a variety of fields, such as drought resistance in dry areas, masks in cosmetics, antipyretic stickers, analgesic stickers, agricultural films, new prevention agents in construction, humidity control, water-blocking agents in the petroleum industry, dehydration of crude oil and refined oil, dust suppressants in mining, food preservatives, thickeners, pharmaceutical carriers in medical applications, tissue engineering scaffolds, wound dressings, etc. It is worth noting that different application fields require different polymer materials (Bellet et al., 2021).

Hydrogels can be classified by different methods according to several views (Figure 6.1). Recent literature reports hydrogels can be classified into natural hydrogels and synthetic hydrogels based on their origin. Natural polymers are collagen, gelatin, hyaluronic acid, fibrin, sodium alginate, agarose, chitosan, dextran, cyclodextrin, etc. (F. Zhu et al., 2021). The hydrogels prepared with these abundant polymers inherit these excellent properties and are sensitive to the external environment. Most animals and plants contain a large number of natural polymer gels (Seidi et al., 2020)

The study of the structure and properties of such hydrogels is also an important way to study the physiological mechanisms of tissues for biomimetic applications. These hydrogels are biocompatible, biodegradable, and nontoxic (Larrañeta et al., 2018). However, they have poor mechanical properties and batch variation may lead to poor reproducibility. Synthetic hydrogels are cross-linked polymers prepared by the addition reaction or ring-opening polymerization under artificial conditions (Yuan et al., 2019). Polyacrylic acid and its derivatives, polyvinyl alcohol, polyethylene glycol, its copolymers, polyvinylpyrrolidone (Wakuda et al., 2018) are usually used as the skeletons to prepare synthetic hydrogels. Synthetic hydrogels have the advantages of easy industrial production and chemical modification and precisely controllable properties, yet with poor biocompatibility, bioactivity, and biodegradability, as compared with natural polymer hydrogels (Yuan et al., 2019).

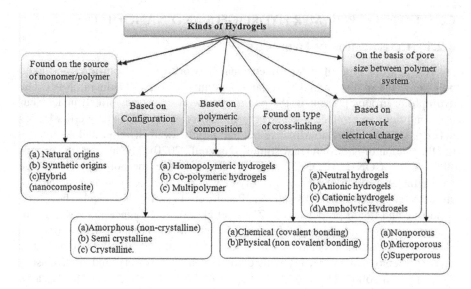

FIGURE 6.1 Classification of hydrogels based on different characteristics. (Adapted from ResearchGate.)

Based on cross-linking, they can be classified as physical and chemical hydrogels (Ciolacu & Suflet, 2018). Chemical hydrogels result in a strong covalent bond between polymer strands and it forms an irreversible gel structure (T. Zhu et al., 2019), which gives degradation resistance compared to other materials. Physical hydrogels have noncovalent bonds (Eshkol-Yogev et al., 2021). Polymers are capable of forming hydrogen bonds, hydrophobic interactions, ionic interactions, and crystallization cross-linking (Narayanan & Han, 2017). These have high biocompatibility and low toxicity and are also easily reversible under specific conditions. Classification according to the degree of swelling can be low swelling, medium swelling, high swelling, and superabsorbent (Anjani et al., 2021). Cross-linked network of polymer hydrogel swells when water or solvent is present. The swelling property of hydrogels depends on the density of the network present in hydrogels, polymer–solvent interactions, nature of solvent present, and act. The hydrophilic polymer can absorb a high amount of water and swell up quickly. Depending on hydrophilic groups present in the backbone structure of the polymer, swelling ability differs. This swelling ability of hydrogels gives raise to different porosity in the polymer hydrogel materials. This leads to nonporous, microporous, and superporous. The swelling of superporous hydrogels is governed by capillary wetting (Guo et al., 2020).

Based on biodegradability, hydrogels can be divided into biodegradable hydrogels and nonbiodegradable hydrogels. Most of the natural polymer hydrogels are biodegradable hydrogels (Ilyas et al., 2018). The three-dimensional structures of these hydrogels can be destroyed by the actions of microorganisms and enzymes under natural conditions (Aranaz et al., 2021). The bonding between the molecular chains and within the molecular chains is then broken, and the strength of the hydrogel is reduced. Eventually, the hydrogel is degraded into small molecules.

b. Nonbiodegradable hydrogels are a class of hydrogels that are insensitive to environmental stimuli and can maintain stable structural, physical, and chemical properties for a long time (Saba et al., 2018). Most synthetic hydrogels were prepared by chemical cross-linking (Dilks, 1980).

Classification according to polymeric composition can be homopolymer, co-polymeric, and semi-interpenetrating network, and interpenetrating network hydrogels (Giri et al., 2019). Polymer network derived from single species of monomer is known as a homopolymer. Hydrogels are composed of two types of monomers referred to as copolymers (El-Zawawy, 2005). In this type, at least one monomer is hydrophilic in nature. A semi-interpenetrating network forms when linear polymer penetrates into the cross-linked polymer network, without forming any other chemical bonds in-between them. An interpenetrating polymer network forms by the joining of two polymers (Pacelli et al., 2018).

Some hydrogels are sensitive to environmental stimuli. These hydrogels are named smart hydrogels. They can be characterized into pH-sensitive hydrogels, temperature-sensitive hydrogels, electrosensitive hydrogels, light-responsive hydrogels, antigen-sensitive hydrogels, and pressure-sensitive. pH-sensitive polymers will contain acid or base hangings on their structures (Kaniewska et al., 2020). These will respond to the pH change by the gain or loss of a proton. Temperature-sensitive hydrogels will form a gel-like structure under temperature changes. These hydrogels will have hydrophobic nature due to the presence of hydrogen bond-making substances (methyl, ethyl, and propane and act.) on their structures. Hydrogen bonds between these hydrogels are formed to swell (Bellet et al., 2021). Formed hydrogen bonds are consonant to the temperature. Electrosensitive hydrogels will swell or shrink when an applied electric field is present (Ruggiero et al., 2017). These have the ability to convert electrical energy to mechanical energy. Due to this characteristic, these can be used in vast applications. In light-sensitive hydrogels, when irradiated with light sources they will undergo structural transformation (Malagurski et al., 2017). These changes result in expansion and contract. At a specific wavelength of light source chromophores in the hydrogel, the structure will absorb and increase the temperature of the hydrogels. This will lead to altering the swelling character of thermosensitive hydrogels (Aung et al., 2020).

6.2.2 TYPES OF NANOFILLERS

Nanofillers are usually classified as organic and inorganic nanofillers (Seidi et al., 2020), which are discussed in Figure 6.2. Nanophase can be categorized based on nanosized dimensions Figure 6.3. They are 3D nanofiller, 2D nanofillers, and 1D nanofillers. 3D nanofillers include nanosphere, nanocrystals/nanoparticles (all dimensions < 100 nm). 2D nanofillers include nanofibrils, nanowhiskers, and nanotubes (diameter < 100 nm). The 1D nanofillers include nanosheets or nanoplatelets (thickness < 100 nm) (H. Liu et al., 2019). Based on sources used for the synthesis of nanofillers, they categorized into organic and inorganic nanofillers. Inorganic nanofillers mainly consist of metal oxides and clay nanoparticles. Nanofillers can improve the material property by varying the mixture of ratios between organic matrixes and nanofillers (Wei et al., 2010). Nanofillers were incorporated into polymer matrixes at

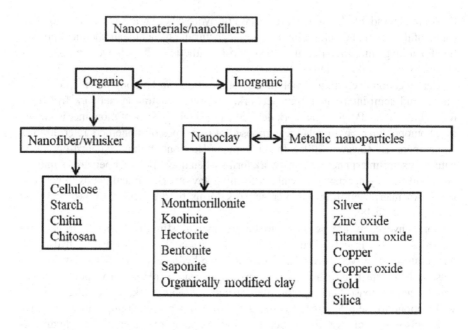

FIGURE 6.2 Different types of nanofillers. (Adapted from ResearchGate.)

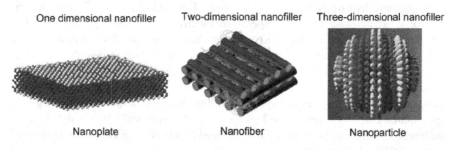

FIGURE 6.3 Nanofillers based on dimensions (1D, 2D, and 3D). (Adapted from the ResearchGate.)

1%–10% by mass. By incorporating various nanofiller types into different materials, we can improve their physical and chemical properties according to our needs (Dai et al., 2019).

Polymer nanocomposites, a class of polymers reinforced with low quantities of well-dispersed nanoparticles, offer advantages over conventional composites (Tie et al., 2021). It has been proved that when the sizes of nanofillers are very small (at least one of their dimensions is under 100 nm), the interface regions are so large that they start to interact at a very low level of loadings. Nanoparticles have a great effect on the properties and morphology of polymeric nanocomposites due to their large specific surface and high surface energy (Qian et al., 2020). The interactions between polymer matrix and nanoparticles alter polymer chemistry, that is, chain mobility and degree of cure, and generate new trap centers in the composite which brings about a significant change in electrical properties (Sun et al., 2020). Where the size of the particle is close to that of the polymer chain length, the particles do not behave like foreign inclusions and space charge densities are small.

6.2.3 TYPES OF ORGANIC NANOFILLERS

Naturally occurring polysaccharides can be classified according to their origin. Plant-based polysaccharides are starch, cellulose, pectin, gum, and act. Algae-based polysaccharides are agar, alginate, and Galatian's act. Animal-based polysaccharides are chitin, chitosan, hyaluronic acid, and cellulose. Bacteria-based polysaccharides are xanthan, cellulose, dextran, and levan. Fungi-based polysaccharides are chitin, chitosan, and yeast glucan (Seidi et al., 2020). Organic nanofillers mainly consist of naturally occurring cellulose, starch, chitosan, and chitin in the form of nanocrystals, nanofibrils, nanowhiskers, or NPS based on the method of extraction and source. Nanofillers have a greater surface area per mass. The nanometric size effect leads to the intensive topic in the research area of nanosized particles. The high surface area of nanosize particles opens various routes for novel interactive behavior with polymers/hydrogel materials (Tomczykowa & Plonska-Brzezinska, 2019).

Chitin (CTN) is the richest natural polysaccharide in the world. It is considered to be one of the most important renewable resources. It is found in cuticles or exoskeletons of invertebrates and in the cell walls of algae and fungi species. The dry weight of chitin obtained from crustaceans is highest when compared with the other sources (Tovar et al., 2020). CTN is a large, structural homopolymer of β-(1–4)-N-acetyl D-glucosamine (β-NAG) linkages. Based on the source of origin, it occurs in two forms, namely, α- and β-forms (Vashist et al., 2019). The chains of chitin are interconnected through hydrogen bonding, which impacts its solubility, reactivity, and swelling. The α chitin form is abundant in nature, as compared to the β form (Ali & Ahmed, 2018). The crystallographic measures of the two forms clearly display the fact that per unit cell in α-chitin (α-CTN), there are two antiparallel molecules, but only one in β-chitin (β-CTN), which are arranged in parallel. In both cases, chains in the sheets are held through hydrogen bonds, which are assumed to account for the resistance with respect to swelling of α-CTN in water, which is white in color, hard, and stiff. Unlike other natural polysaccharides, which are either neutral or acidic, CTN is highly basic. Typically, being hydrophobic and insoluble in water, has a chelating property, the ability to form polyoxysalt and films. It is soluble in certain organic solvents. CTN does not have well-characterized melting points, but they show the excessive application of heat. The water molecules are tightly bound with chitin, and hence, water loss generally occurs in the temperature range of 53°C–100°C. Both CTN and chitosan have numerous advantageous biological properties. They are hemostatic, fungistatic, and spermicidal; they also bind well to microbial and mammalian cells, regenerate gum tissue, and accelerate the synthesis of osteoblasts that is responsible for bone formation. They exhibit antitumor and antioxidant properties (Ahmadi et al., 2015). They can also act as central nervous system depressants and immunologic adjuvants. They have applications in multiple streams because of their fiber- and film-forming properties. Natural polymers are important in the development of biocomposite materials (Eshkol-Yogev et al., 2021). Composite materials are made of two immiscible components (Giri et al., 2019). The main function of reinforcement is to improve the mechanical, thermal, and biological properties of polymer hydrogels. The unique properties of chitin are utilized in packaging, coating, water treatment, biomedicine, and in other fields in various areas. In order to increase the mechanical and water-resistant properties of polymer hydrogel materials, certain organic nanofillers are incorporated into the polymer hydrogel matrix (Qian et al., 2020). Recycling of thermoplastic polymer hydrogel is done by the molding technique. Melting temperature is one of the restrictive conditions in the recycling of chitin-based composites, and it should not increase beyond 300°C, at which point it begins to degrade. This technique does not damage the reinforcement in the polymer hydrogel material (C. T. G. V. M. T. Pires et al., 2014). CTN has attracted significant and chitin-based bionanocomposite's notable attention in the last 6 years as a biodegradable material with a satisfactory way of waste management of the crustacean shells (Antonino et al., 2017).

Cellulose is the most abundant renewable polymer which takes 1st place on earth and is responsible for parallel in a highly ordered manner (Han et al., 2021). Cellulose is mainly synthesized by plants but it is also synthesized by bacteria (Figure 6.4). Cellulose is a homopolymer of glucose. Glucose monomers are joined by β-1,4 linkage. Cellulose chains are ordered in a manner to form microfibrils that build up the plant cell wall. Toughness, water insolubility, and fibrous polysaccharides play important role in keeping the structure of plant cell walls stable (Ilyas et al., 2018). The morphology, size, and other characteristics depend on the origin of the cellulose. the nanoform of cellulose exists mainly as crystalline cellulose (CNC) and cellulose nanofibers (CNFs). Nanoforms of cellulose have been proven for many green materials in recent times (Seidi et al., 2020). Nanocellulose has excellent properties such as abundance, better mechanical properties, high aspect ratio, biocompatibility, and renewable (Jiao et al., 2021). In recent studies, bioethanol was used for the isolation of CNC having highly negative surface ζ-potential compared to sludge-based CNC in acidic pH (pH 5.0). Surface modification of nanocellulose is performed using various routes. A green approach using the enzyme hexokinase can be done for the introduction of phosphate (PO_4^{2+}) groups on the surface of cellulose fiber (Saba et al., 2018). The size of nanocellulose depends on the origin of cellulose, isolation method, and processing conditions (Qian et al., 2020).

The polysaccharides are synthesized by green plants. Starch is a polymer carbohydrate that consists of numerous glycosidic bonds between glucose. Pure starch molecule bears properties like white, tasteless, and odorless powder which is insoluble in cold water and alcohols. Starch nanoparticles (St-NPs) have attracted much attention due to their unique behavior that is different from their bulk material (Huang et al., 2019). In a study, St-NPs were extracted from starch granules by means of physical or chemical treatments (González et al., 2018). Modifying the surface of starch nanoparticles can increase surface functionality (Ilyas et al., 2018). Alginate-based nanoparticles have been widely used in target drug delivery and cell transplant in tissue engineering due to their biocompatibility, low toxicity, low cost, and high surface-area-to-mass ratio (T. Zhu et al., 2019). Pectin is polysaccharides rich in galacturonic acids. Nano form of pectin is commonly used as a gelling agent,

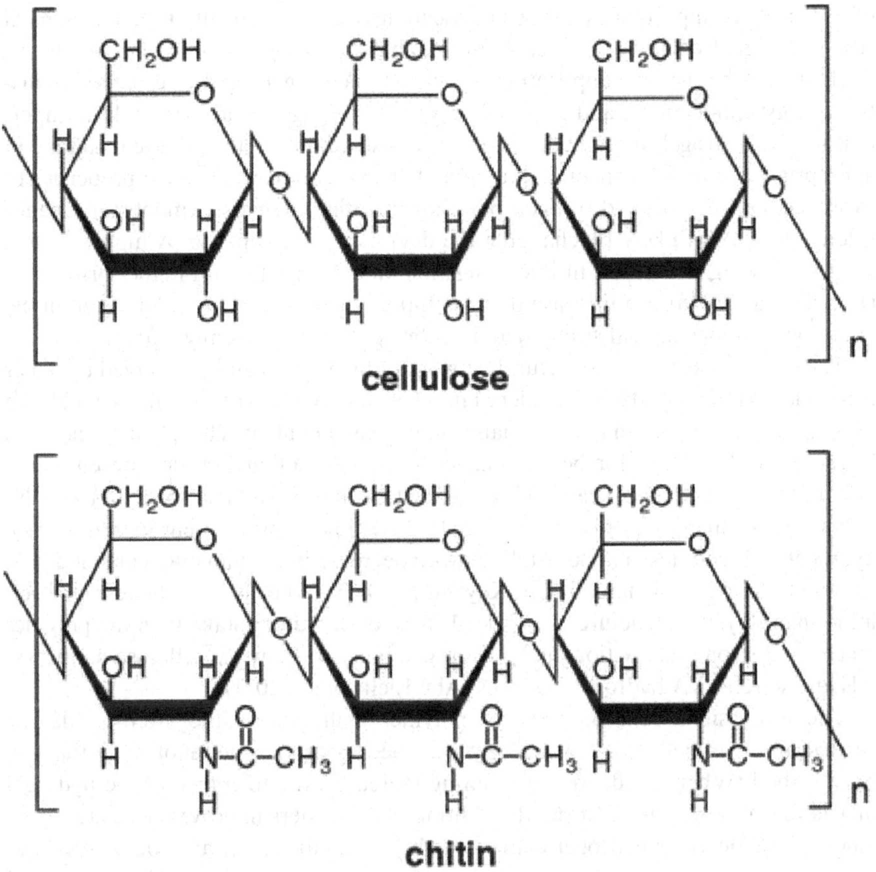

FIGURE 6.4 Structure of cellulose and chitin. (Adapted from Springer link.)

colloidal stabilizers, and thickening agent in the food industries and has also been used for drug-loading vehicles (Martau et al., 2019). In the world, the most abundant natural polymers such as chitosan ranked second place. Organic nanofiller-based hydrogels have been extensively been used in recent years in various fields for several applications (L. Liu et al., 2020).

6.3 EFFECT OF ORGANIC NANOFILLERS ON THE FUNCTIONAL PROPERTIES OF POLYMER HYDROGEL MATERIALS

The use of nonbiodegradable plastic increases with time, and waste produced from that is an increasingly serious problem. Therefore, researches forces on the development of products with biodegradable and eco-friendly properties. The interaction between polymer hydrogels and organic nanofillers leads to the improved functionality of the hydrogel materials. Depending on the properties of organic nanofillers, improved version can be obtained (Mondal et al., 2020). The mechanical strength of hydrogels is important in various applications. The assessment of the mechanical strength of hydrogels is significant for suitable physiological functions in various fields such as biomedical applications, tissue engineering, drug delivery, food industry, energy applications, and gene delivery (Eshkol-Yogev et al., 2021). It is important for the hydrogel to maintain its physical texture during the release or intake of appropriate agents for a specific time period. The required mechanical properties of hydrogels can be attained by incorporating specific polymers comonomers, nano-fillers, and cross-linkers by changing the degree of cross-linking. A higher degree of cross-linking will result in low elongation and electricity with higher brittleness (Pacelli et al., 2018). An increase in cross-linking is important in order to retain the compromised mechanical strength and elasticity (Rallini & Kenny, 2017).

The empty space in the structural network of hydrogels can be occupied by water molecules. This property is considered in the water content of the film. The addition of organic nanofillers can make a change in the water content. This phenomenon can be related to the interaction between nanofillers and functional groups present in the polymer hydrogels, which can lead to a reduction in the available space in the polymer hydrogel matrix (Eshkol-Yogev et al., 2021). The development of starch-PVP hydrogel with cellulose nanocrystals shows a decrease in the moisture content due to the addition of organic nanofillers (Koyani, 2020). This behavior is the result of cellulose nanocrystals structure, which leads to a lower water uptake than the polymer matrix and strong interaction due to hydrogen bonding between cellulose nanocrystals and starch-PVA hydrogel matrixes (Abu Elella et al., 2021).

The most significant property of polymer hydrogels is the swelling degree (Prihatiningtyas et al., 2019), which gives an idea about the amount of water that can be absorbed. When the degree of swelling increases, the tolerance of the hydrogel matrix to water increases. Normally addition of nanofillers improves the water resistance of the polymer hydrogel matrixes (Mohammadinejad et al., 2020). Anyhow, reverse reaction also can be noticed during the addition of fillers. The addition of chitin nanofibers can lead to a higher swelling degree due to the formation of a higher amount of OH^- resulting in increased water absorption (Ahmed, 2015).

The addition of organic nanofillers to polymer hydrogel can affect their mechanical properties (Sánchez et al., 2019). This occurs because nanofillers have a very high specific surface area that can affect the strength and degree of dispersion. Homogeneous dispersion of organic nanofillers within the polymer hydrogels gives rise to a specific transfer of stress via the shear mechanism from the polymer matrix to the nanofillers and can result in increased strength of the polymer hydrogel matrixes (Lee et al., 2020). The strategy of strengthening the tensile strength may also result by the interaction between the nanofillers and polymer hydrogel matrixes. Covalent and hydrogen bonds between hydroxyl groups of hydrogels are formed, which will lead to the strengthening of molecular forces between nanofillers and hydrogel matrixes (Dai et al., 2019).

Antimicrobial packaging system in various fields is important to ensure safety and increase the storage stability of materials by preventing the growth of microorganism (Hassan et al., 2012). The antimicrobial properties of packaging materials can be achieved through the addition of antimicrobial substances to the polymer hydrogels (González et al., 2018). Antimicrobial agents can be categorized into two categories. They are bactericidal and bacteriostatic. Encapsulating organic nanoparticles in polymer hydrogels will increase the antimicrobial property of the materials. The mechanism of antibacterial properties of organic nanofillers is depended on the type of nanofillers as well as on the species and the natural properties of the bacteria (Ali & Ahmed, 2018). Quick-growing bacteria are more sensitive to nanofillers than slow-growing bacteria. Chitin nanofibers can be characterized under bacteriostatic. The addition of chitin nanofillers leads to course blocking the route of access to the nutrient and oxygen, which leads to the inhibition of bacterial growth (Yu Chen, 2019). Chitosan has also been identified as an antibacterial agent to prevent infection in wounds and hasten the wound-healing process by increasing the growth of skin cells. It was also used for food preservative purposes during packaging foods (Armentano et al., 2018).

6.4 FUNCTIONAL APPLICATIONS OF POLYMER HYDROGEL-BASED MATERIALS WITH ORGANIC NANOFILLERS

The idea behind the nanofillers is to use the building blocks with dimensions in nanometer scale to design and create new materials with new properties. Organic nanofillers is the most exploited area in polymer hydrogel science in various fields. Organic nanofillers derived from polymer hydrogels have gained considerable attention in recent years owing to their unique potential via combining the characteristic properties of polymer hydrogels and organic nanofillers. The unique properties of these materials lead research scientists to focus on multiple applications in various fields. Some of those applications are gene delivery, tissue engineering, wound dressing, food packaging, drug delivery, and act (Shah et al., 2020).

6.4.1 DRUG DELIVERY APPLICATIONS

People around the world are connected through the various lines each other than earlier. Globalization is increasing, and the spread of technologies and innovation has led human life to higher standers of living across the globe. The ease of work had made the reduction of human activity (Vashist et al., 2019). The lack of physical exercise leads to face various diseases. Biopolymeric nanoparticles such as polysaccharides and proteins have attracted greater attention in therapeutic applications (Liechty et al., 2010). Chitosan is the most famous polysaccharide, which is ranked 2nd in abundance, the first being cellulose. Chitosan is obtained by the deacetylation reaction of chitin. It is cost-effective, biocompatible, environmentally friendly, and biodegradable (Jamróz et al., 2019). Nanoparticles of the biopolymer chitosan have gained great attention for application in drug delivery agents due to their high stability in aqueous solution, lower toxicity, and providing several routes of administration (Figure 6.5). Therefore, chitosan nanoparticles can be used as promising drug delivery vehicles for several diseases (Zeng et al., 2012).

Chitosan nanoparticles are defined as solid particles with a size in the range of 1–100 nm. There are different methods available for the synthesis of chitosan NP such as emulsion, ionic gelation, reverse micellar, and self-assembling methods

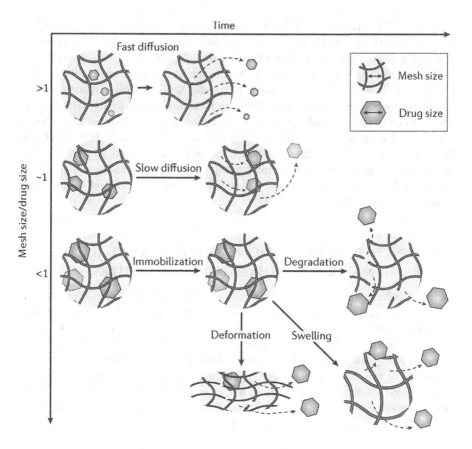

FIGURE 6.5   Drug delivery mechanism. (Adapted from Europe PMC.)

(Hossain & Iqbal, 2014). A popular method used to synthesize a chitosan nanoparticle is reverse micellar and ionic gelation methods. In the ionic gelation method,
chitosan NP can be synthesized by the interaction of oppositely charged macromolecules (Manickam et al., 2020). Tripolyphosphate is extensively used in the preparation and assembling of the NP of chitosan. Tripolyphosphate has less toxicity and
has the ability to form gels through ionic interactions. The interaction of oppositely
charged molecules can be controlled by the charge density of chitosan and tripolyphosphate by changing the pH of the solution. These interactions can be facilitated
by various factors, such as pH, method of mixing, and the ratio of the component
(Vashist et al., 2019). Recently, a very easy method to synthesize Fe_3O_4-chitosan
NP (magnetic) by cross-linking with tripolyphosphate, precipitation with NaOH, and
oxidation with O_2 in aqueous HCl containing chitosan and $Fe(OH)_2$ was reported
(He et al., 2020). Polymeric NP with controlled shape and size can be synthesized
by reverse micellar method. In the preparation of reverse micelles, the organic solvent was used to dissolve the surfactant (Sun et al., 2020). The solution of chitosan
is added with constant shaking to avoid any turbidity. The aqueous phase should be

controlled in such a way as to maintain the whole solution in an optically transparent microemulsion phase. To control the shape and size of chitosan NP, lowering the molar mass of chitosan would be advantageous due to either lowering the viscosity of the internal aqueous phase or an increase in the separation of the polymer chains during the process (Ng et al., 2020).

Chitin and chitosan nanoparticles derived polymer hydrogels and films have also found applications in wound healing. Chitosan is also being used to control natural clotting of blood, block nerve endings, and reduce pain (Kamoun et al., 2015). It expedites wound healing and prevents scar formation. Chitosan films in combination with silver and zinc oxide NP have shown good antibacterial behavior against E. coli, S. aureus, and B. subtilis. Copper NP modified chitosan. Chitosan NP is used as an adjuvant for different vaccines such as hepatitis B and influenza. These NP have better antigen uptake ability by mucosal lymphoid tissues and lead to induction of strong immune responses against antigens (Aranaz et al., 2021).

6.4.2 GENE DELIVERY APPLICATIONS

The polysaccharides, importantly chitosan, are extensively used in biomedical and pharmaceutical industries due to their capacity to regulate the release of drugs, DNA, proteins, vaccines, and genes. Gene is the unit cell of heredity and sequence of nucleotides in DNA and RNA which contain information to code their respective proteins (Massaro et al., 2018), which is located on chromosomes. Chromosomes are located inside the cells. Gene therapy is a technique used to treat genetic disorders at their genetic roots (Saba et al., 2018), by injecting genes into the cells. In gene therapy, an active vector is used that is biocompatible at productive in the most significant barriers. The virus has limitations in cell transduction, due to its extreme adverse effects. So, there is a need to develop nonviral gene delivery systems. Organic nanofillers can enhance the therapeutic effectiveness of the drug by monitoring their efficient delivery time and circulation time. Chitosan-based nanofillers have been established due to their high biocompatibility, low toxicity, high positive charge, inexpensive, low immunogenicity, and ability to form polyelectrolyte complexes with DNA (Zhou et al., 2020).

Partially diacetyl chitosan consists of repeating units of glucosamine and N-acetyl glucosamine, which is soluble in an acidic medium and positively charged. In neutral pH, it does not dissolve. The solubility of chitosan can be modified by enhancing surface functionalization to produce chitosan derivatives. Various types of chitosan-based derivatives are made which include trimethyl, and dimethyl chitosan by alkylation of the amino groups of chitosan (Ali & Ahmed, 2018). This process is aided with an appropriate aldehyde with the help of a reducing agent and then applied in various fields (Ningaraju et al., 2018). New properties of chitosan can be achieved by alteration of hydroxyl functional groups present in chitosan, thiolates, hydrophobic, and by chemically grafting by deriving chitosan, which will improve the properties of chitosan such as drug permeation efficiency, being a barrier for acid-sensitive biomolecules and drugs, easy solubility at neutral pH, the enhanced release of drugs and reaching target sites (Aranaz et al., 2021). Chitosan-based gene delivery systems should form a stable complex with genetic materials, targeting the cells, being a barrier for nuclear degradation, and ease cellular transportation (Kumar Teli et al., 2010).

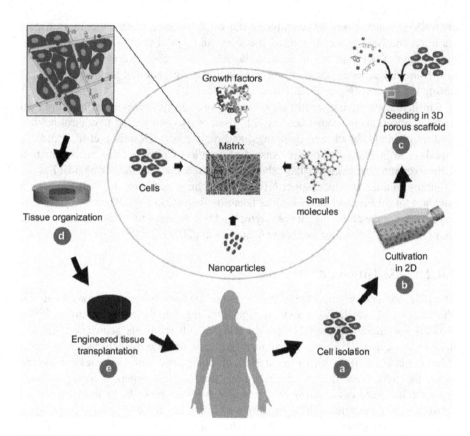

FIGURE 6.6   Steps involved in the tissue engineering. (Adapted from ResearchGate.)

6.4.3   TISSUE ENGINEERING APPLICATIONS

Biomedical engineering involves and uses a combination of cell, engineering, materials, methods, and new technologies to restore physiochemical properties and replace the biological tissues (Figure 6.6). Various naturally occurring biomaterials have been suggested as a templating material, as they have the biocompatibility, degradability, desired mechanical strength, and biological activity in vivo. Alginates are salts of algin, which is a polysaccharide derived from the cell wall of brown algae (F. Zhu et al., 2021). Alginate is extracted by initially treating the algae with aqueous NaOH. Alginate hydrogels are a network of water-insoluble polymeric chains. They are a type of colloidal gel having water as its dispersion medium (Mohamadali et al., 2017). The gels can contain over 99% water of their total volume; regardless of the high water content, they also retain their stiffness. The high water content and the desired stiffness render the gel an extracellular matrix-like property, which makes them suitable to provide mechanical support for an injured tissue and serve as a platform for delivering biomolecules to the injured tissue for its repair and regeneration (Ahmadi et al., 2015). Alginate hydrogels are used in combination with cells; the hydrogels are modified to make them compatible for interaction with the cells.

An increased cell–hydrogel matrix interaction is observed upon covalently conjugating the hydrogel with heparin-binding peptides and related peptide sequences (Mohamadali et al., 2017). Sodium alginate hydrogels have a lot of potential in cell and tissue engineering applications and have been studied extensively in the field of regenerative medicine. The modulation of the physico-chemical characteristics of these alginate-based hydrogels is usually achieved by selecting the appropriate divalent cation employed in the gel. Peptide-loaded modified alginate hydrogels act as extracellular matrix providing adhesion and attachment to cell for proliferation and differentiation. The use of calcium ions for cross-linking sodium alginate hydrogels has been shown to significantly increase the proliferation and differentiation of osteoblasts (Mohamadali et al., 2017; Piantanida et al., 2019). Using barium and strontium as a divalent cation in place of calcium leads to more rigid hydrogels. Drug-encapsulated alginate hydrogels can be modified to act as an extracellular matrix and accelerate the healing process. Wound-healing properties of sodium alginate films have been well established with findings that suggest rapid and efficient healing with faster wound closure (Udayakumar et al., 2021).

The injectable polymer hydrogels based on chitosan, cellulose nanocrystals, and pectin were designed with compatible internal structures that are prominent for tissue engineering (Ali & Ahmed, 2018). Cartilage's extracellular matrix is made up of a collagen mesh containing slowly dividing chondrocytes and negatively charged proteoglycans (Kamoun et al., 2015). Any injury and trauma to the cartilage lead to defects, which do not heal properly. The insufficient cartilage regeneration causes severe pain, rendering the person disabled, ultimately requiring a total joint replacement to treat the condition. Alginate–polyacrylamide-based hybrid hydrogel with organic nanofillers has shown immense potential as a substitute for cartilage (Mohammadinejad et al., 2020). This alginate–polyacrylamide–silica hydrogel consists of short chains of alginate that dissipate the strain energy via ionic cross-linking, providing the structure healing properties by reforming the ionic cross-links.

6.4.4 ENERGY APPLICATION

Nanostructured materials are critically important in many areas of technology because of their unusual physical and chemical properties due to their limited dimensions (Ranjan et al., 2018). Micro- or nanostructures owing to unique chemical or physical properties, tunable functionalities, hydrogels, and derivatives have emerged as an important class of functional materials and receive increasing interest from research scientists (Kausar, 2020a). Bottom-up synthetic methods to rationally design and modify their molecular architectures enable nanostructured functional hydrogels to give several critical challenges in advanced energy technologies (S. Kumar et al., 2018). Integrating the intrinsic or extrinsic properties of various functional materials, nanostructured functional hydrogels hold the promise to break the limitations of current materials, improving the device performance of energy storage and conversion. The focus is on the fundamentals and applications of nanostructured functional hydrogels in energy conversion and storage. Various organic nanofillers derived hydrogel works as core components in batteries, supercapacitors, and various other energy-providing materials (F. Zhao et al., 2018).

As a promising functional material, hydrogels have attracted extensive attention, especially in flexible wearable sensor fields, but it remains a great challenge to facilely integrate excellent mechanical properties, self-adhesion, and strain sensitivity into a single hydrogel (Tie et al., 2021). In this work, we present high in strength, stretchable, conformable, and self-adhesive chitosan/poly (acrylic acid) double-network organic nanofiller-derived hydrogels for application in epidermal strain sensor via in situ polymerization of acrylic acid in chitosan acid aqueous solution with tannic acid-coated cellulose nanocrystal (TA-CNC) acting as nanofillers to reinforce tensile properties, followed by a soaking process in a saturated NaCl solution to cross-link chitosan chains. With the addition of a small amount of TA-CNC, the double-network nanocomposite hydrogels became highly adhesive and mechanically compliant, which were critical factors for the development of conformable and resilient wearable epidermal sensors (Khan et al., 2018). The salt-soaking process was applied to cross-link chitosan chains by shielding electrostatic repulsions between positively charged amino groups, drastically enhancing the mechanical properties of the hydrogels (Cui et al., 2019). The obtained double-network nanocomposite hydrogels exhibited excellent tunable mechanical properties that could be conveniently tailored with fracture stress and fracture strain ranging from 0.39 to 1.2 MPa and 370% to 800%, respectively (Massaro et al., 2018). Besides, the hydrogels could be tightly attached onto diverse substrates, including wood, glass, plastic, polytetrafluoroethylene, glass, metal, and skin, demonstrating high adhesion strength and compliant adhesion behavior (Ali & Ahmed, 2018; Piantanida et al., 2019). In addition, benefiting from the abundant free ions from strong electrolytes, the flexible hydrogel sensors demonstrated stable conductivity and strain sensitivity, which could monitor both large human motions and subtle motions (González et al., 2018). Furthermore, the antibacterial property originating from chitosan made the hydrogels suitable for wearable epidermal sensors (González et al., 2018). The facile soaking strategy proposed in this work would be promising in fabricating high-strength multi-functional conductive hydrogels used for wearable epidermal devices (Cui et al., 2019).

6.4.5 FOOD AND ENVIRONMENT APPLICATIONS

In recent years, a huge development in packaging materials based on proteins and polysaccharides has been developed, which is caused by environmental problems resulting from the presence of an increasing number of packaging based on petroleum (Tomono, 2019). Biocompatibility, biodegradability, and antimicrobial properties of chitosan indicate that this polysaccharide can be used as a packaging material (Seidi et al., 2020). Enriched with various additives such as nanofillers, it generates completely new active properties. Despite the fact that there is a huge number of studies regarding the production and characteristics of various types of chitosan-based films, its industrial use is still limited. Currently, research needs to be focused on developing a scheme for obtaining cheap chitosan packaging material that would eliminate the use of the expensive and environmentally unfriendly chemicals that are used in the production of synthetic films (Jamróz et al., 2019).

Chitin biopolymers are biodegradable and renewable (Ali & Ahmed, 2018). To achieve the preferred functionalities to improve product safety, an active food packaging system is employed in which active components are loaded in the chitin matrix. These active agents can act as absorbers, releasers, or reacting agents. Biobased plastics were less than 1% in 2009 and they started to increase to 20% by the end of 2020. Solubility of chitin in acidic media is a valuable characteristic since it can be modified as fibers, films, sponges, beads, gels, or solutions and blended with polymers of both natural and synthetic origin (Jeon et al., 2000). Nanofillers can be fabricated by using electrospraying techniques, which produce ultrathin solids with health-promoting active packaging strategies (Akpan et al., 2018). There are numerous challenges as well as opportunities involved in commercializing biodegradable packaging. An extra major requirement for such biocomposite films used in packaging applications is that they comply with government legislation. To be more specific, such packaging material should exhibit minimal or no migration effect and not interact with the product. Biopackaging is expected to offer a clean and pollution-free environment in the future (Yang Chen et al., 2019; Fertahi et al., 2021).

Chitin nanofillers are both biodegradable and renewable. To achieve the preferred functionalities to improve product safety, an active packaging system is employed in which active components are loaded in the polymer hydrogels (Selim et al., 2017). These active agents can act as absorbers, releasers, or reacting agents. Solubility of chitin in acidic media is a valuable characteristic since it can be modified as fibers, films, sponges, beads, gels, or solutions and blended with polymers of both natural and synthetic origin. Nanofillers can be generated by using electrospraying techniques, which produce ultrathin solids with health-promoting active packaging art (Hu et al., 2021; Rallini & Kenny, 2017). The CTN composite-based packaging system showed enhanced improvement in wettability, as well as in mechanical characteristics, and helped in preventing grapes from microbial attack. There are numerous challenges as well as opportunities involved in commercializing biodegradable packaging (Aljbour et al., 2019; J. Pires et al., 2021). To be more specific, such packaging material should exhibit minimal or no migration effect and not interact with the product. Biopackaging is expected to offer a clean and pollution-free environment in the future.

6.5 OTHER APPLICATIONS

Cellulose nanocrystals (CNCs) with >2000 photoactive groups on each can act as highly efficient initiators for radical polymerizations, cross-linkers, as well as covalently embedded nanofillers for nanocomposite hydrogels. This is achieved by a simple and reliable method for surface modification of CNCs with a photoactive bis(acyl) phosphate oxide derivative. Shape-persistent and free-standing 3D structured objects were printed with a monofunctional methacrylate, showing a superior swelling capacity and improved mechanical properties (González et al., 2018).

Since corrosion has tremendous economic effects, academics and industries are in need to develop more effective coatings (Hettiarachchi et al., 2019). These efforts have led to the profound importance of nanocomposite coatings based on polymers and carbon nanostructures. It is shown that good reinforcement, advanced mechanical properties, and high corrosion resistance are only found at relatively low levels of nanocarbon (i.e., fullerene, carbon black, carbon nanotubes, graphene, graphene oxide, and carbon dots) loadings in coating compositions (Pourhashem et al., 2020). Herein, a survey of breakthrough scientific studies on the application of carbon nanostructures in corrosion-resistant organic coatings is carried out to pave the way for future developments in novel nanocoatings. The chitosan NP has also been used as an anti-aging skin agent and has shown the skin regenerative properties when materials were tested against skin cell fibroblasts and keratinocytes in the laboratory (Arshad et al., 2020). Infections caused by bacterial wounds are common infections. Excess use of antibiotics leads to raising the growth of antibiotic-resistant bacteria. Using antibiotics to prevent and treat wound infections is not successful with the spread of antibiotic-resistant bacteria. Antibacterial wound dressing involves encapsulating an antiseptic agent into polymer hydrogel matrixes to avoid microbial contamination. The study of poly (acrylic acid) (PAA) and bacterial cellulose nanofiber shows effectiveness in wound dressing applications (Eshkol-Yogev et al., 2021; Jao et al., 2017; T. Zhu et al., 2019).

6.6 THE FUTURE OF NANOMATERIALS

Several research articles and magazines on organic nanofillers derived from polymer hydrogels reflect the significant value of organic nanofillers. Many important technological developments have been used to increase the abilities and advance the consistency of organic-based nanocomposites for various applications, particularly in gene delivery and the medical field. Organic nanofillers are biocompatible, natural, biodegradable, toxin-free, which has ideal properties for biomedicine applications. In addition, nanofillers and nanocomposites are resources that have an extensive range of useful applications (Tekade et al., 2017; Yanat & Schroën, 2021). These have some drawbacks that limit its application, such as its basic and insoluble nature in neutral media. Several nanoparticles and other polymers, such as carbon nanotubes, have served as supportive candidates for the emerging and enormously varied applications of chitosan nanocomposites. Furthermore, chitosan can be improved with polysaccharide-based nanocomposites for gene delivery and tissue engineering various functional groups to regulate the hydrophobic, anionic, and cationic parameters (Nalwa, 2014). This is why chitosan is comparatively unique among biopolymers. The most unusual properties of chitosan arise from primary amines groups with their structure. These types of structures communicate to the polysaccharides, which have highly encouraged physical and chemical properties, but also have specific contacts with proteins, cells, and living creatures. In this chapter, the importance of chitosan nanocomposites and the basic properties of chitosan have also been studied. Making an accurately biodegradable product provides both environmental and humanoid protection (Tao et al., 2019). At the least, production must be nontoxic. Other eco-friendly qualities are the use of sustainable ingredients, produced in ways that will not damage the bionetwork. Polysaccharide-based nanofillers displays a capacity for gentle applications in healthcare products, which are not toxic and are safe materials. However, care is necessary to ensure that the material remains pure, as metal, or many supplementary impurities, could produce harmful properties, individually or in derivative combinations and various quantities (Kausar, 2020b). The cross-linking or derivatization and unreacted substances must be carefully eliminated to avoid unintended consequences, as many substances are toxic. It was shown that for superior applications such as biomedical gene delivery and food packing, the thermal and mechanical aspects of pure nanofillers are not. Additionally, several studies focus on schemes with organic nanocomposites. The addition of a small number of clay nanomaterials improved thermal or mechanical properties, overcoming the restrictions of organic nanofiller-derived hydrogels alone being functionalized in fields that need a definite thermal and strength stability. The polymeric nanocomposites and polymeric system can offer a good result in terms of gene delivery, as well as drug discharge enzyme immobilization, high biocompatibility, and toxin-free properties offering a broad diversity of applications such as medical, material engineering, and pharmaceuticals in the future.

REFERENCES

Abu Elella, M. H., Goda, E. S., Gab-Allah, M. A., Hong, S. E., Pandit, B., Lee, S., Gamal, H., Rehman, A. U., & Yoon, K. R. (2021). Xanthan gum-derived materials for applications in environment and eco-friendly materials: A review. *Journal of Environmental Chemical Engineering*, *9*(1). https://doi.org/10.1016/j.jece.2020.104702

Ahmadi, F., Oveisi, Z., Samani, M., & Amoozgar, Z. (2015). Chitosan based hydrogels: Characteristics and pharmaceutical applications. *Research in Pharmaceutical Sciences*, *10*(1), 1–16.

Ahmed, E. M. (2015). Hydrogel: Preparation, characterization, and applications: A review. *Journal of Advanced Research*, *6*(2), 105–121. https://doi.org/10.1016/j.jare.2013.07.006

Akpan, E. I., Shen, X., Wetzel, B., & Friedrich, K. (2018). Design and Synthesis of Polymer Nanocomposites. In Pielichowski, Krzysztof & Majka, Tomasz M. (eds) *Polymer Composites with Functionalized Nanoparticles: Synthesis, Properties, and Applications* (pp. 47–83). Elsevier Inc. https://doi.org/10.1016/B978-0-12-814064-2.00002-0

Ali, A., & Ahmed, S. (2018). A review on chitosan and its nanocomposites in drug delivery. *International Journal of Biological Macromolecules*, *109*, 273–286. https://doi.org/10.1016/j.ijbiomac.2017.12.078

Aljbour, N. D., Beg, M. D. H., & Gimbun, J. (2019). Acid hydrolysis of chitosan to oligomers using hydrochloric acid. *Chemical Engineering and Technology*, *9*, 1741–1746. https://doi.org/10.1002/ceat.201800527

Anjani, Q. K., Permana, A. D., Cárcamo-Martínez, Á., Domínguez-Robles, J., Tekko, I. A., Larrañeta, E., Vora, L. K., Ramadon, D., & Donnelly, R. F. (2021). Versatility of hydrogel-forming microneedles in in vitro transdermal delivery of tuberculosis drugs. *European Journal of Pharmaceutics and Biopharmaceutics*, *158*(December 2020), 294–312. https://doi.org/10.1016/j.ejpb.2020.12.003

Antonino, R. S. C. M. D. Q., Fook, B. R. P. L., Lima, V. A. D. O., Rached, R. Í. D. F., Lima, E. P. N., Lima, R. J. D. S., Covas, C. A. P., & Fook, M. V. L. (2017). Preparation and characterization of chitosan obtained from shells of shrimp (Litopenaeus vannamei Boone). *Marine Drugs*, *15*(5), 1–12. https://doi.org/10.3390/md15050141

Aranaz, I., Alcántara, A. R., Civera, M. C., Arias, C., Elorza, B., Caballero, A. H., & Acosta, N. (2021). Chitosan: An overview of its properties and applications. *Polymers*, *13*(19). https://doi.org/10.3390/polym13193256

Armentano, I., Puglia, D., Luzi, F., Arciola, C. R., Morena, F., Martino, S., & Torre, L. (2018). Nanocomposites based on biodegradable polymers. *Materials*, *11*(5). https://doi.org/10.3390/ma11050795

Arshad, M. S., Fatima, S., Nazari, K., Ali, R., Farhan, M., Muhammad, S. A., Abbas, N., Hussain, A., Kucuk, I., Chang, M. W., Mehta, P., & Ahmad, Z. (2020). Engineering and characterisation of BCG-loaded polymeric microneedles. *Journal of Drug Targeting*, *28*(5), 525–532. https://doi.org/10.1080/1061186X.2019.1693577

Aung, N. N., Ngawhirunpat, T., Rojanarata, T., Patrojanasophon, P., Pamornpathomkul, B., & Opanasopit, P. (2020). Fabrication, characterization and comparison of α-arbutin loaded dissolving and hydrogel forming microneedles. *International Journal of Pharmaceutics*, *586*, 119508. https://doi.org/10.1016/j.ijpharm.2020.119508

Bellet, P., Gasparotto, M., Pressi, S., Fortunato, A., Scapin, G., Mba, M., Menna, E., & Filippini, F. (2021). Graphene-based scaffolds for regenerative medicine. *Nanomaterials*, *11*(2), 1–41. https://doi.org/10.3390/nano11020404

Chen, Yang, Huang, A., Zhang, Y., & Bie, Z. (2019). Recent advances of boronate affinity materials in sample preparation. *Analytica Chimica Acta*, *1076*, 1–17. https://doi.org/10.1016/j.aca.2019.04.050

Chen, Yu. (2019). *Hydrogels Based on Natural Polymers*. Elsevier B.V. https://doi.org/10.1016/C2018-0-00171-1

Ciolacu, D. E., & Suflet, D. M. (2018). Cellulose-Based Hydrogels for Medical/Pharmaceutical Applications. In Popa, Valentin & Volf, Irina (eds) *Biomass as Renewable Raw Material to Obtain Bioproducts of High-Tech Value* (pp. 401–439). Elsevier B.V. https://doi.org/10.1016/B978-0-444-63774-1.00011-9

Cui, C., Shao, C., Meng, L., & Yang, J. (2019). High-strength, self-adhesive, and strain-sensitive chitosan/poly(acrylic acid) double-network nanocomposite hydrogels fabricated by salt-soaking strategy for flexible sensors [research-article]. *ACS Applied Materials and Interfaces*, *11*(42), 39228–39237. https://doi.org/10.1021/acsami.9b15817

Dai, X., Du, Y., Yang, J., Wang, D., Gu, J., Li, Y., Wang, S., Xu, B. B., & Kong, J. (2019). Recoverable and self-healing electromagnetic wave absorbing nanocomposites. *Composites Science and Technology*, *174*(February), 27–32. https://doi.org/10.1016/j.compscitech.2019.02.018

Dilks, A. (1980). Characterisation of Polymers By Esca. In Dawkins, J. V. (ed) *Developments in Polymer Characterisation* (Vol. 1, pp. 145–182). Springer. https://doi.org/10.1007/978-94-010-9237-1_4

El-Zawawy, W. K. (2005). Preparation of hydrogel from green polymer. *Polymers for Advanced Technologies*, *16*(1), 48–54. https://doi.org/10.1002/pat.537

Eshkol-Yogev, I., Gilboa, E., Giladi, S., & Zilberman, M. (2021). Formulation - Properties effects of novel dual composite hydrogels for use as medical sealants. *European Polymer Journal*, *152*(January), 110470. https://doi.org/10.1016/j.eurpolymj.2021.110470

Feldman, D. (2019). Polymers and polymer nanocomposites for cancer therapy. *Applied Sciences (Switzerland)*, *9*(18). https://doi.org/10.3390/app9183899

Fertahi, S., Ilsouk, M., Zeroual, Y., Oukarroum, A., & Barakat, A. (2021). Recent trends in organic coating based on biopolymers and biomass for controlled and slow release fertilizers. *Journal of Controlled Release*, *330*(December 2020), 341–361. https://doi.org/10.1016/j.jconrel.2020.12.026

Giri, A., Bhowmick, R., Prodhan, C., Majumder, D., Bhattacharya, S. K., & Ali, M. (2019). Synthesis and characterization of biopolymer based hybrid hydrogel nanocomposite and study of their electrochemical efficacy. *International Journal of Biological Macromolecules*, *123*, 228–238. https://doi.org/10.1016/j.ijbiomac.2018.11.010

González, K., García-Astrain, C., Santamaria-Echart, A., Ugarte, L., Avérous, L., Eceiza, A., & Gabilondo, N. (2018). Starch/graphene hydrogels via click chemistry with relevant electrical and antibacterial properties. *Carbohydrate Polymers*, *202*(April), 372–381. https://doi.org/10.1016/j.carbpol.2018.09.007

Guo, Y., Bae, J., Fang, Z., Li, P., Zhao, F., & Yu, G. (2020). Hydrogels and hydrogel-derived materials for energy and water sustainability. *Chemical Reviews*, *120*(15), 7642–7707. https://doi.org/10.1021/acs.chemrev.0c00345

Han, L., Zhang, H., Yu, H. Y., Ouyang, Z., Yao, J., Krucinska, I., Kim, D., & Tam, K. C. (2021). Highly sensitive self-healable strain biosensors based on robust transparent conductive nanocellulose nanocomposites: Relationship between percolated network and sensing mechanism. *Biosensors and Bioelectronics*, *191*(February), 113467. https://doi.org/10.1016/j.bios.2021.113467

Hassan, B. A. R., Yusoff, Z. B. M., Hassali, M. A., & Othman, B. S. (2012). *Supportive and Palliative Care in Solid Cancer Patients*. IntechOpen. http://dx.doi.org/10.5772/55358

He, R., Niu, Y., Li, Z., Li, A., Yang, H., Xu, F., & Li, F. (2020). A hydrogel microneedle patch for point-of-care testing based on skin interstitial fluid. *Advanced Healthcare Materials*, *9*(4), 1–11. https://doi.org/10.1002/adhm.201901201

Heifler, O., Borberg, E., Harpak, N., Zverzhinetsky, M., Krivitsky, V., Gabriel, I., Fourman, V., Sherman, D., & Patolsky, F. (2021). Clinic-on-a-needle array toward future minimally invasive wearable artificial pancreas applications. *ACS Nano*, *15*(7), 12019–12033. https://doi.org/10.1021/acsnano.1c03310

Hettiarachchi, N. M., De Silva, R. T., Prasanga Gayanath Mantilaka, M. M. M. G., Pasbakhsh, P., Nalin De Silva, K. M., & Amaratunga, G. A. J. (2019). Synthesis of calcium carbonate microcapsules as self-healing containers. *RSC Advances*, *9*(41), 23666–23677. https://doi.org/10.1039/c9ra03804c

Hossain, M. S., & Iqbal, A. (2014). Production and characterization of chitosan from shrimp waste. *Journal of the Bangladesh Agricultural University*, *12*(1), 153–160.

Hu, H., Teng, X., Zhang, S., Liu, T., Li, X., & Wang, D. (2021). Structural characteristics, rheological properties, and antioxidant activity of novel polysaccharides from "deer tripe mushroom." *Journal of Food Quality*, *2021*. https://doi.org/10.1155/2021/6593293

Huang, Y., Kormakov, S., He, X., Gao, X., Zheng, X., Liu, Y., Sun, J., & Wu, D. (2019). Conductive polymer composites from renewable resources: An overview of preparation, properties, and applications. *Polymers*, *11*(2). https://doi.org/10.3390/polym11020187

Ilyas, R. A., Sapuan, S. M., Ishak, M. R., & Zainudin, E. S. (2018). Development and characterization of sugar palm nanocrystalline cellulose reinforced sugar palm starch bionanocomposites. *Carbohydrate Polymers*, *202*, 186–202. https://doi.org/10.1016/j.carbpol.2018.09.002

Jamróz, E., Kulawik, P., & Kopel, P. (2019). The effect of nanofillers on the functional properties of biopolymer-based films: A review. *Polymers*, *11*(4), 1–42. https://doi.org/10.3390/polym11040675

Jao, D., Xue, Y., Medina, J., & Hu, X. (2017). Protein-based drug-delivery materials. *Materials*, *10*(5), 1–24. https://doi.org/10.3390/ma10050517

Jeon, Y. J., Shahidi, F., & Kim, S. K. (2000). Preparation of chitin and chitosan oligomers and their applications in physiological functional foods. *Food Reviews International*, *16*(2), 159–176. https://doi.org/10.1081/FRI-100100286

Jiao, Y., Lu, K., Lu, Y., Yue, Y., Xu, X., Xiao, H., Li, J., & Han, J. (2021). Highly viscoelastic, stretchable, conductive, and self-healing strain sensors based on cellulose nanofiber-reinforced polyacrylic acid hydrogel. *Cellulose*, *28*(7), 4295–4311. https://doi.org/10.1007/s10570-021-03782-1

Kamoun, E. A., Chen, X., Mohy Eldin, M. S., & Kenawy, E. R. S. (2015). Crosslinked poly(vinyl alcohol) hydrogels for wound dressing applications: A review of remarkably blended polymers. *Arabian Journal of Chemistry*, *8*(1), 1–14. https://doi.org/10.1016/j.arabjc.2014.07.005

Kaniewska, K., Karbarz, M., & Katz, E. (2020). Nanocomposite hydrogel films and coatings – Features and applications. *Applied Materials Today*, *20*, 100776. https://doi.org/10.1016/j.apmt.2020.100776

Kausar, A. (2020a). Nanocarbon in polymeric nanocomposite hydrogel—Design and multifunctional tendencies. *Polymer-Plastics Technology and Materials*, *59*(14), 1505–1521. https://doi.org/10.1080/25740881.2020.1757106

Kausar, A. (2020b). Shape memory polyester-based nanomaterials: Cutting-edge advancements. *Polymer-Plastics Technology and Materials*, *59*(7), 765–779. https://doi.org/10.1080/25740881.2019.1695268

Khan, M., Khurram, A. A., Li, T., Zhao, T., Śubhani, T., Gul, I. H., Ali, Z., & Patel, V. (2018). Synergistic effect of organic and inorganic nano fillers on the dielectric and mechanical properties of epoxy composites. *Journal of Materials Science and Technology*, *34*(12), 2424–2430. https://doi.org/10.1016/j.jmst.2018.06.014

Koyani, R. D. (2020). Journal of drug delivery science and technology synthetic polymers for microneedle synthesis: From then to now. *Journal of Drug Delivery Science and Technology*, *60*(September), 102071. https://doi.org/10.1016/j.jddst.2020.102071

Kumar, A., Zo, S. M., Kim, J. H., Kim, S. C., & Han, S. S. (2019). Enhanced physical, mechanical, and cytocompatibility behavior of polyelectrolyte complex hydrogels by reinforcing halloysite nanotubes and graphene oxide. *Composites Science and Technology*, *175*(March), 35–45. https://doi.org/10.1016/j.compscitech.2019.03.008

Kumar, S., Sarita, Nehra, M., Dilbaghi, N., Tankeshwar, K., & Kim, K. H. (2018). Recent advances and remaining challenges for polymeric nanocomposites in healthcare applications. *Progress in Polymer Science*, *80*, 1–38. https://doi.org/10.1016/j.progpolymsci.2018.03.001

Kumar Teli, M., Mutalik, S., & Rajanikant, G. K. (2010). Nanotechnology and nanomedicine: Going small means aiming big. *Current Pharmaceutical Design*, *16*(16), 1882–1892. https://doi.org/10.2174/138161210791208992

Larrañeta, E., Henry, M., Irwin, N. J., Trotter, J., Perminova, A. A., & Donnelly, R. F. (2018). Synthesis and characterization of hyaluronic acid hydrogels crosslinked using a solvent-free process for potential biomedical applications. *Carbohydrate Polymers*, *181*(December), 1194–1205. https://doi.org/10.1016/j.carbpol.2017.12.015

Lee, T. H., Oh, J. Y., Jang, J. K., Moghadam, F., Roh, J. S., Yoo, S. Y., Kim, Y. J., Choi, T. H., Lin, H., Kim, H. W., & Park, H. B. (2020). Elucidating the role of embedded metal-organic frameworks in water and ion transport properties in polymer nanocomposite membranes. *Chemistry of Materials*, *32*(23), 10165–10175. https://doi.org/10.1021/acs.chemmater.0c03692

Liechty, W. B., Kryscio, D. R., Slaughter, B. V., & Peppas, N. A. (2010). *Polymers for Drug Delivery Systems*. Annual Review of Chemical and Biomolecular Engineering. https://doi.org/10.1146/annurev-chembioeng-073009-100847

Liu, H., Peng, H., Xin, Y., & Zhang, J. (2019). Metal-organic frameworks: A universal strategy towards super-elastic hydrogels. *Polymer Chemistry*, *10*(18), 2263–2272. https://doi.org/10.1039/c9py00085b

Liu, L., Lyu, J., Mo, J., Yan, H., Xu, L., Peng, P., Li, J., Jiang, B., Chu, L., & Li, M. (2020). Comprehensively-upgraded polymer electrolytes by multifunctional aramid nanofibers for stable all-solid-state Li-ion batteries. *Nano Energy*, *69*, 104398. https://doi.org/10.1016/j.nanoen.2019.104398

Malagurski, I., Levic, S., Nesic, A., Mitric, M., Pavlovic, V., & Dimitrijevic-Brankovic, S. (2017). Mineralized agar-based nanocomposite films: Potential food packaging materials with antimicrobial properties. *Carbohydrate Polymers*, *175*, 55–62. https://doi.org/10.1016/j.carbpol.2017.07.064

Manickam, P., Vashist, A., Madhu, S., Sadasivam, M., Sakthivel, A., Kaushik, A., & Nair, M. (2020). Gold nanocubes embedded biocompatible hybrid hydrogels for electrochemical detection of H2O2. *Bioelectrochemistry*, *131*, 107373. https://doi.org/10.1016/j.bioelechem.2019.107373

Martau, G. A., Mihai, M., & Vodnar, D. C. (2019). The use of chitosan, alginate, and pectin in the biomedical and food sector-biocompatibility, bioadhesiveness, and biodegradability. *Polymers*, *11*(11). https://doi.org/10.3390/polym11111837

Massaro, M., Cavallaro, G., Colletti, C. G., Lazzara, G., Milioto, S., Noto, R., & Riela, S. (2018). Chemical modification of halloysite nanotubes for controlled loading and release. *Journal of Materials Chemistry B*, *6*(21), 3415–3433. https://doi.org/10.1039/c8tb00543e

Mohamadali, M., Irani, S., & Soleimani, M. (2017). *PANi/PAN copolymer as scaffolds for the muscle cell-like differentiation of mesenchymal stem cells.* December 2016. https://doi.org/10.1002/pat.4000

Mohammadinejad, R., Kumar, A., Ranjbar-Mohammadi, M., Ashrafizadeh, M., Han, S. S., Khang, G., & Roveimiab, Z. (2020). Recent advances in natural gum-based biomaterials for tissue engineering and regenerative medicine: A review. *Polymers, 12*(1). https://doi.org/10.3390/polym12010176

Mondal, S., Das, S., & Nandi, A. K. (2020). A review on recent advances in polymer and peptide hydrogels. *Soft Matter, 16*(6), 1404–1454. https://doi.org/10.1039/c9sm02127b

Nalwa, H. S. (2014). A special issue on reviews in nanomedicine, drug delivery and vaccine development. *Journal of Biomedical Nanotechnology, 10*(9), 1635–1640. https://doi.org/10.1166/jbn.2014.2033

Narayanan, K. B., & Han, S. S. (2017). Dual-crosslinked poly(vinyl alcohol)/sodium alginate/silver nanocomposite beads – A promising antimicrobial material. *Food Chemistry, 234*, 103–110. https://doi.org/10.1016/j.foodchem.2017.04.173

Ng, J. Y., Obuobi, S., Chua, M. L., Zhang, C., Hong, S., Kumar, Y., Gokhale, R., & Ee, P. L. R. (2020). Biomimicry of microbial polysaccharide hydrogels for tissue engineering and regenerative medicine – A review. *Carbohydrate Polymers, 241*, 116345. https://doi.org/10.1016/j.carbpol.2020.116345

Ningaraju, S., Hegde, V. N., Prakash, A. P. G., & Ravikumar, H. B. (2018). Free volume dependence on electrical properties of poly (styrene co-acrylonitrile)/nickel oxide polymer nanocomposites. *Chemical Physics Letters, 698*, 24–35. https://doi.org/10.1016/j.cplett.2018.03.002

Pacelli, S., Paolicelli, P., Avitabile, M., Varani, G., Di Muzio, L., Cesa, S., Tirillò, J., Bartuli, C., Nardoni, M., Petralito, S., Adrover, A., & Casadei, M. A. (2018). Design of a tunable nanocomposite double network hydrogel based on gellan gum for drug delivery applications. *European Polymer Journal, 104*, 184–193. https://doi.org/10.1016/j.eurpolymj.2018.04.034

Park, K. (2007). Nanotechnology: What it can do for drug delivery. *Journal of Controlled Release, 120*(1–2), 1–3. https://doi.org/10.1016/j.jconrel.2007.05.003

Piantanida, E., Alonci, G., Bertucci, A., & De Cola, L. (2019). Design of nanocomposite injectable hydrogels for minimally invasive surgery. *Accounts of Chemical Research, 52*(8), 2101–2112. https://doi.org/10.1021/acs.accounts.9b00114

Pires, C. T. G. V. M. T., Vilela, J. A. P., & Airoldi, C. (2014). The effect of chitin alkaline deacetylation at different condition on particle properties. *Procedia Chemistry, 9*, 220–225. https://doi.org/10.1016/j.proche.2014.05.026

Pires, J., de Paula, C. D., Souza, V. G. L., Fernando, A. L., & Coelhoso, I. (2021). Understanding the barrier and mechanical behavior of different nanofillers in chitosan films for food packaging. *Polymers, 13*(5), 1–29. https://doi.org/10.3390/polym13050721

Polarz, S., & Smarsly, B. (2002). Nanoporous materials. *Journal of Nanoscience and Nanotechnology, 2*(6), 581–612. https://doi.org/10.1166/jnn.2002.151

Pourhashem, S., Ghasemy, E., Rashidi, A., & Vaezi, M. R. (2020). A review on application of carbon nanostructures as nanofiller in corrosion-resistant organic coatings. *Journal of Coatings Technology and Research, 17*(1). https://doi.org/10.1007/s11998-019-00275-6

Prihatiningtyas, I., Volodin, A., & Van Der Bruggen, B. (2019). 110th anniversary: Cellulose nanocrystals as organic nanofillers for cellulose triacetate membranes used for desalination by pervaporation. *Industrial and Engineering Chemistry Research, 58*(31), 14340–14349. https://doi.org/10.1021/acs.iecr.9b02106

Qian, C., Higashigaki, T., Asoh, T. A., & Uyama, H. (2020). Anisotropic conductive hydrogels with high water content. *ACS Applied Materials and Interfaces, 12*(24), 27518–27525. https://doi.org/10.1021/acsami.0c06853

Rallini, M., & Kenny, J. M. (2017). Nanofillers in Polymers. In Jasso-Gastinel, Carlos F. & Kenny, José M. (eds) *Modification of Polymer Properties* (pp. 47–86). Elsevier Inc. https://doi.org/10.1016/B978-0-323-44353-1.00003-8

Ranjan, R., Narnaware, S. D., & Patil, N. V. (2018). A novel technique for synthesis of calcium carbonate nanoparticles. *National Academy Science Letters*, *41*(6), 403–406. https://doi.org/10.1007/s40009-018-0704-4

Ruggiero, F., Vecchione, R., Bhowmick, S., Coppola, G., Coppola, S., Esposito, E., Lettera, V., Ferraro, P., & Netti, P. A. (2017). Electro-drawn polymer microneedle arrays with controlled shape and dimension. *Sensors & Actuators: B. Chemical*. https://doi.org/10.1016/j.snb.2017.08.165

Saba, N., Jawaid, M., & Asim, M. (2018). Nanocomposites with Nanofibers and Fillers from Renewable Resources. In Koronis, Georgios & Silva, Arlindo (eds) *Green Composites for Automotive Applications* (pp. 145–170). Elsevier Ltd. https://doi.org/10.1016/B978-0-08-102177-4.00007-0

Sánchez, J. A. L., Díez-Pascual, A. M., Capilla, R. P., & Díaz, P. G. (2019). The effect of hexamethylene diisocyanate-modified graphene oxide as a nanofiller material on the properties of conductive polyaniline. *Polymers*, *11*(6). https://doi.org/10.3390/polym11061032

Seidi, F., Jin, Y., Han, J., Saeb, M. R., Akbari, A., Hosseini, S. H., Shabanian, M., & Xiao, H. (2020). Self-healing polyol/borax hydrogels: Fabrications, properties and applications. *Chemical Record*, *20*(10), 1142–1162. https://doi.org/10.1002/tcr.202000060

Selim, M. S., Shenashen, M. A., El-Safty, S. A., Higazy, S. A., Selim, M. M., Isago, H., & Elmarakbi, A. (2017). Recent progress in marine foul-release polymeric nanocomposite coatings. *Progress in Materials Science*, *87*, 1–32. https://doi.org/10.1016/j.pmatsci.2017.02.001

Shah, S. A., Kulhanek, D., Sun, W., Zhao, X., Yu, S., Parviz, D., Lutkenhaus, J. L., & Green, M. J. (2020). Aramid nanofiber-reinforced three-dimensional graphene hydrogels for supercapacitor electrodes. *Journal of Colloid and Interface Science*, *560*, 581–588. https://doi.org/10.1016/j.jcis.2019.10.066

Shin, J. E., Kim, H. W., Yoo, B. M., & Park, H. B. (2018). Graphene oxide nanosheet-embedded crosslinked poly(ethylene oxide) hydrogel. *Journal of Applied Polymer Science*, *135*(24), 1–11. https://doi.org/10.1002/app.45417

Sun, X., Yao, F., & Li, J. (2020). Nanocomposite hydrogel-based strain and pressure sensors: A review. *Journal of Materials Chemistry A*, *8*(36), 18605–18623. https://doi.org/10.1039/d0ta06965e

Tao, Y., Wei, C., Liu, J., Deng, C., Cai, S., & Xiong, W. (2019). Nanostructured electrically conductive hydrogels obtained: Via ultrafast laser processing and self-assembly. *Nanoscale*, *11*(18), 9176–9184. https://doi.org/10.1039/c9nr01230c

Tekade, R. K., Maheshwari, R., Soni, N., Tekade, M., & Chougule, M. B. (2017). Nanotechnology for the Development of Nanomedicine. In Mishra, Vijay, Kesharwani, Prashant, Amin, Mohd Cairul Mohd, & Iyer, Arun (eds) *Nanotechnology-Based Approaches for Targeting and Delivery of Drugs and Genes* (pp. 3–61). Elsevier Inc. https://doi.org/10.1016/B978-0-12-809717-5.00001-4

Tie, J., Chai, H., Mao, Z., Zhang, L., Zhong, Y., Sui, X., & Xu, H. (2021). Nanocellulose-mediated transparent high strength conductive hydrogel based on in-situ formed polypyrrole nanofibrils as a multimodal sensor. *Carbohydrate Polymers*, *273*(2999), 118600. https://doi.org/10.1016/j.carbpol.2021.118600

Tomczykowa, M., & Plonska-Brzezinska, M. E. (2019). Conducting polymers, hydrogels and their composites: Preparation, properties and bioapplications. *Polymers*, *11*(2), 1–36. https://doi.org/10.3390/polym11020350

Tomono, T. (2019). A new way to control the internal structure of microneedles: A case of chitosan lactate. *Materials Today Chemistry*, *13*, 79–87. https://doi.org/10.1016/j.mtchem.2019.04.009

Tovar, G. I., Fernández de Luis, R., Arriortua, M. I., Wolman, F. J., & Copello, G. J. (2020). Enhanced chitin gel with magnetic nanofiller for lysozyme purification. *Journal of Industrial and Engineering Chemistry*, *88*, 90–98. https://doi.org/10.1016/j.jiec.2020.03.026

Udayakumar, G. P., Muthusamy, S., Selvaganesh, B., Sivarajasekar, N., Rambabu, K., Banat, F., Sivamani, S., Sivakumar, N., Hosseini-Bandegharaei, A., & Show, P. L. (2021). Biopolymers and composites: Properties, characterization and their applications in food, medical and pharmaceutical industries. *Journal of Environmental Chemical Engineering*, *9*(4), 105322. https://doi.org/10.1016/j.jece.2021.105322

Vashist, A., Ghosal, A., Vashist, A., Kaushik, A., Gupta, Y. K., Nair, M., & Ahmad, S. (2019). Impact of nanoclay on the pH-responsiveness and biodegradable behavior of biopolymer-based nanocomposite hydrogels. *Gels*, *5*(4), 44.

Vashist, A., Kaushik, A., Ghosal, A., Bala, J., Nikkhah-Moshaie, R., Wani, W. A., Manickam, P., & Nair, M. (2018). Nanocomposite hydrogels: Advances in nanofillers used for nanomedicine. *Gels*, *4*(3), 1–15. https://doi.org/10.3390/gels4030075

Wakuda, Y., Nishimoto, S., Suye, S. I., & Fujita, S. (2018). Native collagen hydrogel nanofibres with anisotropic structure using core-shell electrospinning. *Scientific Reports*, *8*(1), 1–10. https://doi.org/10.1038/s41598-018-24700-9

Wei, L., Hu, N., & Zhang, Y. (2010). Synthesis of polymer—Mesoporous silica nanocomposites. *Materials*, *3*(7), 4066–4079. https://doi.org/10.3390/ma3074066

Xing, W., & Tang, Y. (2021). On mechanical properties of nanocomposite hydrogels: Searching for superior properties. *Nano Materials Science*, *May*. https://doi.org/10.1016/j.nanoms.2021.07.004

Yanat, M., & Schroën, K. (2021). Preparation methods and applications of chitosan nanoparticles; with an outlook toward reinforcement of biodegradable packaging. *Reactive and Functional Polymers*, *161*(December 2020). https://doi.org/10.1016/j.reactfunctpolym.2021.104849

Yuan, T., Cui, X., Liu, X., Qu, X., & Sun, J. (2019). Highly tough, stretchable, self-healing, and recyclable hydrogels reinforced by in situ-formed polyelectrolyte complex nanoparticles. *Macromolecules*, *52*(8), 3141–3149. https://doi.org/10.1021/acs.macromol.9b00053

Zeng, J. B., He, Y. S., Li, S. L., & Wang, Y. Z. (2012). Chitin whiskers: An overview. *Biomacromolecules*, *13*(1), 1–11. https://doi.org/10.1021/bm201564a

Zhang, C., Wu, B., Zhou, Y., Zhou, F., Liu, W., & Wang, Z. (2020). Mussel-inspired hydrogels: From design principles to promising applications. *Chemical Society Reviews*, *49*(11), 3605–3637. https://doi.org/10.1039/c9cs00849g

Zhang, L., Jin, L., Liu, B., He, J., & Liu, B. (2019). Templated growth of crystalline mesoporous materials : From soft/hard templates to colloidal templates. *Frontiers in Chemistry*, *7*(January), 1–13. https://doi.org/10.3389/fchem.2019.00022

Zhao, C., Zhang, P., Zhou, J., Qi, S., Yamauchi, Y., Shi, R., Fang, R., Ishida, Y., Wang, S., Tomsia, A. P., Jiang, L., & Liu, M. (2020). Layered nanocomposites by shear-flow-induced alignment of nanosheets. *Nature*, *580*(7802), 210–215. https://doi.org/10.1038/s41586-020-2161-8

Zhao, F., Bae, J., Zhou, X., Guo, Y., & Yu, G. (2018). Nanostructured functional hydrogels as an emerging platform for advanced energy technologies. *Advanced Materials*, *30*(48), 1–16. https://doi.org/10.1002/adma.201801796

Zhou, L., Ramezani, H., Sun, M., Xie, M., Nie, J., Lv, S., Cai, J., Fu, J., & He, Y. (2020). 3D printing of high-strength chitosan hydrogel scaffolds without any organic solvents. *Biomaterials Science*, *8*(18), 5020–5028. https://doi.org/10.1039/d0bm00896f

Zhu, F., Du, B., Xu, B., Science, F., & Program, T. (2021). Polysaccharides. Springer, 1–16. https://doi.org/10.1007/978-3-319-03751-6

Zhu, T., Mao, J., Cheng, Y., Liu, H., Lv, L., Ge, M., Li, S., Huang, J., Chen, Z., Li, H., Yang, L., & Lai, Y. (2019). Recent progress of polysaccharide-based hydrogel interfaces for wound healing and tissue engineering. *Advanced Materials Interfaces*, *6*(17), 1–22. https://doi.org/10.1002/admi.201900761

7 Environmental Impact in Terms of Nanotoxicity and Limitations of Employing Organic Nanofillers in Polymers

Habibul Islam and Md Enamul Hoque
Military Institute of Science and Technology

Shek Md Atiqure Rahman
University of Sharjah

Faris M. Al-Oqla
Hashemite University

CONTENTS

7.1   Introduction 200
7.2   Nanotoxicity 201
    7.2.1   Definition 201
    7.2.2   Factors 201
        7.2.2.1   Size 201
        7.2.2.2   Particle Shape, Surface Area, and Surface Charge 203
7.3   Exposure Routes 203
    7.3.1   Dermal Route 204
    7.3.2   Respiratory Route 204
    7.3.3   Gastrointestinal Tracts 206
7.4   Nanoparticle Emission in the Environment 207
    7.4.1   Emission in Air 207
    7.4.2   Emission in Soil 209
    7.4.3   Emission in Water 209
7.5   Detection of Nanoparticles in the Environment 209
7.6   Nanoparticles' Interaction with the Environment 210
7.7   Toxicity Mechanisms of Nanoparticles 213

DOI: 10.1201/9781003279372-7

7.8   Environmental Impact of Nanoparticles 213
          7.8.1   Positive Impacts...213
          7.8.2   Negative Impacts 215
7.9   Limitations of Organic Nanofillers..218
7.10   Future Aspects and Conclusion..219
References...220

7.1   INTRODUCTION

Although American physicist Richard P. Feynman is considered to be the introducer of the concept of nanomaterial manipulation and thus nanotechnology (Hulla et al., 2015), earlier Indian potteries from the 6th century BC were reported to have used carbon nanotubes in coating (Kokarneswaran et al., 2020). Damascus steel, a material from the 9th century, was reported to have cementite nanowires (Sanderson, 2006). From ancient use of nanotechnology to modern-day applications, the field has come a long way and has become one of the most important sections of science and engineering today.

Nanoparticles are materials that are on the nanoscale. Because of their smaller size, these materials have a greater surface area-to-volume ratio, which allows these particles to interact with different systems differently than conventional materials. As the size decreases, the physiochemical properties take an interesting turn. Due to their excellent and exceptional properties, nanomaterials have been used in applications as follows:

Lightweight transport system can significantly reduce fuel consumption: Carbon nanotubes have been experimented to be used in next-generation aircraft systems, which will make the systems lighter, and conductivity of carbon nanotubes will also improve their electromagnetic shielding and thermal management (Gohardani et al., 2014).

Biomedical imaging contrast agents: Contrast agents for different diagnostic methods such as magnetic field-based imaging techniques like Magnetic Resonance Imaging (MRI) and Magnetic Particle Imaging (MPI). Recent advances have been made to use nanoscale ferrites as contrast agents, instead of conventional gadolinium-based contrast agents (Ravichandran & Velumani, 2020).

Drug delivery and tissue engineering applications: Biocompatibility and bioactivity of bionanomaterials have made these materials ideal candidates for biomedical applications. Many pieces of research have been carried out to use nanomaterials for tissue regeneration and biosensing applications (Hoque et al., 2018; Kim et al., 2013; Mahbub & Hoque, 2020; Padmanabhan et al., 2019; Rabbani et al., 2020; Suri et al., 2007).

Construction materials: The invention of the crystal nanostructure of calcium silicate hydrate has opened new doors in the construction sector. These materials have water and wear resistance, corrosion protection.

Energy Applications: Nanomaterials have been reported to be used in developing more efficient and environment-friendly batteries. A different example of energy applications where nanomaterials have been extensively researched is solar cells, hydrogen fuel cells, and nanographene batteries (Ghernaout et al., 2018; Y. Li et al., 2017; Sarno, 2020).

Although there is a constant increase of nanotechnology in different applications, there is also a rising question regarding the ecotoxicity of nanomaterials. How these nanomaterials interact with human bodies has been the most common topic in nano-toxicology, but their interaction with the environment and other living things has got limited attention. In this era of the nanotechnology boom, the limited number of research and investments in the assessment of ecotoxicity due to nanoparticles is a great concern. This chapter tries to give an overview of how nanoparticles can enter the environment and create adverse effects on different components of the environment.

7.2 NANOTOXICITY

In this growing era of nanomaterial usage, concerns about the negative effects of nanoscale particles on the human body and environment have been a topic of discussion. Quantum size effects that come with a great reduction in material sizes and a larger surface area-to-volume ratio guarantee exceptional properties of these materials compared to larger materials (Khan et al., 2019; Monticone et al., 2000). With these excellent properties, there also comes an issue with toxicity. Although there are specific protocols to use these nanoparticles as human bodies can be directly exposed to these nanoparticles via various routes, there are still risks involved with these materials (Chakraborty et al., 2011).

7.2.1 DEFINITION

Nanotoxicology can be defined as the evaluation of the toxicity of nanoparticles, especially to the human body. Engineered and modified nanomaterials are used in different sectors such as regenerative medicine, tissue engineering applications, automation industry, space research, construction, electronics industries, security and defense industries, textiles, and cosmetics. These nanoparticles can get exposed to human bodies via different routes. There are mainly five forms of in vitro nanotoxicity, as shown in Figure 7.1 (Aillon et al., 2009).

7.2.2 FACTORS

7.2.2.1 Size

One of the main factors of toxicity of nanoparticles is their size. With the reduction in the size of nanoparticles, the surface area increases, which leads to a greater number of molecules binding to the surface area, resulting in more toxic effects. Some studies have shown that macrophage clearance mechanisms in the lungs work less efficiently on nanoparticles than other larger-size molecules with adverse effects (E. Fröhlich & Salar-Behzadi, 2014; Nho, 2020; Schraufnagel, 2020). Figure 7.2 shows how the size of nanoparticles can affect toxicity (Akçan et al., 2020; Chithrani et al., 2006; Elechiguerra et al., 2005; Jiang et al., 2008; Osaki et al., 2004).

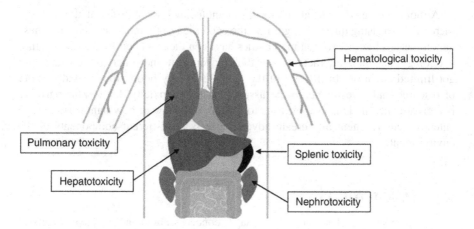

FIGURE 7.1   Forms of in vitro nanotoxicity. (Reprinted with permission from Aillon et al. (2009). Copyright (2009) Elsevier.)

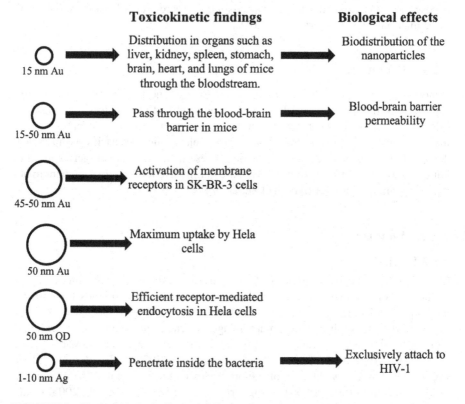

FIGURE 7.2   Toxicokinetic findings of different sizes of nanoparticles.

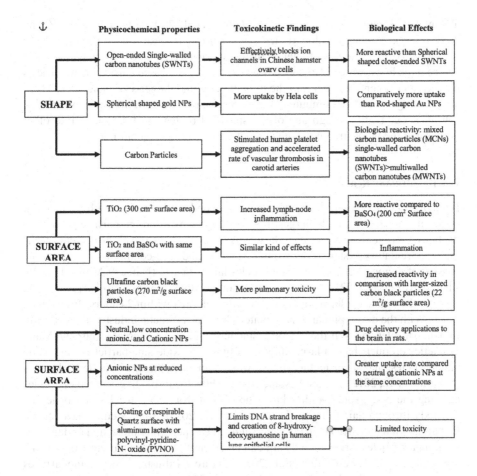

FIGURE 7.3   Effects of shape, surface area, and charge on toxicity.

7.2.2.2   Particle Shape, Surface Area, and Surface Charge

The surface chemistry of nanoparticles can also affect the toxicity of materials by influencing the interaction between the environment of the human or animal model and nanoparticles. Figure 7.3 shows toxicokinetic findings and biological effects for nanoparticles of different shapes, charges, and surface areas (Chithrani et al., 2006; Driscoll et al., 1996; Jiang et al., 2008; Lockman et al., 2004; Nigavekar et al., 2004; Nikula et al., 1995; Park et al., 2003; Radomski et al., 2005; Schins et al., 2002).

7.3   EXPOSURE ROUTES

Due to the increase in the use of nanotechnology in different sectors of health science and engineering, it has become more likely to get exposed to these nanoparticles and their potential risk elements. Toxic nanoparticles can enter the human body through various routes, such as skin contact, respiratory, dental, and ocular routes, and intestinal tract (Davoren et al., 2007; Günter et al., 2005; Yah et al., 2012; Zheng et al., 2007). Although nanoparticles can get through these routes, the main issue with toxicity relies on how efficiently different organs can remove these nanomaterials from the body, which also depends on the factors discussed earlier. Nanoparticles have greater potential to get into the human body than larger-size molecules. There are seemingly numerous amounts of airborne nanoparticles in the environment that can enter our body through any of the specified routes and translocate into other organs and the bloodstream, which can put adverse effects on our body.

7.3.1 DERMAL ROUTE

Being the largest organ of the body and with immediate interaction with the external environment, the skin is the largest route through which foreign particles can enter the body. Once toxic particles pass the dermal layer, they can travel to different organs via the circulatory system. Any wound, scrapes, and skin conditions can allow the passing of nanoparticles as well as particles larger in size (Nafisi & Maibach, 2018; Thanigaivel et al., 2021). In one study, quantum dots were seen to be penetrating through the viable epidermis and into the upper dermis within 24 hours (Prow et al., 2012). With the development of nanomaterials in the cosmetic industry, the exposure to nanoparticles through the use of cosmetic products has increased largely. Many cosmetics products have a large amount of titanium oxide nanoparticles, zinc oxide nanoparticles, and silver nanoparticles, which can enter the body via dermal routes (Piccinno et al., 2012). Although some studies show the unlikeliness of nanoparticles entering the body, some studies have shown that iron nanoparticles can penetrate the body through hair follicles (Watkinson et al., 2013). Some studies have shown that although nanoparticles in the range of 4 nm can penetrate intact and undamaged skin, nanoparticles larger than 45 nm can enter through the dermal layer only if the skin is damaged (Larese Filon et al., 2015). Figure 7.4 illustrates how nanoparticles can enter through the skin (Palmer & DeLouise, 2016).

7.3.2 RESPIRATORY ROUTE

The respiratory tract is one of the most common routes of nanomaterial exposure as there are countless airborne nanoparticles surrounding our environment. Some amount of nanoparticles entering the body through the respiratory tract is usually discharged by various methods. According to Sajid et al, one-third of the nanoparticles inhaled are removed by various defense mechanisms of the respiratory system (Sajid et al., 2015). The size of nanoparticles affects their capacity to enter the human respiratory tract (Bakand & Hayes, 2016). Large nanoparticles that have a size range between 5 and 30 usually reside in the nasopharyngeal region, which is a part of the upper respiratory tract. Smaller nanoparticles with a size range of 1–5 deposit in the tracheobronchial region of the lower respiratory tract. Further smaller nanoparticles (0.1–1) reside in the deepest part of the lower respiratory region, which is the alveolar region. This segmentation happens due to size differences of the nanoparticles as well as gravitational sedimentation and Brownian diffusion. Figure 7.5 shows

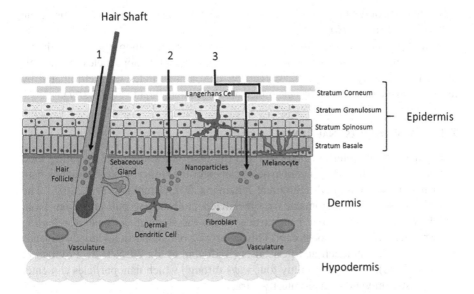

FIGURE 7.4 Illustration of nanoparticle skin penetration pathways. (Reprinted with permission from Palmer and DeLouise (2016). Copyright (2016) MDPI.)

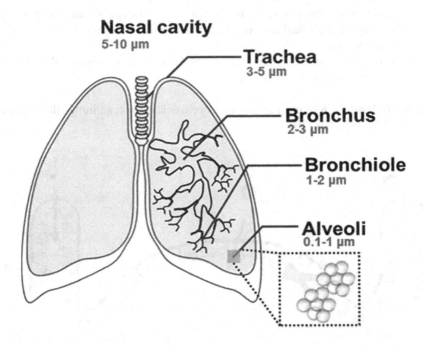

FIGURE 7.5 Size-dependent hosting of nanoparticles inside the lungs. (Reprinted with Permission from Yhee et al. (2016). Copyright (2016) MDPI.)

different parts of the lungs and the size-dependent hosting of nanoparticles inside the lungs (Yhee et al., 2016).

Male Sprague–Dawley rats exposed to titanium oxide nanoparticles for 6 hours via the inhalation route were found to have traces of TiO_2 nanoparticles inside their lungs, liver, kidney, and spleen, suggesting that nanoparticles were eliminated by feces and urine by mucociliary clearance and ingestion (Pujalté et al., 2017).

7.3.3 Gastrointestinal Tracts

Nanoparticles can enter the body through the gastrointestinal tract by eating food and drinking. Additionally, nanoparticles can enter the stomach via the trachea with the help of mucociliary cells (Bergin & Witzmann, 2013; Braakhuis et al., 2015). The epithelial tissues in the stomach are specialized in absorbing nutrients from food, which can also absorb toxic nanoparticles from the stomach, ultimately releasing those particles into the bloodstream. This absorption of nanoparticles largely depends on the pH of the stomach and peristalsis (Bellmann et al., 2015; E. E. Fröhlich & Fröhlich, 2016). There are mainly four ways through which nanoparticles can enter the bloodstream via the gastrointestinal tract.

1. Nanoparticles of less than 100 nm in diameter can be absorbed by endocytosis in epithelial cells.
2. Large nanoparticles and microparticles can pass through the intestinal epithelium via transcytosis and perception by M-cells.
3. Nanoparticles can also change the morphology of the epithelium, which can form gaps in the apical zone of villi through which nanoparticles can cross the epithelium.
4. In disease conditions, paracellular uptake can also happen by which nanoparticles can exit the stomach.

Figure 7.6 shows the four mechanisms of nanoparticles crossing through the stomach discussed (Powell et al., 2010).

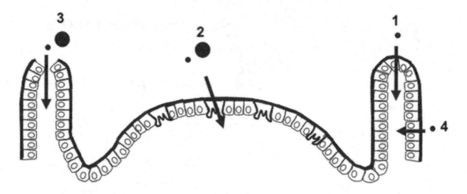

FIGURE 7.6 Nanoparticle crossing through the gastrointestinal route. (Reprinted with permission from Powell et al. (2010). Copyright(2010) Elsevier.)

With the increasing use of nanoparticles in food and pharmaceutical industries as additives, the risk of getting exposed to an excessive amount of nanoparticles has increased (Kaida Takahiro et al., 2004). Titanium oxide, which is commercially known as E171, is extensively used in sweet foods (Weir et al., 2012). Ag nanoparticles and SiO_2 nanoparticles have been used in food industries in recent times. Some studies have shown that the ideal amount of consumption of these nanoparticles is 31.5 mg/day for TiO_2 nanoparticles, 80 g/day for Ag nanoparticles, and 126 mg/day for SiO_2 nanoparticles for a 70 kg person (Dekkers et al., 2011; Lomer et al., 2000). But due to the increased usage of these nanoparticles, risks of toxicity have been an issue for discussion.

7.4 NANOPARTICLE EMISSION IN THE ENVIRONMENT

With the increased use of nanoparticles in the modern world, the possibility of nanoparticle emission in the environment has also increased. Although nanotechnology has created new possibilities in the field of health and biological sciences, the impact on the environment due to the excessive use of nanoparticles is not necessarily positive. Nanoparticles can be formed in two ways: natural and industrial. Both natural and industrial nanomaterials can be produced either accidentally or artificially. Accidentally and naturally produced nanoparticles are released into the environment where nanoparticles are produced by the industrial process. Figure 7.7 represents different ways of nanoparticle production (Martínez et al., 2021).

Engineered and natural nanoparticles can impact the environment by getting emitted into the air, soil, and water by various means. These nanoparticles can either positively impact the weather and soil quality or can have adverse effects such as pollution, and animal and plant toxicity. The total emission of nanoparticles can be divided into four categories (Keller et al., 2013):

1. Emission into landfills: (63%–91% of total production volumes)
2. Emission into soil: (8%–28% of total production volumes)
3. Emission into the aquatic environment (7% of total production volumes)
4. Emission to air (1.5% of total production volumes)

7.4.1 Emission in Air

Although large amounts of nanoparticles are released into the air, a limited number of research studies have been performed for finding the ultimate stage of those released nanomaterials. Nanoparticles can get released into the air in every stage of production, handling, packaging, and applications (Caballero-Guzman & Nowack, 2016). Airborne nanomaterials are generally produced by traffic exhaust smoke, fire, combustion, explosion, industrial exhaust, and oxidation of atmospheric gases. After the production of nanomaterials, some amounts are released into the air during internal handling and transportation (Brouwer, 2010; Ding et al., 2017). Also, emissions can happen from stacks, during use, industrial continuous-release, waste treatment and disposal, and accidental emission (John et al., 2017). Another way nanoparticles

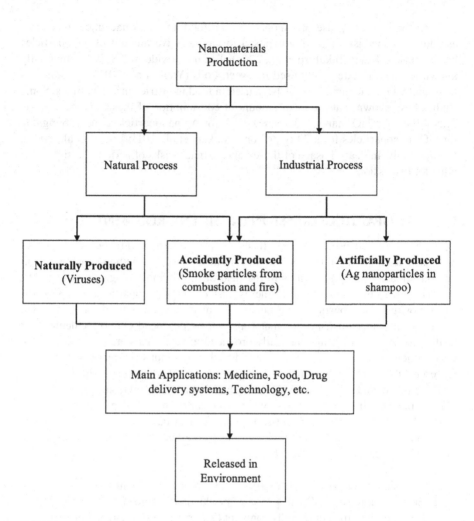

FIGURE 7.7   Classification of nanoparticles by production methods.

get released into the air is by vehicle exhaustion. One study has shown that nanoparticles get emitted from vehicles even when not fueled (Rönkkö et al., 2014). The release of nanoparticles from waste incineration centers mainly depends on the rate of combustion and composition of the nanoparticles. Usually, carbonaceous and organic nanoparticles burn out completely in the incineration center, but nanoclays tend to have residues even after the combustion process (Ounoughene et al., 2015). In many cases, nanomaterials get released into the air as a larger mass of materials by aggregation (Ding et al., 2017). Higher amounts of nanoparticles can also be emitted due to accidents in incineration centers (John et al., 2017). After getting released into the air, nanoparticles can face physiochemical changes in composition due to various factors like sunlight, UV radiation, and other chemical treatments.

7.4.2 Emission in Soil

Emission of nanoparticles into the soil can occur from fertilizers, waste, wastewater, plant protection products, floodplains, biosolids, etc. (Batley et al., 2013). Most of the nanoparticles in the soil come from the use of nanotechnology in agricultural products (Chhipa, 2017). Nanoparticles in these agricultural products and other sources can enter the soil through the matrix pores and accumulate in the soil. These nanoparticles due to their greater surface area hold the soil further aggregating (Mukhopadhyay, 2014). Artificially produced nanomaterials can have adverse effects on the soil as these materials can be resilient to degradation and may also accumulate in the soil. Nanoparticles in the soil affect the fertility of the soil. These nanoparticles which are toxic to plants can influence plant growth and germination (Khodakovskaya et al., 2009). Various studies nanoparticles harm **microorganisms** in soils. Several studies have shown that nanoparticles have different effects on microorganisms present in the soil. Generally, upon exposure to nanoparticles such as gold, copper, silver, and titanium oxide, reduction in the microbial community and biomass in the soil was observed (Asadishad et al., 2017; Hänsch & Emmerling, 2010; Javed et al., 2019; Kumar et al., 2012; Pradhan et al., 2011; Tong et al., 2010). But there was no noticeable difference in the microbial community when the soil was exposed to a mixture of cobalt, iron, nickel, and silver nanoparticles (Kumar et al., 2012).

Along with negative impacts on the microenvironment, nanoparticles can also affect plant or crop growth, pH value of soil, and other characteristics of the soil environment.

7.4.3 Emission in Water

Nanoparticles can enter the water stream through industrial discharge, wastewater treatment effluent dumping, and/or surface runoff from soils (Batley et al., 2013)in water depend on the composition, size, and coating of the materials. Jang et al. showed that silver nanoparticles coated with citrate and polyvinylpyrrolidone showed fate dependency on the coating materials (Jang et al., 2014). How nanomaterials interact with water depends on the size and surface chemistry of nanoparticles. Usually, if nanoparticles aggregate, their surface area gets decreased, making these particles less reactive, ultimately resulting in slow degradation (Baker et al., 2014). Nanomaterials in water can react with other substances and particles in water. Studies have shown genetic damage, death, oxidative stress, and growth inhibition in aquatic organisms due to nanoparticle exposure (Baker et al., 2014; Grillo et al., 2015; Rocha et al., 2015). Brien and Cummins prepared a model to understand the characteristics of nanomaterials in surface water and human exposure (O'Brien & Cummins, 2010).

7.5 DETECTION OF NANOPARTICLES IN THE ENVIRONMENT

Usually, there are some straightforward analytical methods by which nanoparticles can be detected. But, with time, new computational approaches have been developed that give better accuracy than analytical methods. Generally, there are three analytical approaches to detecting and characterizing nanoparticles in the environment (Picó & Andreu, 2014).

Measurements of nanoparticles in environment media: In this method, detection and analysis of nanoparticles are carried out in different environments and ecosystems with different media such as solid, gas, and liquid. The analysis is performed by techniques such as electron microscopy, chromatography, centrifugation, laser light scattering, ultrafiltration, and spectroscopy (Petersen & Henry, 2012).

Detection and size distribution: In this approach, detection and size distribution of nanoparticles by using three techniques.

1. Scanning mobility particle sizer
2. Imaging techniques
3. Light absorbance and emission

Chemical composition and quantification: In this approach, a sample suspension is prepared. After the preparation of samples, separation techniques like field flow fractionation, liquid chromatography, capillary electrophoresis, and hydrodynamic chromatography are performed. Finally, a quantitative determination is performed. Some common quantitative determination techniques are given below:

1. Atomic absorption spectroscopy
2. Inductively coupled plasma atomic emission spectroscopy
3. UV/visible spectroscopy
4. Fluorescence-based techniques
5. Raman spectroscopy
6. Mass spectrometry

Although these analytical techniques give somewhat accurate results, computational approaches have been proved to be more efficient and accurate (Mueller & Nowack, 2008). Material flow models also show limiting accuracy due to an insignificant amount of detail (Caballero-Guzman & Nowack, 2016). Recently, advanced models with probabilistic approaches have been proved to be more accurate (Sun et al., 2014). These models predict the time-dependent material flow of different nanoparticles in different kinds of environments. Although these models have not shown relative success in determining the outcome of nanoparticles in the environment, the accuracy of detecting nanoparticles and measuring their quantitative properties has been a great success (Baalousha et al., 2016; Bäuerlein et al., 2017; Caballero-Guzman & Nowack, 2016; Mitrano et al., 2017). For nanoparticles with more complicated structures, a new multi-element method of detection such as sp-ICP-Time of Flight (ToF)-MS has been prepared and has been used to detect CeO_2 nanoparticles in the soil (Praetorius et al., 2017).

7.6   NANOPARTICLES' INTERACTION WITH THE ENVIRONMENT

The fate of nanoparticles in the environment depends on the size, surface chemistry, surface area, and surface charge of that particular nanoparticle. Figure 7.8 shows nanoparticle pathways to the environment.

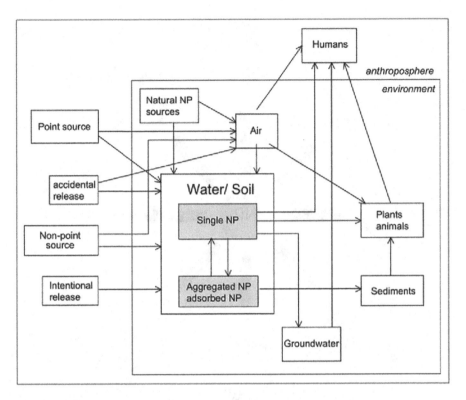

FIGURE 7.8 Nanoparticle pathways to the environment. (Reproduced with permission Nowack and Bucheli (2007). Copyright (2007) Elsevier.)

The type of interaction that happens between the nanoparticles and the environment decides if the material will cause ecotoxicity. There can be changes in chemical properties, degradation rate, dissolution, and surface charge alteration of nanomaterials due to different factors present in media. Many pieces of research have been performed to detect the chemical transformation of nanomaterials in soil and aquatic environments (Baer et al., 2013; Barton et al., 2014; Sekine et al., 2013; Sivry et al., 2014). Figure 7.9 illustrates different types of interaction of nanoparticles with the environment.

Nanoparticles face dissolution in the environment depending on the chemical composition of the materials. In aerobic conditions, an oxide layer can be created around silver nanoparticles, which ultimately results in releasing silver ions (Elzey & Grassian, 2010). This happens not only due to chemical characteristics like surface charge and coating but also because of environmental factors such as pH and temperature (Brunetti et al., 2015; Khaksar et al., 2015; Metreveli et al., 2016). Nanoparticles in the environment can also face sulfidation. Due to sulfidation, silver nanoparticles turn into hollow silver sulfide nanoparticles, which can be toxic to many organisms (Kraas et al., 2017; Levard et al., 2012; Thalmann et al., 2016). Homo-aggregation, hetero-aggregation, and disaggregation are also possible outcomes of nanomaterials'

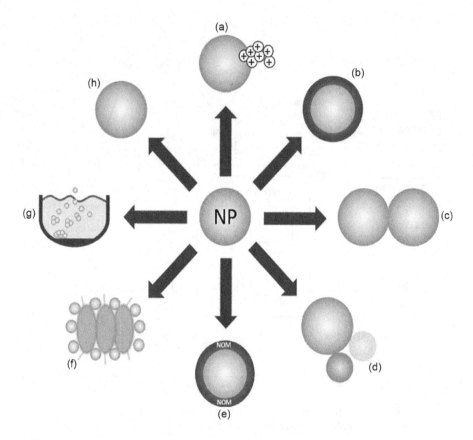

FIGURE 7.9 Interaction and the results of nanoparticles in the environment. (a) Dissolution, (b) sulfidation, (c) homo-aggregation, (d) hetero-aggregation, (e) coating with natural organic matter, (f) nanoparticle adsorption on biological surfaces, (g) sedimentation/deposition, and (h) persistence. (Reprinted with permission from Bundschuh et al. (2018). Copyright (2018) Springer.)

interaction with the environment. These outcomes depend mainly on the colloidal stability of the nanoparticles (Schaumann et al., 2015; H. Wang et al., 2015).

In addition to engineered nanoparticles, environmental nanoparticles also interact with other factors and elements of the environment. After emission from the source, nanoparticles are called primary nanoparticles, which in the presence of H_2SO_4, HNO_3, and other organic chemicals become secondary or volatile nanoparticles under different conditions of temperature, residence time, dilution ratio, and concentration of carbon compounds (Morawska et al., 2008). These increasing number of particles can lead to atmospheric optical effects, building soiling, and climate change (Nowack & Bucheli, 2007). Nanomaterials' interaction with ocean water changes with depth. Nanoparticles can accumulate on the surface microlayers of the ocean, creating a risk of aerosol exposure to marine animals and organisms residing under the surface microlayers (Kennedy et al., 2004). Oberdörster et al. performed acute toxicity assays to assess the toxicity of C60. They exposed three different aquatic and marine creatures to fullerenes and found toxicity of the nanomaterial in the specimens (Oberdörster et al., 2006).

7.7 TOXICITY MECHANISMS OF NANOPARTICLES

Nanoparticles even if not proven toxic to humans can induce cytotoxicity and genotoxicity in plants and other animals. The main way nanoparticles induce toxicity to organisms is by producing reactive oxygen species. Due to their reactivity and surface charge, nanomaterials react with organelles containing O_2^-, OH radicals, and H_2O_2 and create reactive oxygen species. These reactive particles can then interact with proteins, molecules, and other parts of tissues, which can result in cell death (Lojk et al., 2020). Although oxidative stress is known as the most common toxicity mechanism of nanoparticles, some reports showed modification of hormones, which resulted in reproduction system complications (Muller et al., 2015). These mechanisms for toxicity can be classified as transgenerational effects (Jacobasch et al., 2014). Recent research has proved the accumulation of nanoparticles in terrestrial plants results in biochemical and physiochemical changes (Du et al., 2017; Marslin et al., 2017; Tripathi, Shweta, et al., 2017). Cao et al. showed negative impacts on carbon fixation and water usage during photosynthesis when soybean was exposed to cerium oxide nanoparticles (Cao et al., 2017). Table 7.1 shows exposure of different nanoparticles to different plants and their biological effects on the plants.

7.8 ENVIRONMENTAL IMPACT OF NANOPARTICLES

With the increasing use of nanotechnology and the possibility of further expansion of the sector, the environment is filled with nanomaterials. Nanomaterials can affect the environment in both positive and negative ways. Figure 7.10 provides an overview of the positive and negative impacts of nanoparticles on the environment.

7.8.1 POSITIVE IMPACTS

Nanotechnology can significantly influence their surroundings by creating more efficient solutions. With their wide range of applications, nanoparticles can significantly decrease pollution and energy stress on the environment. Using nanoparticles in aircraft can reduce the weight of the aircraft significantly, which can result in less fuel use (Kausar et al., 2017). Another way nanoparticles are working to make energy more efficient is through wind turbines. By using nanoparticles in turbine blades, the turbines get lighter but stronger, which means better energy efficiency (Patel & Mahajan, 2017). As the demand for green energy is increasing day by day, the use of photovoltaic cells that utilize solar radiation to convert that energy into electricity has been on the rise. But due to their low energy absorption efficiency and high production cost, alternative materials have been searched to use within these cells. Some studies have shown that carbon nanotubes can be an alternate for this application as carbon nanotubes are easier to synthesize and have better energy absorption quality

TABLE 7.1
Different Nanoparticle Exposure to Different Plants and Their Biological Effects on the Plant

Nanoparticle	Plant	Biological Effects	Reference
Ag	*Triticum aestivum* L.	Negatively affect seedling growth	Vannini et al. (2014)
	Pisum sativum L.	Chlorophyll fluorescence	Tripathi, Singh, et al. (2017)
	Allium cepa	Induced oxidative stress and toxicity	Cvjetko et al. (2017)
	Phaseolus radiatus	Greater accumulation in roots and shoots.	W.-M. Lee et al. (2012)
Au	*Arabidopsis thaliana*	Root length was decreased by 75%. Accumulation	Taylor et al. (2014)
CeO$_2$, ZnO	Glycine max	Greater reactive oxygen species and noticeable lipid peroxidation	Priester et al. (2017)
CuO	Ipt-cotton	No reduction in height and root length	le Van et al. (2016)
Fe$_3$O$_4$, TiO$_2$, and C nanoparticles	Cucumber plants	Negatively affect seed germination rate, root elongation, and germination index	Mushtaq (2011)
MWCNTs, Ag, Cu, ZnO, and Si	*Cucurbita pepo*	Adverse effects on seed germination, root elongation, and biomass	Stampoulis et al. (2009)
ZnO	*Allium cepa*	Cytotoxic and genotoxic effects include lipid peroxidation, a reduction of the mitotic index, and an increase in chromosomal aberration indexes	Kumari et al. (2011)
Al$_2$O$_3$	*Arabidopsis thaliana*	Toxic effects on seed germination, root elongation, and number of leaves	C. W. Lee et al. (2010)
Carbon nanotubes	Tomato plants	Increased seed germination and growth	Khodakovskaya et al. (2009)
TiO$_2$	Riticum aestivum, *Brassica napus*, and *Arabidopsis thaliana*	No effect on the plants	Larue et al. (2011)
TiO$_2$	D1/D2/Cyt b559 complex of spinach	Better energy utilization and conversion	Su et al. (2009)

(Ong et al., 2010). Due to their photocatalytic activity, titanium oxide nanoparticles have also been considered in some research for this application (Abdel-Mottaleb et al., 2011).

Purification of air and water by nanoparticles with greater efficiency was reported in some studies (Kunduru et al., 2017). Although some studies proved the toxicity of nanoparticles to specific plants, using nanoparticles in fertilizers, herbicides, and pesticides can lead to improved plant growth with a limited amount of these

1. Light and Stringer Structure
2. CO_2 Isolation
3. Self-Cleaning
4. More Efficient Solar Cells
5. Disaster Management
6. Precision Manufacturing
7. Purification and Filtration

Positive Impacts

1. Cytotoxicity and Genotoxicity
2. DNA Damage
3. Reduction in Plant Growth
4. Adverse Health effects
5. Carrier of Pollutants

Negative Impacts

FIGURE 7.10 An overview of the benefits and risks of nanoparticles on the environment.

substances as some nanoparticles even in smaller quantities can ultimately enhance the activity of these growth promoter compounds (Meier et al., 2020). By using less amounts of these compounds, the rate of soil pollution and water pollution decreases by a significant amount. Compounds used in pesticides and fertilizers can sometimes be subjected to photodegradation. Coating these compounds with zinc and aluminum nanoparticles has been proved to protect active components of pesticides against photodegradation (Medina-Pérez et al., 2019). In recent times, the effectiveness of nanomaterials for soil remediation has been studied extensively. Zinc nanoparticles have shown efficacy in the degradation of compounds in such a way that the soil does not become toxic and polluted (Pandey, 2018). Fe nanoparticles were shown to remove heavy metal and organochlorine compounds from the soil via the electronic donation process (Guerra et al., 2018). Saif et al. showed that iron oxide nanoparticles can remove chromium from the soil, with an efficiency of more than 90% (Saif et al., 2016). Carbon nanotubes are porous, which is why their surface-to-volume ratio is high. This property of carbon nanotubes can be used to filer inorganic and organic compounds from water, which makes them an appropriate candidate for nanofilters (Ong et al., 2010). Graphene oxide nanoparticles have also been tested to filter polycyclic aromatic hydrocarbons, gasoline, dyes, heavy metals, etc. (Kemp et al., 2013). Metallic iron nanoparticles have also been proven to be effective in filtering organochlorine compounds, arsenic, or petroleum derivatives (Saif et al., 2016).

Some recent studies reported the antibacterial effect of silver nanoparticles against different types of bacteria (Durán et al., 2005). Silver nanoparticles have also shown antifungal activities against different fungi species (Bratovcic, 2019). However, in their early stages, nanoparticles have shown promises in sensing different contaminants, pollutants, pesticides, antibiotics, and microorganisms due to their larger surface area and high reactivity (Z. Li et al., 2020; R. P. Singh, 2011; Song et al., 2020; Zhu et al., 2020).

7.8.2 NEGATIVE IMPACTS

Although there are not enough studies to back up the claim that nanoparticles have a more negative than positive impact, the issue of nanotoxicity needs to be addressed. With the increasing use of nanotechnology in almost every sector of science and engineering, the aspect of toxicity is needed to be assessed with more importance than ever. Although limited, several studies proved toxicity issues and negative impacts of nanoparticles on the environment, animals, and plants as well as humans. Due to less efficient detection methods, low detection limits, and other environmental constraints, determining the impacts of nanoparticles on the environment is difficult. The United States Environmental Protection Agency stated that because of their exceptional chemical characteristics, high reactivity, and dissolubility in water, the toxicity of nanoparticles is very difficult to measure. Because of these properties, the characteristics of specimens can change even when collecting and analyzing the samples (U.S. Environmental Protection Agency, 2007).

The main issue with nanoparticles' negative impact is the absorption and distribution of nanoparticles by microorganisms in the environment. Microorganisms absorb nanoparticles through their cell surface, but for other complex living things, nanoparticles enter through exposure routes discussed in this chapter before. Carbon nanotubes can enter the human body through the pulmonary epithelium and can cause toxicity, but fullerenes are non-toxic (Garner & Keller, 2014). Nanoparticles cause more toxicity to eukaryotic cells than to prokaryotic cells as prokaryotic cell walls can stop the transfer of nanoparticles inside the cell, whereas eukaryotic cells do not have this type of mechanism (Taghavi et al., 2013). Wang et al. showed that porous cell walls of seaweeds allow nanoparticle entrance inside the cell cytoplasm (Y. Wang & Xia, 2019). After the entrance into the cell cytoplasm, the severity and type of toxicity will depend on the particular material. For example, carbon nanotubes induce toxicity to mitochondria (Samiei et al., 2020). Silver nanoparticles stay at the cell membrane affecting the permeability of the cell membrane, which can disrupt the transport mechanisms of the cell (Xiang et al., 2020). In some cases, when the nuclear membrane becomes vulnerable, nanoparticles can enter the nuclei and disrupt the normal function of the Golgi complex (Elsaesser & Howard, 2012). The main mechanism by which nanoparticles can induce toxicity at the cellular level is discussed in previous sections. Figure 7.11 shows the toxic effects of nanoparticles on different intracellular components (Martínez et al., 2021).

Bacteria, as an essential component of the ecosystem, nanoparticles' effects on these microorganisms have been a subject of research. Although the amounts of research are limited, several studies assessed toxicity induced by different nanomaterials in both gram-positive and gram-negative strains. The kind and scale of toxicity depend on the cell structure and components of microorganisms. A limited number of studies, in this case, have shown that nanoparticles can induce oxidative stress, metal ion release, and other non-oxidative mechanisms in bacterial cells (Prajitha et al., 2019). In one study, Ge et al. showed a decrease in the growth rate of *Rhizobiales* or *Methylobacteriaceae* and an increase in the growth rate of *Streptomyces* when introduced to TiO_2 and ZnO nanoparticles at different doses (Yuan et al., 2012). Singh et al. performed genotoxicity and cytotoxicity assays and found that zinc oxide nanoparticles introduced cell membrane damage, and morphological changes to *Deinococcus radiodurans* by producing reactive oxygen species and altering energy routes inside the cells (R. Singh et al., 2020). Most of the research regarding this area has been performed with silver nanoparticles. Silver nanoparticles showed toxicity in five different kinds of soils. Another study by Beddow et al. reported a reduction in the nitrification potential in three ammonia-oxidizing bacteria

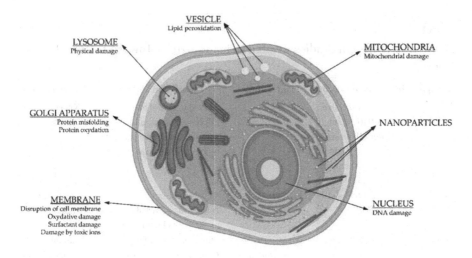

FIGURE 7.11 Toxic effects of nanoparticles on different intracellular components. (Reprinted with Permission from Martínez et al. (2021). Copyright (2021) MDPI.)

(*Nitrosomonas europaea, Nitrosospira multiformis*, and *Nitrosococcus oceani*). Ag nanoparticles were also found to cause cell lysis and cell membrane disruption for *Microcystis aeruginosa*. Although this can be toxic for the microorganism, using this method bloom-forming bacteria can be removed from water. Two separate studies on polymeric nanomaterials against *Vibrio fischeri* showed different results. Silicon nanoparticles and polyethyleneimine polystyrene nanoparticles showed no toxicity to the mentioned strain, but poly (*N*-isopropyl acrylamide) and *N*-isopropylacrylamide/ *N*-tert-butylacrylamide polymeric nanomaterials showed toxicity after limited exposure (Casado et al., 2013; Naha et al., 2009). Many nanotoxicity studies have been performed for plants, specifically plants related to the food chain. Table 7.2 demonstrates the findings of different studies related to nanotoxicity toward different plants.

For proper assessment of nanotoxicity and its impact on the environment, nanotoxicity studies on mammals and marine animals are needed to be performed. Some studies have tried to determine probable toxicity levels of nanoparticles in different animal models. Unirne et al. examined the exposure of Cu, Ag, and Au nanoparticles exposure to earthworms. The effect of these nanoparticles depended on the kinds of soil media (Unrine et al., 2010). When *Daphnia magna* was directly exposed to carbon nanotubes, intestinal toxicity and movement impairment were visible, but when carbon nanotubes were given with food, no toxicity was observed (FraseR et al., 2011; Petersen et al., 2011). This indicates that nanotoxicity induced by carbon nanotubes is due to interaction with organism surfaces. A similar kind of research has been performed for different species such as *Tigriopus japonicus* and *Elasmopus rapax* larvae (Wong et al., 2010), Mediterranean mussel hemocytes (Canesi et al., 2015), zebrafish (Lehner et al., 2019), *Oreochromis niloticus*, and tilapia *zillii* fishes (Saddick et al., 2017), Wistar rats (Thakur et al., 2014), etc.

TABLE 7.2

Findings of Different Studies Related to Nanotoxicity Toward Different Plants

Nanoparticle	Concentrations	Plant Samples	Experiment Findings	References
TiO$_2$	For 21 nm size –0 to 100 mg/L For 60 nm size –0 to 360 mg/L For 400 nm size –0 to 500 mg/L	*Nitzschia closterium*	Increase in Half maximal effective concentration with size which shows a decrease in toxicity with size.	Xia et al. (2015)
TiO$_2$	0,2,4,6,8, and 10 mM	*Allium cepa*	• DNA folding • Chromosomal aberrations • ROS production	Ghosh et al. (2010)
ZnO	10, 20, 40 and 80 mg/L	*Phaeodactylum tricornutum, Cylindrotheca gracilis, Thalassiosira pseudonana*	Growth inhibition for *Cylindrotheca gracilis* and *Thalassiosira pseudonana* and concentration-based growth reduction for *Phaeodactylum tricornutum*	Matranga and Corsi (2012)
ZnO	0–300 mg/L	*Arabidopsis thaliana*	Oxidative stress produced by Zn^{2+} results in cellular toxicity, disruption in gene expression, photosynthesis, and growth	X. Wang et al. (2016)
Ag	55 mg/L	*Ulva lactuca*	Reduction in photosynthesis	Turner et al. (2012)
Ag	20, 200, and 2000 mg/kg	*Triticum aestivum L.*	• Dose-dependent toxicity • Severe phytotoxicity • Short height • Lower grain weight • Reduced nutrients	Yang et al. (2018)
Fe$_2$O$_3$, CuO	50 and 500 mg/kg	*Arachis hypogaea*	Reduced amino acid concentration	Rui et al. (2017, 2018)

7.9 LIMITATIONS OF ORGANIC NANOFILLERS

With increasing concern in green nanotechnology and green synthesis of polymer materials, researchers have been trying solutions in organic nanofillers such as starch, cellulose, chitosan, and proteins. Although many studies have suggested that by incorporating these organic nanofillers into polymers, non-toxic and biodegradable materials can be produced, seemingly, there have been some limitations regarding these nanofillers. Inorganic and organic nanofillers have been studied to be used in specific drug delivery applications (Hoque et al., 2015; Majumder et al., 2020). Organic nanofillers have been reported to be respiratory irritants in some

TABLE 7.3

Weight Loss and Strength Reduction in Different Organic Nanofillers Incorporated with Polymer Composite Materials

Filler Material	Weight Loss (%)	Reduction in Tensile Strength (%)	Reduction in Tensile Modulus (%)
Palm	15.4	49	54.6
Banana	18.6	86.1	83
Bagasse	20.3	88	94
Flax	20	87.2	84

studies (Alves et al., 2010). Polymer materials have been reported to have excellent mechanical and thermal properties, but adding these organic nanofillers can ultimately impact the usual properties of polymer composites.

As organic nanofillers like cellulose, chitosan, and starch have lower degradation temperatures than conventional polymers, there can be a decrease in thermal stability for polymer materials. Different environmental conditions such as temperature, chemical compounds, and UV radiation can degrade organic nanofillers faster than polymers, ultimately resulting in less durability. Due to these reasons, it is difficult to accurately determine the lifetime of these nanofillers. Also, organic nanofillers are subjected to microbial attack more commonly. For example, flax fibers have been reported to be under microbial attack within 3 days (Yan et al., 2021). Ibrahim et al. examined different organic nanofillers and their effect on polymer composites. The research reported a reduction in tensile strength and modulus due reinforcement of different nanofillers. Table 7.3 illustrates weight loss and strength reduction in different organic nanofillers incorporated with polymer composite materials (Ibrahim et al., 2018).

7.10 FUTURE ASPECTS AND CONCLUSION

In a technologically expanding world, the upcoming era is the era of nanotechnology where different industries and sectors are coming together on the importance of the nanotechnology issue. Although different sectors intend to give importance to different properties of nanomaterials, the end goal is to create a technological system that depends on nanomaterials. For instance, for medical sectors, the most importance is given to biocompatibility, bioactivity, and non-toxicity of the materials, whereas industrial sectors dive deep into creating a more efficient manufacturing system. If these approaches can be combined wholeheartedly, a future with safe and effective nanotechnology can evolve.

The use of nanotechnology includes both benefits and risks. To properly use nanotechnology for humans as well as the environment, a proper balance between these benefits and risks should be established. Many regulatory bodies have been formed to properly evaluate any use and research of nanomaterials. In the United States, the use of nanomaterials in food, drugs, medicine, and cosmetics is regulated by the Food and Drugs Administration (FDA) (*U.S. Food and Drug Administration*, n.d.), whereas the emission of nanoparticles in the environment is regulated by the United States Environmental Protection Agency (*U.S. Environmental Protection Agency | US EPA*, n.d.). The European Union developed the Biomaterial Risk Management (BIORIMA) project to establish a framework to assess bio-nanomaterials used in medical applications (*BIORIMA | Risk Management of Biomaterials*, n.d.). Nanodatabase is an initiative created with funds from the European Research Council and Danish Consumer Council. Nanodatabase categorizes different nanomaterials based on toxicity to human health and the environment. Currently, there are more than 5000 products in the database (*Welcome to The Nanodatabase*, n.d.).

Nanomaterial emissions into the environment can happen any time during their life cycle, which is why determining the timeline of exposure is difficult. Although many studies reported the biocompatibility, biodegradation, bioactivity, barrier properties, reactive and thermal properties, and electrical characteristics of nanomaterials and nanomaterials incorporated with polymer composites, the study of ecotoxicity induced by nanoparticles is fairly limited. But in those limited number of studies, the hint of dose and concentration-dependent ecotoxicity was reported by exposing different nanoparticles to different plants, microorganisms, mammals, and aquatic animals.

REFERENCES

Abdel-Mottaleb, M. S. A., Byrne, J. A., & Chakarov, D. (2011). Nanotechnology and solar energy. *International Journal of Photoenergy*, *2011*, 194146. https://doi.org/10.1155/2011/194146

Aillon, K. L., Xie, Y., El-Gendy, N., Berkland, C. J., & Forrest, M. L. (2009). Effects of nanomaterial physicochemical properties on in vivo toxicity. *Advanced Drug Delivery Reviews*, *61*(6), 457–466. https://doi.org/10.1016/j.addr.2009.03.010

Akçan, R., Aydogan, H. C., Yildirim, M. Ş., Taştekin, B., & Sağlam, N. (2020). Nanotoxicity: A challenge for future medicine. *Turkish Journal of Medical Sciences*, *50*(4), 1180–1196. https://doi.org/10.3906/sag-1912-209

Alves, C., Ferrão, P. M. C., Silva, A. J., Reis, L. G., Freitas, M., Rodrigues, L. B., & Alves, D. E. (2010). Ecodesign of automotive components making use of natural jute fiber composites. *Journal of Cleaner Production*, *18*(4), 313–327. https://doi.org/10.1016/j.jclepro.2009.10.022

Asadishad, B., Chahal, S., Cianciarelli, V., Zhou, K., & Tufenkji, N. (2017). Effect of gold nanoparticles on extracellular nutrient-cycling enzyme activity and bacterial community in soil slurries: role of nanoparticle size and surface coating. *Environmental Science: Nano*, *4*(4), 907–918. https://doi.org/10.1039/C6EN00567E

Baalousha, M., Yang, Y., Vance, M. E., Colman, B. P., McNeal, S., Xu, J., Blaszczak, J., Steele, M., Bernhardt, E., & Hochella, M. F. (2016). Outdoor urban nanomaterials: The emergence of a new, integrated, and critical field of study. *Science of The Total Environment*, *557–558*, 740–753. https://doi.org/10.1016/j.scitotenv.2016.03.132

Baer, D. R., Engelhard, M. H., Johnson, G. E., Laskin, J., Lai, J., Mueller, K., Munusamy, P., Thevuthasan, S., Wang, H., Washton, N., Elder, A., Baisch, B. L., Karakoti, A., Kuchibhatla, S. V. N. T., & Moon, D. (2013). Surface characterization of nanomaterials and nanoparticles: Important needs and challenging opportunities. *Journal of Vacuum Science & Technology A*, *31*(5), 050820. https://doi.org/10.1116/1.4818423

Bakand, S., & Hayes, A. (2016). Toxicological considerations, toxicity assessment, and risk management of inhaled nanoparticles. *International Journal of Molecular Sciences*, *17*(6). https://doi.org/10.3390/ijms17060929

Baker, T. J., Tyler, C. R., & Galloway, T. S. (2014). Impacts of metal and metal oxide nanoparticles on marine organisms. *Environmental Pollution*, *186*, 257–271. https://doi.org/10.1016/j.envpol.2013.11.014

Barton, L. E., Auffan, M., Bertrand, M., Barakat, M., Santaella, C., Masion, A., Borschneck, D., Olivi, L., Roche, N., Wiesner, M. R., & Bottero, J.-Y. (2014). Transformation of pristine and citrate-functionalized CeO2 nanoparticles in a laboratory-scale activated sludge reactor. *Environmental Science & Technology*, *48*(13), 7289–7296. https://doi.org/10.1021/es404946y

Batley, G. E., Kirby, J. K., & McLaughlin, M. J. (2013). Fate and risks of nanomaterials in aquatic and terrestrial environments. *Accounts of Chemical Research*, *46*(3), 854–862. https://doi.org/10.1021/ar2003368

Bäuerlein, P. S., Emke, E., Tromp, P., Hofman, J. A. M. H., Carboni, A., Schooneman, F., de Voogt, P., & van Wezel, A. P. (2017). Is there evidence for man-made nanoparticles in the Dutch environment? *Science of The Total Environment*, *576*, 273–283. https://doi.org/10.1016/j.scitotenv.2016.09.206

Bellmann, S., Carlander, D., Fasano, A., Momcilovic, D., Scimeca, J. A., Waldman, W. J., Gombau, L., Tsytsikova, L., Canady, R., Pereira, D. I. A., & Lefebvre, D. E. (2015). Mammalian gastrointestinal tract parameters modulating the integrity, surface properties, and absorption of food-relevant nanomaterials. *WIREs Nanomedicine and Nanobiotechnology*, *7*(5), 609–622. https://doi.org/10.1002/wnan.1333

Bergin, I. L., & Witzmann, F. A. (2013). Nanoparticle toxicity by the gastrointestinal route: evidence and knowledge gaps. *International Journal of Biomedical Nanoscience and Nanotechnology*, *3*(1/2), 163–210.

BIORIMA | Risk Management of Biomaterials. (n.d.). Retrieved May 20, 2022, from https://www.biorima.eu/

Braakhuis, H. M., Kloet, S. K., Kezic, S., Kuper, F., Park, M. V. D. Z., Bellmann, S., van der Zande, M., le Gac, S., Krystek, P., Peters, R. J. B., Rietjens, I. M. C. M., & Bouwmeester, H. (2015). Progress and future of in vitro models to study translocation of nanoparticles. *Archives of Toxicology*, *89*(9), 1469–1495. https://doi.org/10.1007/s00204-015-1518-5

Bratovcic, A. (2019). Different applications of nanomaterials and their impact on the environment. *International Journal of Material Science and Engineering*, *5*(1), 1–7.

Brouwer, D. (2010). Exposure to manufactured nanoparticles in different workplaces. *Toxicology*, *269*(2), 120–127. https://doi.org/10.1016/j.tox.2009.11.017

Brunetti, G., Donner, E., Laera, G., Sekine, R., Scheckel, K. G., Khaksar, M., Vasilev, K., de Mastro, G., & Lombi, E. (2015). Fate of zinc and silver engineered nanoparticles in sewerage networks. *Water Research*, *77*, 72–84. https://doi.org/10.1016/j.watres.2015.03.003

Bundschuh, M., Filser, J., Lüderwald, S., McKee, M. S., Metreveli, G., Schaumann, G. E., Schulz, R., & Wagner, S. (2018). Nanoparticles in the environment: Where do we come from, where do we go to? *Environmental Sciences Europe*, *30*(1), 6. https://doi.org/10.1186/s12302-018-0132-6

Caballero-Guzman, A., & Nowack, B. (2016). A critical review of engineered nanomaterial release data: Are current data useful for material flow modeling? *Environmental Pollution*, *213*, 502–517. https://doi.org/10.1016/j.envpol.2016.02.028

Canesi, L., Ciacci, C., Bergami, E., Monopoli, M. P., Dawson, K. A., Papa, S., Canonico, B., & Corsi, I. (2015). Evidence for immunomodulation and apoptotic processes induced by cationic polystyrene nanoparticles in the hemocytes of the marine bivalve Mytilus. *Marine Environmental Research*, *111*, 34–40. https://doi.org/10.1016/j.marenvres.2015.06.008

Cao, Z., Stowers, C., Rossi, L., Zhang, W., Lombardini, L., & Ma, X. (2017). Physiological effects of cerium oxide nanoparticles on the photosynthesis and water use efficiency of soybean (Glycine max (L.) Merr.). *Environmental Science: Nano, 4*(5), 1086–1094. https://doi.org/10.1039/C7EN00015D

Casado, M. P., Macken, A., & Byrne, H. J. (2013). Ecotoxicological assessment of silica and polystyrene nanoparticles assessed by a multitrophic test battery. *Environment International, 51,* 97–105. https://doi.org/10.1016/j.envint.2012.11.001

Chakraborty, M., Jain, S., & Rani, V. (2011). Nanotechnology: Emerging tool for diagnostics and therapeutics. *Applied Biochemistry and Biotechnology, 165*(5), 1178–1187. https://doi.org/10.1007/s12010-011-9336-6

Chhipa, H. (2017). Nanofertilizers and nanopesticides for agriculture. *Environmental Chemistry Letters, 15*(1), 15–22. https://doi.org/10.1007/s10311-016-0600-4

Chithrani, B. D., Ghazani, A. A., & Chan, W. C. W. (2006). Determining the size and shape dependence of gold nanoparticle uptake into mammalian cells. *Nano Letters, 6*(4), 662–668. https://doi.org/10.1021/nl052396o

Cvjetko, P., Milošić, A., Domijan, A.-M., Vinković Vrček, I., Tolić, S., Peharec Štefanić, P., Letofsky-Papst, I., Tkalec, M., & Balen, B. (2017). Toxicity of silver ions and differently coated silver nanoparticles in Allium cepa roots. *Ecotoxicology and Environmental Safety, 137,* 18–28. https://doi.org/10.1016/j.ecoenv.2016.11.009

Davoren, M., Herzog, E., Casey, A., Cottineau, B., Chambers, G., Byrne, H. J., & Lyng, F. M. (2007). In vitro toxicity evaluation of single walled carbon nanotubes on human A549 lung cells. *Toxicology in Vitro, 21*(3), 438–448. https://doi.org/10.1016/j.tiv.2006.10.007

Dekkers, S., Krystek, P., Peters, R. J. B., Lankveld, D. P. K., Bokkers, B. G. H., van Hoeven-Arentzen, P. H., Bouwmeester, H., & Oomen, A. G. (2011). Presence and risks of nano-silica in food products. *Nanotoxicology, 5*(3), 393–405. https://doi.org/10.3109/17435390.2010.519836

Ding, Y., Kuhlbusch, T. A. J., van Tongeren, M., Jiménez, A. S., Tuinman, I., Chen, R., Alvarez, I. L., Mikolajczyk, U., Nickel, C., Meyer, J., Kaminski, H., Wohlleben, W., Stahlmecke, B., Clavaguera, S., & Riediker, M. (2017). Airborne engineered nanomaterials in the workplace—A review of release and worker exposure during nanomaterial production and handling processes. *Journal of Hazardous Materials, 322,* 17–28. https://doi.org/10.1016/j.jhazmat.2016.04.075

Driscoll, K. E., Carter, J. M., Howard, B. W., Hassenbein, D. G., Pepelko, W., Baggs, R. B., & Oberdörster, G. (1996). Pulmonary inflammatory, chemokine, and mutagenic responses in rats after subchronic inhalation of carbon black. *Toxicology and Applied Pharmacology, 136*(2), 372–380. https://doi.org/10.1006/taap.1996.0045

Du, W., Tan, W., Peralta-Videa, J. R., Gardea-Torresdey, J. L., Ji, R., Yin, Y., & Guo, H. (2017). Interaction of metal oxide nanoparticles with higher terrestrial plants: Physiological and biochemical aspects. *Plant Physiology and Biochemistry, 110,* 210–225. https://doi.org/10.1016/j.plaphy.2016.04.024

Durán, N., Marcato, P. D., Alves, O. L., de Souza, G. I. H., & Esposito, E. (2005). Mechanistic aspects of biosynthesis of silver nanoparticles by several Fusarium oxysporum strains. *Journal of Nanobiotechnology, 3*(1), 8. https://doi.org/10.1186/1477-3155-3-8

Elechiguerra, J. L., Burt, J. L., Morones, J. R., Camacho-Bragado, A., Gao, X., Lara, H. H., & Yacaman, M. J. (2005). Interaction of silver nanoparticles with HIV-1. *Journal of Nanobiotechnology, 3*(1), 6. https://doi.org/10.1186/1477-3155-3-6

Elsaesser, A., & Howard, C. V. (2012). Toxicology of nanoparticles. *Advanced Drug Delivery Reviews, 64*(2), 129–137. https://doi.org/10.1016/j.addr.2011.09.001

Elzey, S., & Grassian, V. H. (2010). Agglomeration, isolation and dissolution of commercially manufactured silver nanoparticles in aqueous environments. *Journal of Nanoparticle Research, 12*(5), 1945–1958. https://doi.org/10.1007/s11051-009-9783-y

FraseR, T. W. K., Reinardy, H. C., Shaw, B. J., Henry, T. B., & Handy, R. D. (2011). Dietary toxicity of single-walled carbon nanotubes and fullerenes (C60) in rainbow trout (Oncorhynchus mykiss). *Nanotoxicology*, *5*(1), 98–108. https://doi.org/10.3109/17435 390.2010.502978

Fröhlich, E. E., & Fröhlich, E. (2016). Cytotoxicity of nanoparticles contained in food on intestinal cells and the gut microbiota. *International Journal of Molecular Sciences*, *17*(4). https://doi.org/10.3390/ijms17040509

Fröhlich, E., & Salar-Behzadi, S. (2014). Toxicological assessment of inhaled nanoparticles: Role of in vivo, ex vivo, in vitro, and in silico studies. *International Journal of Molecular Sciences*, *15*(3), 4795–4822. https://doi.org/10.3390/ijms15034795

Garner, K. L., & Keller, A. A. (2014). Emerging patterns for engineered nanomaterials in the environment: a review of fate and toxicity studies. *Journal of Nanoparticle Research*, *16*(8), 2503. https://doi.org/10.1007/s11051-014-2503-2

Ghernaout, D., Alghamdi, A., Touahmia, M., Aichouni, M., Aït-Messaoudène, N., Ait, N., & Nanotechnology, M. (2018). Nanotechnology phenomena in the light of the solar energy. *Journal of Energy, Environmental & Chemical Engineering*, *3*, 1–8. https://doi.org/10.11648/j.jeece.20180301.11

Ghosh, M., Bandyopadhyay, M., & Mukherjee, A. (2010). Genotoxicity of titanium dioxide (TiO2) nanoparticles at two trophic levels: Plant and human lymphocytes. *Chemosphere*, *81*(10), 1253–1262. https://doi.org/10.1016/j.chemosphere.2010.09.022

Gohardani, O., Elola, M. C., & Elizetxea, C. (2014). Potential and prospective implementation of carbon nanotubes on next generation aircraft and space vehicles: A review of current and expected applications in aerospace sciences. *Progress in Aerospace Sciences*, *70*, 42–68. https://doi.org/10.1016/j.paerosci.2014.05.002

Grillo, R., Rosa, A. H., & Fraceto, L. F. (2015). Engineered nanoparticles and organic matter: A review of the state-of-the-art. *Chemosphere*, *119*, 608–619. https://doi.org/10.1016/j.chemosphere.2014.07.049

Guerra, F. D., Attia, M. F., Whitehead, D. C., & Alexis, F. (2018). Nanotechnology for environmental remediation: Materials and applications. *Molecules*, *23*(7). https://doi.org/10.3390/molecules23071760

Günter, O., Eva, O., & Jan, O. (2005). Nanotoxicology: An emerging discipline evolving from studies of ultrafine particles. *Environmental Health Perspectives*, *113*(7), 823–839. https://doi.org/10.1289/ehp.7339

Hänsch, M., & Emmerling, C. (2010). Effects of silver nanoparticles on the microbiota and enzyme activity in soil. *Journal of Plant Nutrition and Soil Science*, *173*(4), 554–558. https://doi.org/10.1002/jpln.200900358

Hoque, M. E., Daei, J. M. G., & Khalid, M. (2018). Next generation biomimetic bone tissue engineering matrix from poly (L-lactic acid) pla/calcium carbonate composites doped with silver nanoparticles. *Current Analytical Chemistry*, *14*(3), 268–277.

Hoque, M. E., Prasad, R. G. S. V., Aparna, R. S. L., & Sapuan, S. M. (2015). Nanofibers: Drug Delivery. In M. Mishra (Ed.), *Encyclopedia of Biomedical Polymers and Polymeric Biomaterials* (1st ed., Vol. 11). Taylor and Francis.

Hulla, J. E., Sahu, S. C., & Hayes, A. W. (2015). Nanotechnology: History and future. *Human & Experimental Toxicology*, *34*(12), 1318–1321. https://doi.org/10.1177/0960327115603588

Ibrahim, H., Mehanny, S., Darwish, L., & Farag, M. (2018). A comparative study on the mechanical and biodegradation characteristics of starch-based composites reinforced with different lignocellulosic fibers. *Journal of Polymers and the Environment*, *26*(6), 2434–2447. https://doi.org/10.1007/s10924-017-1143-x

Jacobasch, C., Völker, C., Giebner, S., Völker, J., Alsenz, H., Potouridis, T., Heidenreich, H., Kayser, G., Oehlmann, J., & Oetken, M. (2014). Long-term effects of nanoscaled titanium dioxide on the cladoceran Daphnia magna over six generations. *Environmental Pollution*, *186*, 180–186. https://doi.org/10.1016/j.envpol.2013.12.008

Jang, M.-H., Bae, S.-J., Lee, S.-K., Lee, Y.-J., & Hwang, Y. S. (2014). Effect of material properties on stability of silver nanoparticles in water. *Journal of Nanoscience and Nanotechnology*, *14*(12), 9665–9669.

Javed, Z., Dashora, K., Mishra, M., Fasake, V., & Srivastava, A. (2019). Effect of accumulation of nanoparticles in soil health-a concern on future. *Frontiers in Nanoscience and Nanotechnology*, *5*, 1–9. https://doi.org/10.15761/FNN.1000181

Jiang, W., Kim, B. Y. S., Rutka, J. T., & Chan, W. C. W. (2008a). Nanoparticle-mediated cellular response is size-dependent. *Nature Nanotechnology*, *3*(3), 145–150. https://doi.org/10.1038/nnano.2008.30

John, A. C., Küpper, M., Manders-Groot, A. M. M., Debray, B., Lacome, J.-M., & Kuhlbusch, T. A. J. (2017). Emissions and possible environmental implication of engineered nanomaterials (ENMs) in the atmosphere. *Atmosphere*, *8*(5). https://doi.org/10.3390/atmos8050084

Kaida, T., Kobayashi, K., Adachi, M., & Suzuki, F. (2004). Optical characteristics of titanium oxide interference film and the film laminated with oxides and their applications for cosmetics. *Journal of Cosmetic Science*, *55*(2), 219–220.

Kausar, A., Rafique, I., & Muhammad, B. (2017). Aerospace application of polymer nanocomposite with carbon nanotube, graphite, graphene oxide, and nanoclay. *Polymer-Plastics Technology and Engineering*, *56*(13), 1438–1456. https://doi.org/10.1080/03602559.2016.1276594

Keller, A. A., McFerran, S., Lazareva, A., & Suh, S. (2013). Global life cycle releases of engineered nanomaterials. *Journal of Nanoparticle Research*, *15*(6), 1692. https://doi.org/10.1007/s11051-013-1692-4

Kemp, K. C., Seema, H., Saleh, M., Le, N. H., Mahesh, K., Chandra, V., & Kim, K. S. (2013). Environmental applications using graphene composites: Water remediation and gas adsorption. *Nanoscale*, *5*(8), 3149–3171. https://doi.org/10.1039/C3NR33708A

Kennedy, C. B., Scott, S. D., & Ferris, F. G. (2004). Hydrothermal phase stabilization of 2-line ferrihydrite by bacteria. *Chemical Geology*, *212*(3), 269–277. https://doi.org/10.1016/j.chemgeo.2004.08.017

Khaksar, M., Jolley, D. F., Sekine, R., Vasilev, K., Johannessen, B., Donner, E., & Lombi, E. (2015). In situ chemical transformations of silver nanoparticles along the water–sediment continuum. *Environmental Science & Technology*, *49*(1), 318–325. https://doi.org/10.1021/es504395m

Khan, I., Saeed, K., & Khan, I. (2019). Nanoparticles: Properties, applications and toxicities. *Arabian Journal of Chemistry*, *12*(7), 908–931. https://doi.org/10.1016/j.arabjc.2017.05.011

Khodakovskaya, M., Dervishi, E., Mahmood, M., Xu, Y., Li, Z., Watanabe, F., & Biris, A. S. (2009). Carbon nanotubes are able to penetrate plant seed coat and dramatically affect seed germination and plant growth. *ACS Nano*, *3*(10), 3221–3227. https://doi.org/10.1021/nn900887m

Kim, N. J., Lee, S. J., & Atala, A. (2013). 1- Biomedical Nanomaterials in Tissue Engineering. In A. K. Gaharwar, S. Sant, M. J. Hancock, & S. A. Hacking (Eds.), *Nanomaterials in Tissue Engineering* (pp. 1–25e). Woodhead Publishing. https://doi.org/10.1533/9780857097231.1

Kokarneswaran, M., Selvaraj, P., Ashokan, T., Perumal, S., Sellappan, P., Murugan, K. D., Ramalingam, S., Mohan, N., & Chandrasekaran, V. (2020). Discovery of carbon nanotubes in sixth century BC potteries from Keeladi, India. *Scientific Reports*, *10*(1), 19786. https://doi.org/10.1038/s41598-020-76720-z

Kraas, M., Schlich, K., Knopf, B., Wege, F., Kägi, R., Terytze, K., & Hund-Rinke, K. (2017). Long-term effects of sulfidized silver nanoparticles in sewage sludge on soil microflora. *Environmental Toxicology and Chemistry*, *36*(12), 3305–3313. https://doi.org/10.1002/etc.3904

Kumar, N., Shah, V., & Walker, V. K. (2012). Influence of a nanoparticle mixture on an arctic soil community. *Environmental Toxicology and Chemistry*, *31*(1), 131–135. https://doi.org/10.1002/etc.721

Kumari, M., Khan, S. S., Pakrashi, S., Mukherjee, A., & Chandrasekaran, N. (2011). Cytogenetic and genotoxic effects of zinc oxide nanoparticles on root cells of Allium cepa. *Journal of Hazardous Materials*, *190*(1), 613–621. https://doi.org/10.1016/j.jhazmat.2011.03.095

Kunduru, K. R., Nazarkovsky, M., Farah, S., Pawar, R. P., Basu, A., & Domb, A. J. (2017). 2-Nanotechnology for Water Purification: Applications of Nanotechnology Methods in Wastewater Treatment. In A. M. Grumezescu (Ed.), *Water Purification* (pp. 33–74). Academic Press. https://doi.org/10.1016/B978-0-12-804300-4.00002-2

Larese Filon, F., Mauro, M., Adami, G., Bovenzi, M., & Crosera, M. (2015). Nanoparticles skin absorption: New aspects for a safety profile evaluation. *Regulatory Toxicology and Pharmacology*, *72*(2), 310–322. https://doi.org/10.1016/j.yrtph.2015.05.005

Larue, C., Khodja, H., Herlin-Boime, N., Brisset, F., Flank, A. M., Fayard, B., Chaillou, S., & Carrière, M. (2011). Investigation of titanium dioxide nanoparticles toxicity and uptake by plants. *Journal of Physics: Conference Series*, *304*, 012057. https://doi.org/10.1088/1742-6596/304/1/012057

le Van, N., Rui, Y., Cao, W., Shang, J., Liu, S., Nguyen Quang, T., & Liu, L. (2016). Toxicity and bio-effects of CuO nanoparticles on transgenic Ipt-cotton. *Journal of Plant Interactions*, *11*(1), 108–116. https://doi.org/10.1080/17429145.2016.1217434

Lee, C. W., Mahendra, S., Zodrow, K., Li, D., Tsai, Y.-C., Braam, J., & Alvarez, P. J. J. (2010). Developmental phytotoxicity of metal oxide nanoparticles to Arabidopsis thaliana. *Environmental Toxicology and Chemistry*, *29*(3), 669–675. https://doi.org/10.1002/etc.58

Lee, W.-M., Kwak, J. I., & An, Y.-J. (2012). Effect of silver nanoparticles in crop plants Phaseolus radiatus and Sorghum bicolor: Media effect on phytotoxicity. *Chemosphere*, *86*(5), 491–499. https://doi.org/10.1016/j.chemosphere.2011.10.013

Lehner, R., Weder, C., Petri-Fink, A., & Rothen-Rutishauser, B. (2019). Emergence of nano-plastic in the environment and possible impact on human health. *Environmental Science & Technology*, *53*(4), 1748–1765. https://doi.org/10.1021/acs.est.8b05512

Levard, C., Hotze, E. M., Lowry, G. V., & Brown, G. E. (2012). Environmental transformations of silver nanoparticles: Impact on stability and toxicity. *Environmental Science & Technology*, *46*(13), 6900–6914. https://doi.org/10.1021/es2037405

Li, Y., Yang, J., & Song, J. (2017). Nano energy system model and nanoscale effect of graphene battery in renewable energy electric vehicle. *Renewable and Sustainable Energy Reviews*, *69*, 652–663. https://doi.org/10.1016/j.rser.2016.11.118

Li, Z., Wang, Z., Khan, J., LaGasse, M. K., & Suslick, K. S. (2020). Ultrasensitive monitoring of museum airborne pollutants using a silver nanoparticle sensor array. *ACS Sensors*, *5*(9), 2783–2791. https://doi.org/10.1021/acssensors.0c00583

Lockman, P. R., Koziara, J. M., Mumper, R. J., & Allen, D. D. (2004). Nanoparticle surface charges alter blood–brain barrier integrity and permeability. *Journal of Drug Targeting*, *12*(9–10), 635–641. https://doi.org/10.1080/10611860400015936

Lojk, J., Repas, J., Veranič, P., Bregar, V. B., & Pavlin, M. (2020). Toxicity mechanisms of selected engineered nanoparticles on human neural cells in vitro. *Toxicology*, *432*, 152364. https://doi.org/10.1016/j.tox.2020.152364

Lomer, M. C. E., Thompson, R. P. H., Commisso, J., Keen, C. L., & Powell, J. J. (2000). Determination of titanium dioxide in foods using inductively coupled plasma optical emission spectrometry. *Analyst*, *125*(12), 2339–2343. https://doi.org/10.1039/B006285P

Mahbub, T., & Hoque, M. E. (2020). Chapter 1-Introduction to Nanomaterials and Nanomanufacturing for Nanosensors. In K. Pal & F. Gomes (Eds.), *Nanofabrication for Smart Nanosensor Applications* (pp. 1–20). Elsevier. https://doi.org/10.1016/B978-0-12-820702-4.00001-5

Majumder, S., Sharif, A., & Hoque, M. E. (2020). Chapter 9-Electrospun Cellulose Acetate Nanofiber: Characterization and Applications. In F. M. Al-Oqla & S. M. Sapuan (Eds.), *Advanced Processing, Properties, and Applications of Starch and Other Bio-Based Polymers* (pp. 139–155). Elsevier. https://doi.org/10.1016/B978-0-12-819661-8.00009-3

Marslin, G., Sheeba, C. J., & Franklin, G. (2017). Nanoparticles alter secondary metabolism in plants via ROS burst. *Frontiers in Plant Science, 8*. https://doi.org/10.3389/fpls.2017.00832

Martínez, G., Merinero, M., Pérez-Aranda, M., Pérez-Soriano, E. M., Ortiz, T., Villamor, E., Begines, B., & Alcudia, A. (2021). Environmental impact of nanoparticles' application as an emerging technology: A review. *Materials, 14*(1). https://doi.org/10.3390/ma14010166

Matranga, V., & Corsi, I. (2012). Toxic effects of engineered nanoparticles in the marine environment: Model organisms and molecular approaches. *Marine Environmental Research, 76*, 32–40. https://doi.org/10.1016/j.marenvres.2012.01.006

Medina-Pérez, G., Fernández-Luqueño, F., Vazquez-Nuñez, E., López-Valdez, F., Prieto-Mendez, J., Madariaga-Navarrete, A., & Miranda-Arámbula, M. (2019). Remediating polluted soils using nanotechnologies: Environmental benefits and risks. *Polish Journal of Environmental Studies, 28*(3), 1013–1030. https://doi.org/10.15244/pjoes/87099

Meier, S., Moore, F., Morales, A., González, M.-E., Seguel, A., Meriño-Gergichevich, C., Rubilar, O., Cumming, J., Aponte, H., Alarcón, D., & Mejías, J. (2020). Synthesis of calcium borate nanoparticles and its use as a potential foliar fertilizer in lettuce (Lactuca sativa) and zucchini (Cucurbita pepo). *Plant Physiology and Biochemistry, 151*, 673–680. https://doi.org/10.1016/j.plaphy.2020.04.025

Metreveli, G., Frombold, B., Seitz, F., Grün, A., Philippe, A., Rosenfeldt, R. R., Bundschuh, M., Schulz, R., Manz, W., & Schaumann, G. E. (2016). Impact of chemical composition of ecotoxicological test media on the stability and aggregation status of silver nanoparticles. *Environmental Science: Nano, 3*(2), 418–433. https://doi.org/10.1039/C5EN00152H

Mitrano, D. M., Mehrabi, K., Dasilva, Y. A. R., & Nowack, B. (2017). Mobility of metallic (nano)particles in leachates from landfills containing waste incineration residues. *Environmental Science: Nano, 4*(2), 480–492. https://doi.org/10.1039/C6EN00565A

Monticone, S., Tufeu, R., Kanaev, A. V., Scolan, E., & Sanchez, C. (2000). Quantum size effect in TiO2 nanoparticles: Does it exist? *Applied Surface Science, 162–163*, 565–570. https://doi.org/10.1016/S0169-4332(00)00251-8

Morawska, L., Ristovski, Z., Jayaratne, E. R., Keogh, D. U., & Ling, X. (2008). Ambient nano and ultrafine particles from motor vehicle emissions: Characteristics, ambient processing and implications on human exposure. *Atmospheric Environment, 42*(35), 8113–8138. https://doi.org/10.1016/j.atmosenv.2008.07.050

Mueller, N. C., & Nowack, B. (2008). Exposure modeling of engineered nanoparticles in the environment. *Environmental Science & Technology, 42*(12), 4447–4453. https://doi.org/10.1021/es7029637

Mukhopadhyay, S. S. (2014). Nanotechnology in agriculture: Prospects and constraints. *Nanotechnology, Science and Applications, 7*(2), 63–71. https://doi.org/10.2147/NSA.S39409

Muller, E. B., Lin, S., & Nisbet, R. M. (2015). Quantitative adverse outcome pathway analysis of hatching in zebrafish with CuO nanoparticles. *Environmental Science & Technology, 49*(19), 11817–11824. https://doi.org/10.1021/acs.est.5b01837

Mushtaq, Y. K. (2011). Effect of nanoscale Fe3O4, TiO2 and carbon particles on cucumber seed germination. *Journal of Environmental Science and Health, Part A, 46*(14), 1732–1735. https://doi.org/10.1080/10934529.2011.633403

Nafisi, S., & Maibach, H. I. (2018). Chapter 3-Skin Penetration of Nanoparticles. In R. Shegokar & E. B. Souto (Eds.), *Emerging Nanotechnologies in Immunology* (pp. 47–88). Elsevier. https://doi.org/10.1016/B978-0-323-40016-9.00003-8

Naha, P. C., Casey, A., Tenuta, T., Lynch, I., Dawson, K. A., Byrne, H. J., & Davoren, M. (2009). Preparation, characterization of NIPAM and NIPAM/BAM copolymer nanoparticles and their acute toxicity testing using an aquatic test battery. *Aquatic Toxicology*, *92*(3), 146–154. https://doi.org/10.1016/j.aquatox.2009.02.001

Nho, R. (2020). Pathological effects of nano-sized particles on the respiratory system. *Nanomedicine: Nanotechnology, Biology and Medicine*, *29*, 102242. https://doi.org/10.1016/j.nano.2020.102242

Nigavekar, S. S., Sung, L. Y., Llanes, M., El-Jawahri, A., Lawrence, T. S., Becker, C. W., Balogh, L., & Khan, M. K. (2004). 3H Dendrimer nanoparticle organ/tumor distribution. *Pharmaceutical Research*, *21*(3), 476–483. https://doi.org/10.1023/B:PHAM.0000019302.26097.cc

Nikula, K. J., Snipes, M. B., Barr, E. B., Griffith, W. C., Henderson, R. F., & Mauderly, J. L. (1995). Comparative pulmonary toxicities and carcinogenicities of chronically inhaled diesel exhaust and carbon black in F344 rats. *Fundamental and Applied Toxicology*, *25*(1), 80–94. https://doi.org/10.1006/faat.1995.1042

Nowack, B., & Bucheli, T. D. (2007). Occurrence, behavior and effects of nanoparticles in the environment. *Environmental Pollution*, *150*(1), 5–22. https://doi.org/10.1016/j.envpol.2007.06.006

Oberdörster, E., Zhu, S., Blickley, T. M., McClellan-Green, P., & Haasch, M. L. (2006). Ecotoxicology of carbon-based engineered nanoparticles: Effects of fullerene (C60) on aquatic organisms. *Carbon*, *44*(6), 1112–1120. https://doi.org/10.1016/j.carbon.2005.11.008

O'Brien, N., & Cummins, E. (2010). Nano-scale pollutants: Fate in Irish surface and drinking water regulatory systems. *Human and Ecological Risk Assessment: An International Journal*, *16*(4), 847–872. https://doi.org/10.1080/10807039.2010.501270

Ong, Y. T., Ahmad, A. L., Zein, S. H. S., & Tan, S. H. (2010). A review on carbon nanotubes in an environmental protection and green engineering perspective. *Brazilian Journal of Chemical Engineering*, *27*(2), 227–242. https://doi.org/10.1590/S0104-66322010000200002

Osaki, F., Kanamori, T., Sando, S., Sera, T., & Aoyama, Y. (2004). A quantum dot conjugated sugar ball and its cellular uptake. On the size effects of endocytosis in the subviral region. *Journal of the American Chemical Society*, *126*(21), 6520–6521. https://doi.org/10.1021/ja048792a

Ounoughene, G., le Bihan, O., Chivas-Joly, C., Motzkus, C., Longuet, C., Debray, B., Joubert, A., le Coq, L., & Lopez-Cuesta, J.-M. (2015). Behavior and fate of halloysite nanotubes (HNTs) when incinerating PA6/HNTs nanocomposite. *Environmental Science & Technology*, *49*(9), 5450–5457. https://doi.org/10.1021/es505674j

Padmanabhan, V. P., Sankara Narayanan, T. S. N., Sagadevan, S., Hoque, M. E., & Kulandaivelu, R. (2019). Advanced lithium substituted hydroxyapatite nanoparticles for antimicrobial and hemolytic studies. *New J. Chem.*, *43*(47), 18484–18494. https://doi.org/10.1039/C9NJ03735G

Palmer, B. C., & DeLouise, L. A. (2016). Nanoparticle-enabled transdermal drug delivery systems for enhanced dose control and tissue targeting. *Molecules*, *21*(12). https://doi.org/10.3390/molecules21121719

Pandey, G. (2018). Prospects of nanobioremediation in environmental cleanup. *Oriental Journal of Chemistry*, *34*(6), 2838–2850.

Park, K. H., Chhowalla, M., Iqbal, Z., & Sesti, F. (2003). Single-walled carbon nanotubes are a new class of ion channel blockers*. *Journal of Biological Chemistry*, *278*(50), 50212–50216. https://doi.org/10.1074/jbc.M310216200

Patel, V., & Mahajan, Y. R. (2017). Techno-Commercial Opportunities of Nanotechnology in Wind Energy. In *Nanotechnology for Energy Sustainability* (pp. 1079–1106). John Wiley & Sons, Ltd. https://doi.org/10.1002/9783527696109.ch43

Petersen, E. J., & Henry, T. B. (2012). Methodological considerations for testing the eco-toxicity of carbon nanotubes and fullerenes: Review. *Environmental Toxicology and Chemistry*, *31*(1), 60–72. https://doi.org/10.1002/etc.710

Petersen, E. J., Zhang, L., Mattison, N. T., O'Carroll, D. M., Whelton, A. J., Uddin, N., Nguyen, T., Huang, Q., Henry, T. B., Holbrook, R. D., & Chen, K. L. (2011). Potential release pathways, environmental fate, and ecological risks of carbon nanotubes. *Environmental Science & Technology*, *45*(23), 9837–9856. https://doi.org/10.1021/es201579y

Piccinno, F., Gottschalk, F., Seeger, S., & Nowack, B. (2012). Industrial production quantities and uses of ten engineered nanomaterials in Europe and the world. *Journal of Nanoparticle Research*, *14*(9), 1109. https://doi.org/10.1007/s11051-012-1109-9

Picó, Y., & Andreu, V. (2014). 12-Nanosensors and other techniques for detecting nanoparticles in the environment. In K. C. Honeychurch (Ed.), *Nanosensors for Chemical and Biological Applications* (pp. 295–338). Woodhead Publishing. https://doi.org/10.1533/9780857096722.2.295

Powell, J. J., Faria, N., Thomas-McKay, E., & Pele, L. C. (2010). Origin and fate of dietary nanoparticles and microparticles in the gastrointestinal tract. *Journal of Autoimmunity*, *34*(3), J226–J233. https://doi.org/10.1016/j.jaut.2009.11.006

Pradhan, A., Seena, S., Pascoal, C., & Cássio, F. (2011). Can metal nanoparticles be a threat to microbial decomposers of plant litter in streams? *Microbial Ecology*, *62*(1), 58–68. https://doi.org/10.1007/s00248-011-9861-4

Praetorius, A., Gundlach-Graham, A., Goldberg, E., Fabienke, W., Navratilova, J., Gondikas, A., Kaegi, R., Günther, D., Hofmann, T., & von der Kammer, F. (2017). Single-particle multi-element fingerprinting (spMEF) using inductively-coupled plasma time-of-flight mass spectrometry (ICP-TOFMS) to identify engineered nanoparticles against the elevated natural background in soils. *Environmental Science: Nano*, *4*(2), 307–314. https://doi.org/10.1039/C6EN00455E

Prajitha, N., Athira, S. S., & Mohanan, P. v. (2019). Bio-interactions and risks of engineered nanoparticles. *Environmental Research*, *172*, 98–108. https://doi.org/10.1016/j.envres.2019.02.003

Priester, J. H., Moritz, S. C., Espinosa, K., Ge, Y., Wang, Y., Nisbet, R. M., Schimel, J. P., Susana Goggi, A., Gardea-Torresdey, J. L., & Holden, P. A. (2017). Damage assessment for soybean cultivated in soil with either CeO2 or ZnO manufactured nanomaterials. *Science of The Total Environment*, *579*, 1756–1768. https://doi.org/10.1016/j.scitotenv.2016.11.149

Prow, T. W., Monteiro-Riviere, N. A., Inman, A. O., Grice, J. E., Chen, X., Zhao, X., Sanchez, W. H., Gierden, A., Kendall, M. A. F., Zvyagin, A. v, Erdmann, D., Riviere, J. E., & Roberts, M. S. (2012). Quantum dot penetration into viable human skin. *Nanotoxicology*, *6*(2), 173–185. https://doi.org/10.3109/17435390.2011.569092

Pujalté, I., Dieme, D., Haddad, S., Serventi, A. M., & Bouchard, M. (2017). Toxicokinetics of titanium dioxide (TiO2) nanoparticles after inhalation in rats. *Toxicology Letters*, *265*, 77–85. https://doi.org/10.1016/j.toxlet.2016.11.014

Rabbani, M., Hoque, M. E., & Mahbub, Z. Bin. (2020). Chapter 7-Nanosensors in Biomedical and Environmental Applications: Perspectives and Prospects. In K. Pal & F. Gomes (Eds.), *Nanofabrication for Smart Nanosensor Applications* (pp. 163–186). Elsevier. https://doi.org/10.1016/B978-0-12-820702-4.00007-6

Radomski, A., Jurasz, P., Alonso-Escolano, D., Drews, M., Morandi, M., Malinski, T., & Radomski, M. W. (2005). Nanoparticle-induced platelet aggregation and vascular thrombosis. *British Journal of Pharmacology*, *146*(6), 882–893. https://doi.org/10.1038/sj.bjp.0706386

Ravichandran, M., & Velumani, S. (2020). Manganese ferrite nanocubes as an MRI contrast agent. *Materials Research Express*, *7*(1), 016107. https://doi.org/10.1088/2053-1591/ab66a4

Rocha, T. L., Gomes, T., Sousa, V. S., Mestre, N. C., & Bebianno, M. J. (2015). Ecotoxicological impact of engineered nanomaterials in bivalve molluscs: An overview. *Marine Environmental Research*, *111*, 74–88. https://doi.org/10.1016/j.marenvres.2015.06.013

Rönkkö, T., Pirjola, L., Ntziachristos, L., Heikkilä, J., Karjalainen, P., Hillamo, R., & Keskinen, J. (2014). Vehicle engines produce exhaust nanoparticles even when not fueled. *Environmental Science & Technology*, *48*(3), 2043–2050. https://doi.org/10.1021/es405687m

Rui, M., Ma, C., Tang, X., Yang, J., Jiang, F., Pan, Y., Xiang, Z., Hao, Y., Rui, Y., Cao, W., & Xing, B. (2017). Phytotoxicity of silver nanoparticles to peanut (Arachis hypogaea L.): Physiological responses and food safety. *ACS Sustainable Chemistry & Engineering*, *5*(8), 6557–6567. https://doi.org/10.1021/acssuschemeng.7b00736

Rui, M., Ma, C., White, J. C., Hao, Y., Wang, Y., Tang, X., Yang, J., Jiang, F., Ali, A., Rui, Y., Cao, W., Chen, G., & Xing, B. (2018). Metal oxide nanoparticles alter peanut (Arachis hypogaea L.) physiological response and reduce nutritional quality: A life cycle study. *Environ. Sci.: Nano*, *5*(9), 2088–2102. https://doi.org/10.1039/C8EN00436F

Saddick, S., Afifi, M., & Abu Zinada, O. A. (2017). Effect of zinc nanoparticles on oxidative stress-related genes and antioxidant enzymes activity in the brain of Oreochromis niloticus and Tilapia zillii. *Saudi Journal of Biological Sciences*, *24*(7), 1672–1678. https://doi.org/10.1016/j.sjbs.2015.10.021

Saif, S., Tahir, A., & Chen, Y. (2016). Green synthesis of iron nanoparticles and their environmental applications and implications. *Nanomaterials*, *6*(11). https://doi.org/10.3390/nano6110209

Sajid, M., Ilyas, M., Basheer, C., Tariq, M., Daud, M., Baig, N., & Shehzad, F. (2015). Impact of nanoparticles on human and environment: review of toxicity factors, exposures, control strategies, and future prospects. *Environmental Science and Pollution Research*, *22*(6), 4122–4143. https://doi.org/10.1007/s11356-014-3994-1

Samiei, F., Shirazi, F. H., Naserzadeh, P., Dousti, F., Seydi, E., & Pourahmad, J. (2020). Correction to: Toxicity of multi-wall carbon nanotubes inhalation on the brain of rats. *Environmental Science and Pollution Research*, *27*(23), 29699. https://doi.org/10.1007/s11356-020-09667-3

Sanderson, K. (2006). Sharpest cut from nanotube sword. *Nature*. https://doi.org/10.1038/news061113-11

Sarno, M. (2020). Chapter 22-Nanotechnology in Energy Storage: The Supercapacitors. In A. Basile, G. Centi, M. de Falco, & G. Iaquaniello (Eds.), *Studies in Surface Science and Catalysis* (Vol. 179, pp. 431–458). Elsevier. https://doi.org/10.1016/B978-0-444-64337-7.00022-7

Schaumann, G. E., Philippe, A., Bundschuh, M., Metreveli, G., Klitzke, S., Rakcheev, D., Grün, A., Kumahor, S. K., Kühn, M., Baumann, T., Lang, F., Manz, W., Schulz, R., & Vogel, H.-J. (2015). Understanding the fate and biological effects of Ag- and TiO2-nanoparticles in the environment: The quest for advanced analytics and interdisciplinary concepts. *Science of the Total Environment*, *535*, 3–19. https://doi.org/10.1016/j.scitotenv.2014.10.035

Schins, R. P. F., Duffin, R., Höhr, D., Knaapen, A. M., Shi, T., Weishaupt, C., Stone, V., Donaldson, K., & Borm, P. J. A. (2002). Surface modification of quartz inhibits toxicity, particle uptake, and oxidative DNA damage in human lung epithelial cells. *Chemical Research in Toxicology*, *15*(9), 1166–1173. https://doi.org/10.1021/tx025558u

Schraufnagel, D. E. (2020). The health effects of ultrafine particles. *Experimental & Molecular Medicine*, *52*(3), 311–317. https://doi.org/10.1038/s12276-020-0403-3

Sekine, R., Khaksar, M., Brunetti, G., Donner, E., Scheckel, K. G., Lombi, E., & Vasilev, K. (2013). Surface immobilization of engineered nanomaterials for in situ study of their environmental transformations and fate. *Environmental Science & Technology*, *47*(16), 9308–9316. https://doi.org/10.1021/es400839h

Singh, R., Cheng, S., & Singh, S. (2020). Oxidative stress-mediated genotoxic effect of zinc oxide nanoparticles on Deinococcus radiodurans. *3 Biotech*, *10*(2), 66. https://doi.org/10.1007/s13205-020-2054-4

Singh, R. P. (2011). Prospects of nanobiomaterials for biosensing. *International Journal of Electrochemistry*, *2011*, 125487. https://doi.org/10.4061/2011/125487

Sivry, Y., Gelabert, A., Cordier, L., Ferrari, R., Lazar, H., Juillot, F., Menguy, N., & Benedetti, M. F. (2014). Behavior and fate of industrial zinc oxide nanoparticles in a carbonate-rich river water. *Chemosphere*, *95*, 519–526. https://doi.org/10.1016/j.chemosphere.2013.09.110

Song, Y., Zhou, T., Liu, Q., Liu, Z., & Li, D. (2020). Nanoparticle and microorganism detection with a side-micron-orifice-based resistive pulse sensor. *Analyst*, *145*(16), 5466–5474. https://doi.org/10.1039/D0AN00679C

Stampoulis, D., Sinha, S. K., & White, J. C. (2009). Assay-dependent phytotoxicity of nanoparticles to plants. *Environmental Science & Technology*, *43*(24), 9473–9479. https://doi.org/10.1021/es901695c

Su, M., Liu, H., Liu, C., Qu, C., Zheng, L., & Hong, F. (2009). Promotion of nano-anatase TiO2 on the spectral responses and photochemical activities of D1/D2/Cyt b559 complex of spinach. *Spectrochimica Acta Part A: Molecular and Biomolecular Spectroscopy*, *72*(5), 1112–1116. https://doi.org/10.1016/j.saa.2009.01.010

Sun, T. Y., Gottschalk, F., Hungerbühler, K., & Nowack, B. (2014). Comprehensive probabilistic modelling of environmental emissions of engineered nanomaterials. *Environmental Pollution*, *185*, 69–76. https://doi.org/10.1016/j.envpol.2013.10.004

Suri, S. S., Fenniri, H., & Singh, B. (2007). Nanotechnology-based drug delivery systems. *Journal of Occupational Medicine and Toxicology*, *2*(1), 16. https://doi.org/10.1186/1745-6673-2-16

Taghavi, S. M., Momenpour, M., Azarian, M., Ahmadian, M., Souri, F., Taghavi, S. A., Sadeghain, M., & Karchani, M. (2013). Effects of nanoparticles on the environment and outdoor workplaces. *Electronic Physician*, *5*(4), 706–712. https://doi.org/10.14661/2013.706-712

Taylor, A. F., Rylott, E. L., Anderson, C. W. N., & Bruce, N. C. (2014). Investigating the toxicity, uptake, nanoparticle formation and genetic response of plants to gold. *PLOS ONE*, *9*(4), e93793-. https://doi.org/10.1371/journal.pone.0093793

Thakur, M., Gupta, H., Singh, D., Mohanty, I. R., Maheswari, U., Vanage, G., & Joshi, D. S. (2014). Histopathological and ultra structural effects of nanoparticles on rat testis following 90 days (Chronic study) of repeated oral administration. *Journal of Nanobiotechnology*, *12*(1), 42. https://doi.org/10.1186/s12951-014-0042-8

Thalmann, B., Voegelin, A., Morgenroth, E., & Kaegi, R. (2016). Effect of humic acid on the kinetics of silver nanoparticle sulfidation. *Environmental Science: Nano*, *3*(1), 203–212. https://doi.org/10.1039/C5EN00209E

Thanigaivel, S., Vickram, A. S., Anbarasu, K., Gulothungan, G., Nanmaran, R., Vignesh, D., Rohini, K., & Ravichandran, V. (2021). Ecotoxicological assessment and dermal layer interactions of nanoparticle and its routes of penetrations. *Saudi Journal of Biological Sciences*, *28*(9), 5168–5174. https://doi.org/10.1016/j.sjbs.2021.05.048

Tong, M., Ding, J., Shen, Y., & Zhu, P. (2010). Influence of biofilm on the transport of fullerene (C60) nanoparticles in porous media. *Water Research*, *44*(4), 1094–1103. https://doi.org/10.1016/j.watres.2009.09.040

Tripathi, D. K., Shweta, Singh, S., Singh, S., Pandey, R., Singh, V. P., Sharma, N. C., Prasad, S. M., Dubey, N. K., & Chauhan, D. K. (2017). An overview on manufactured nanoparticles in plants: Uptake, translocation, accumulation and phytotoxicity. *Plant Physiology and Biochemistry*, *110*, 2–12. https://doi.org/10.1016/j.plaphy.2016.07.030

Tripathi, D. K., Singh, S., Singh, S., Srivastava, P. K., Singh, V. P., Singh, S., Prasad, S. M., Singh, P. K., Dubey, N. K., Pandey, A. C., & Chauhan, D. K. (2017). Nitric oxide alleviates silver nanoparticles (AgNps)-induced phytotoxicity in Pisum sativum seedlings. *Plant Physiology and Biochemistry*, *110*, 167–177. https://doi.org/10.1016/j.plaphy.2016.06.015

Turner, A., Brice, D., & Brown, M. T. (2012). Interactions of silver nanoparticles with the marine macroalga, Ulva lactuca. *Ecotoxicology, 21*(1), 148–154. https://doi.org/10.1007/s10646-011-0774-2

Unrine, J. M., Tsyusko, O. V., Hunyadi, S. E., Judy, J. D., & Bertsch, P. M. (2010). Effects of particle size on chemical speciation and bioavailability of copper to earthworms (Eisenia fetida) exposed to copper nanoparticles. *Journal of Environmental Quality, 39*(6), 1942–1953. https://doi.org/10.2134/jeq2009.0387

U.S. Environmental Protection Agency. (2007). *Nanotechnology White Paper.* www.epa.gov/osa

U.S. Environmental Protection Agency | US EPA. (n.d.). Retrieved May 20, 2022, from www.epa.gov/

U.S. Food and Drug Administration. (n.d.). Retrieved May 20, 2022, from www.fda.gov/

Vannini, C., Domingo, G., Onelli, E., de Mattia, F., Bruni, I., Marsoni, M., & Bracale, M. (2014). Phytotoxic and genotoxic effects of silver nanoparticles exposure on germinating wheat seedlings. *Journal of Plant Physiology, 171*(13), 1142–1148. https://doi.org/10.1016/j.jplph.2014.05.002

Wang, H., Adeleye, A. S., Huang, Y., Li, F., & Keller, A. A. (2015). Heteroaggregation of nanoparticles with biocolloids and geocolloids. *Advances in Colloid and Interface Science, 226*, 24–36. https://doi.org/10.1016/j.cis.2015.07.002

Wang, X., Yang, X., Chen, S., Li, Q., Wang, W., Hou, C., Gao, X., Wang, L., & Wang, S. (2016). Corrigendum: Zinc oxide nanoparticles affect biomass accumulation and photosynthesis in arabidopsis. *Frontiers in Plant Science, 7*. www.frontiersin.org/article/10.3389/fpls.2016.00559

Wang, Y., & Xia, Y. (2019). Optical, electrochemical and catalytic methods for in-vitro diagnosis using carbonaceous nanoparticles: A review. *Microchimica Acta, 186*(1), 50. https://doi.org/10.1007/s00604-018-3110-1

Watkinson, A. C., Bunge, A. L., Hadgraft, J., & Lane, M. E. (2013). Nanoparticles do not penetrate human skin—A theoretical perspective. *Pharmaceutical Research, 30*(8), 1943–1946. https://doi.org/10.1007/s11095-013-1073-9

Weir, A., Westerhoff, P., Fabricius, L., Hristovski, K., & von Goetz, N. (2012). Titanium dioxide nanoparticles in food and personal care products. *Environmental Science & Technology, 46*(4), 2242–2250. https://doi.org/10.1021/es204168d

Welcome to The Nanodatabase. (n.d.). Retrieved May 20, 2022, from https://nanodb.dk/

Wong, S. W. Y., Leung, P. T. Y., Djurišić, A. B., & Leung, K. M. Y. (2010). Toxicities of nano zinc oxide to five marine organisms: Influences of aggregate size and ion solubility. *Analytical and Bioanalytical Chemistry, 396*(2), 609–618. https://doi.org/10.1007/s00216-009-3249-z

Xia, B., Chen, B., Sun, X., Qu, K., Ma, F., & Du, M. (2015). Interaction of TiO2 nanoparticles with the marine microalga Nitzschia closterium: Growth inhibition, oxidative stress and internalization. *Science of The Total Environment, 508*, 525–533. https://doi.org/10.1016/j.scitotenv.2014.11.066

Xiang, Q.-Q., Wang, D., Zhang, J.-L., Ding, C.-Z., Luo, X., Tao, J., Ling, J., Shea, D., & Chen, L.-Q. (2020). Effect of silver nanoparticles on gill membranes of common carp: Modification of fatty acid profile, lipid peroxidation and membrane fluidity. *Environmental Pollution, 256*, 113504. https://doi.org/10.1016/j.envpol.2019.113504

Yah, C. S., Iyuke, S. E., & Simate, G. S. (2012). A review of nanoparticles toxicity and their routes of exposures. *Iranian Journal of Pharmaceutical Sciences, 8*(1), 299–314. www.ijps.ir/article_2193.html

Yan, W., Riahi, H., Benzarti, K., Chlela, R., Curtil, L., & Bigaud, D. (2021). Durability and reliability estimation of flax fiber reinforced composites using tweedie exponential dispersion degradation process. *Mathematical Problems in Engineering, 2021*, 6629637. https://doi.org/10.1155/2021/6629637

Yang, J., Jiang, F., Ma, C., Rui, Y., Rui, M., Adeel, M., Cao, W., & Xing, B. (2018). Alteration of crop yield and quality of wheat upon exposure to silver nanoparticles in a life cycle study. *Journal of Agricultural and Food Chemistry*, *66*(11), 2589–2597. https://doi.org/10.1021/acs.jafc.7b04904

Yhee, J. Y., Im, J., & Nho, R. S. (2016). Advanced therapeutic strategies for chronic lung disease using nanoparticle-based drug delivery. *Journal of Clinical Medicine*, *5*(9). https://doi.org/10.3390/jcm5090082

Yuan, G., P, S. J., & A, H. P. (2012). Identification of soil bacteria susceptible to TiO_2 and ZnO nanoparticles. *Applied and Environmental Microbiology*, *78*(18), 6749–6758. https://doi.org/10.1128/AEM.00941-12

Zheng, L., Tracy, H., Rebecca, S., Rebecca, C., S, L. S., Shih-Houng, Y., Anna, S., I, L. M., & P, S. P. (2007). Cardiovascular effects of pulmonary exposure to single-wall carbon nanotubes. *Environmental Health Perspectives*, *115*(3), 377–382. https://doi.org/10.1289/ehp.9688

Zhu, C., Zhao, Q., Meng, G., Wang, X., Hu, X., Han, F., & Lei, Y. (2020). Silver nanoparticle-assembled micro-bowl arrays for sensitive SERS detection of pesticide residue. *Nanotechnology*, *31*(20), 205303. https://doi.org/10.1088/1361-6528/ab7100

8 Natural Nanofillers Family, Their Properties, and Applications in Polymers

Aswathy Ajayan and Maida Mary Jude
Sree Sankara College

CONTENTS

8.1 Introduction 233
8.2 Polysaccharide Nanomaterials 235
    8.2.1 Nanocellulose (NC) 235
        8.2.1.1 Preparation of NC 237
        8.2.1.2 Applications of NC 239
    8.2.2 Nanostarch 240
        8.2.2.1 Preparation of Nanostarch 241
        8.2.2.2 Application of Nanostarch 242
    8.2.3 Nanochitin 242
        8.2.3.1 Preparation of Nanochitin 243
        8.2.3.2 Applications of Nanochitin 245
    8.2.4 Nanochitosan 245
        8.2.4.1 Preparation of Nanochitosan 246
        8.2.4.2 Applications of Nanochitosan 247
    8.2.5 Alginate Nanoparticles 247
8.3 Conclusion 249
    8.3.1 Nanomaterials: Future 249
References 250

8.1 INTRODUCTION

The development of materials from renewable, sustainable resources is of great public interest with increasing ecological and environmental concerns nowadays. Nanocomposites have been one of the most promising and developing study fields in recent years. Polymer nanocomposites have several advantages: (i) They are lighter than conventional composites because high degrees of stiffness and strength can be achieved with far less high-density material, (ii) their barrier properties are improved over the neat polymer, (iii) their mechanical and thermal properties are potentially

DOI: 10.1201/9781003279372-8

superior, and (iv) they have excellent flammability properties and increased biodegradability (Dantas de Oliveira & Augusto Gonçalves Beatrice, 2019). In situ polymerisation, solution, and melt mixing are three ways for making polymer nanocomposites. According to the kind of polymeric matrix, nanofiller, and desired qualities for the final products, a suitable procedure is chosen (Mallakpour & Naghdi, 2018).

A filler is a relatively innocuous ingredient added to a plastic to adjust its strength, permanence, working characteristics, or other features, or to reduce costs, according to the definitions proposed by the American Society for Testing and Materials (ASTM-D-883). The type of filler used in nanocomposite formulation is one of the most essential characteristics that define its qualities (Dos Ouros et al., 2014). Before selecting filler that will be included into a polymer matrix to achieve a specific composite property, many factors must be taken into account. Chemical surface composition, nanoparticle size and shape, structure, pore diameters, interlayer distances, hydrophobicity, and mechanical, electrical, and thermal properties are only a few of the filler's critical qualities to consider. Fillers are categorised as either inorganic or organic substances, with further subcategories based on chemical family or form and size (aspect ratio). More than 70 types of particles or flakes, as well as more than 15 types of natural and synthetic fibres, have been employed or assessed as fillers in thermoplastics and thermosets (Wypych, 2016). Nanofillers are fillers with particle sizes ranging from 1 to 100 nm. As nanofillers have a large surface area, they provide a large interphase or border area between the biopolymer matrix and the nanofiller, which is advantageous for nanocomposite materials. The introduction of nanofillers found to modify the aesthetics of the final product. Nanofillers have strong reinforcing effects; they not only reduce the cost of composites, but also impart performance improvements. Even at low filler content, nanocomposites show versatile properties compared to conventional composites due to the nanometric size. Depending on application, nanofillers with suitable physical and dimensional properties can be selected. Nanofillers for polymer composite applications are divided into three categories based on their dimensions: one nanoscale dimension nanoplatelets, two nanoscale dimension nanofibres, and three nanoscale dimension nanoparticulates (Dantas de Oliveira & Augusto Gonçalves Beatrice, 2019). New or improved properties of nanocomposites can be obtained based on the properties of nanofillers, such as barrier qualities, better mechanical strength, antibacterial and antioxidant properties, or thermal stability (Jamróz et al., 2019). The most difficult aspect of working with nanoparticles is achieving homogeneous dispersion within a polymeric matrix.

Nanofillers can be classified according to their origin as natural and synthetic. Natural nanofillers include that originated from animals, plants as well as from minerals (Figure 8.1). Synthetic nanofillers are man-made and can be organic or inorganic. Clays, organic, inorganic, and carbon nanostructure are the four categories of nanofillers based on composition.

Biomass as a source of renewable energy is gaining popularity. Agricultural and forestry wastes, animal residues, sewage, algae, and aquatic crops are the most major biomass sources. Polysaccharides are large, complex macromolecules made up of hexoses and pentoses, with non-sugar components such as short organic acids or sulphates. They are primarily extracted from biomass such as terrestrial plants (starch, cellulose, galactomannan, and pectin), macroalgae (sulphated galactans, alginates, laminarins, or ulvans), animals (chitin, chitosan (CS), and glycosaminoglycans), and bacteria (xanthan, dextran, gellan,

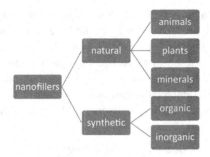

FIGURE 8.1 Classification of nanofillers.

or curdlan) (*AppliedChem|Special Issue : Polysaccharides from Biomass Conversion*, n.d.). Polysaccharides are superior candidates as renewable nanofillers due to the partly crystalline structures conferring interesting properties. Most fillers are considered as additives because of their characteristic features like surface area, geometrical features, and chemical composition. Cellulose and starch are two abundantly occurring polysaccharides in nature. Cellulose is the most ample renewable polymer on our planet and key component of plant cell wall and is a viable biomass source. Nanocellulose (NC) is a cellulose-derived natural fibre. This substance has been labelled a "novel bionanomaterial". Chemical treatments such as alkali extraction and bleaching are used to extract cellulose from plant fibres. Starch is the second most excessive biomass material found in nature. Over long periods, starch is been produced by green plants and algae for energy storage. Owing to their attractive and excellent characteristics such as abundance, high aspect ratio, better mechanical properties, renewability, and biocompatibility, these materials have gained growing interests over the past decades. Chitin (CH, poly-(1–4)-*N*-acetyl-D-glucosamine), obtained mostly from seafood wastes such as shrimp, crab, lobster, and squid pen, is a renewable and plentiful polysaccharide that will soon become one of the most important organic raw resources (Li et al., 2018). Chitin has a micro- and nanofibril structure that is highly structured and incorporates both crystalline and amorphous domains. CS is a linear polysaccharide with inherent antibacterial activity derived from naturally occurring chitin, the most prevalent biopolymer on the planet after cellulose and starch. Pullulan is a neutral polysaccharide derived from the fermentation medium of Aureobasidium pullulans, a fungus-like yeast. It has excellent optical properties, and the use of this non-toxic, biodegradable polymer in innovative optical materials could be promising.

8.2   POLYSACCHARIDE NANOMATERIALS

8.2.1   Nanocellulose (NC)

For thousands of years, cellulose has been used as an engineering material. Furthermore, cellulose found in most lignocellulosic biomass, that is derived from agricultural waste, energy crops, and forestry residues, provides a better option over food crops as a renewable resource for the production of new materials due to its availability in enormous amounts and low cost (Loow et al., 2017). Cellulose is the most abundant component in lignocellulosic biomass, accounting for around 35%–50% of the whole (Figure 8.2). It is made up of a linear homopolysaccharide of 1,4-linked anhydro-d-glucose units as well as a cellobiose repeating unit. A monomer of cellobiose is made up of three hydroxyl groups that form strong hydrogen bonds with neighbouring glucose units in the same chain and with distinct chains, respectively, forming intramolecular and intermolecular hydrogen bonding networks (Figure 8.3). The hydrogen bonding networks in the crystalline sections of cellulose

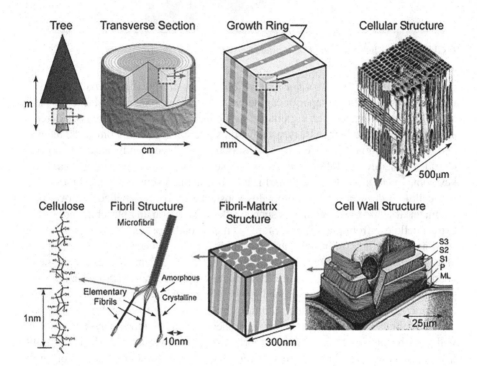

FIGURE 8.2 Wood hierarchical structure: from tree to cellulose (Li et al., 2018).

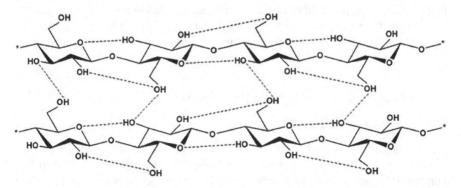

FIGURE 8.3 Intramolecular (---) and intermolecular (---) hydrogen bonding networks in cellulose structure (Phanthong et al., 2018).

fibrils are strongly packed, resulting in a plant cell wall that is stiff, fibrous, insoluble in water, and highly resistant to most organic solvents.

NC has been proven to be one of the most environmentally benign materials in recent history. The word "nanocellulose" refers to a group of compounds made from cellulose that have at least one nanometre dimension. A new bionanomaterial (Dufresne, 2013) has been described for this material. Nanoparticles with one or two dimensions can be extracted from any naturally existing supply of cellulose using appropriate chemical and mechanical treatments. NC is commonly derived by the disintegration of naturally occurring polymers or produced by bacterial action. NC-based materials have a low-carbon footprint and are sustainable, renewable, recyclable, and non-toxic, making them truly green nanomaterials with a wide range of useful and surprising features.

Along with its nanometre size diameter, NC is comprised of different properties such as high surface area, magnificent mechanical performance, low density, and excellent stiffness, which make it attractive for various purposes (Kargarzadeh et al., 2017) with high stiffness up to 220 GPa of elastic modulus which is greater than Kevlar fibre. NC or cellulose nanomaterials can be categorised based on its size and structure into two major groups: nano-objects (10–100 m) and nanostructures (1–50 nm) (Kargarzadeh et al., 2017; Loow et al., 2017).

8.2.1.1   Preparation of NC

NC (Figure 8.4) has been isolated from cellulosic materials using a variety of processes of which the three primary extraction processes are acid hydrolysis, enzymatic hydrolysis, and mechanical process. The most common method for extracting NC from cellulosic materials is acid hydrolysis. Ball milling is another novel and vital mechanical approach for extracting NC from biomass efficiently (Phanthong et al., 2018). Varied extraction procedures resulted in different types and qualities of NC produced (Peng et al., 2011).

As mention above, one of the most common ways to extract NC from cellulosic materials is through acid hydrolysis. Since cellulose chains contain both ordered and disordered sections, the disordered regions can be easily dissolved by acid, leaving just the orderly parts. For acid hydrolysis, sulphuric acid is the most commonly utilised acid (Dong et al., 1998). Due to the esterification of the hydroxyl group by sulphate ions (P. Lu & Hsieh, 2010), it can not only firmly isolate nanocrystalline cellulose, but also disseminate it as a stable colloid system. Reaction time, temperature, and acid content are the primary regulating elements that influence the properties of produced NC. The acid wastewater generated during the washing process to neutralise the pH value (Wang et al., 2007) of the NC suspension is the main disadvantage of

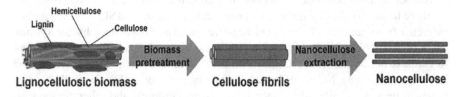

FIGURE 8.4   Schematic of NC extraction from lignocellulosic biomass (Phanthong et al., 2018).

acid hydrolysis. Adding cold water and centrifuging until a neutral pH is obtained is typically how the washing process is carried out. Another approach for washing the collected items is to neutralise the pH value with an alkaline such as sodium hydroxide. Maiti et al. (2013), for example, used 47% sulphuric acid to extract NC from three different types of biomass. After the reaction was completed, the acid was removed by washing with deionised water and centrifugation, and then neutralising the suspension with 0.5 N sodium hydroxide, followed by washing with distilled water. To oxidise the hydroxyl group of cellulose to carboxylates, the 2,2,6,6-tetramethylpiperidine-1-oxyl radical (TEMPO) can be utilised as a catalyst with hypochlorite as the primary oxidant. As a result, the nanofibrillated cellulose is created with a diameter of 3–4 nm and a length of a few microns. TEMPO-oxidised cellulose nanofibres have a uniform width (3–4 nm) and a high aspect ratio, making them suitable for applications such as transparent and flexible displays, gas barrier films for packaging, and nanofibre infill for composite materials (Fukuzumi et al., 2008).

When cellulosic fibres are subjected to several mechanical shearing events, more or fewer individual microfibrils are released. Microfibrillated cellulose is the common name for this material (Dufresne, 2013). Because of their more homogeneous nature, the mechanical characteristics of cellulose microfibrils should be greater and less scattered than those of the lignocellulosic fibres from which they are liberated. To prepare MFC, various mechanical treatment processes have been documented. High-pressure homogenisation and/or grinding are the most common methods (*Nanocellulose: From Nature to High Performance Tailored Materials—Alain Dufresne—Google Books*, n.d.). Pre-treatments such as mechanical cutting (*Microfibrillated Cellulose: Morphology and Accessibility (Conference) | OSTI.GOV*, n.d.), acid hydrolysis (Boldizar et al., 2006), enzymatic pre-treatment (Pääkko et al., 2007; Henriksson et al., 2007), and the introduction of charged groups by carboxymethylation or TEMPO-mediated oxidation (Pääkko et al., 2007) have all been proposed to aid this process. The shape, degree of crystallinity, particle size, and other features of several forms of NC vary. Because of the abundance of interacting surface hydroxyl groups, cellulose nanoparticles have a great tendency for self-association. Various functional nanomaterials with remarkable features, or significantly improved physical, chemical, biological, and electrical properties, can be created by appropriately modifying NCC. The qualities of NCC are affected by a number of parameters, including the type of cellulose utilised, the reaction time and temperature, and the acid employed for hydrolysis (Vaishnav, 2016). Because of the rising potential uses of modified NCC in many industrial sectors, such as personal care, nanocomposites, biomedical, and so on, there is a growing research focus on NCC modification.

Nanocrystalline cellulose, nanofibrillated cellulose, and bacterial NC (BNC) are the three forms of cellulose utilised as a nanoscale component. BNC, which is mainly extracted from cultures of the Gram-negative bacteria, Gluconacetobacter xylinus, has a higher molecular weight and crystallinity than cellulose from plant sources. Regardless of the fact that NC has a number of unique features and is abundant in nature, extracting NC from lignocellulosic biomass or cellulosic products continues to be a major difficulty. Since lignin, hemicelluloses, and other components in plant cell walls clump together, biomass pre-treatment is required to remove all non-cellulosic compounds which always involves multiple procedures, takes a long time, uses harmful chemicals, and produces trash. Furthermore, general methods for extracting NC from cellulosic materials have limitations such as high acid wastewater creation during acid hydrolysis, high energy consumption during the mechanical process, and a long reaction time for enzymatic hydrolysis. The amount of research into cellulose-based materials has expanded significantly; yet, cellulose has some functional constraints. In this regard, NC nanocrystalline cellulose (Echeverria et al., 2015), microfibrillar/nanofibrillar cellulose, and bacterial cellulose (Wiegand et al., 2015; Li et al., 2018) are gaining popularity. Despite the fact that cellulose is the most abundant natural polymer on the planet, it has only lately acquired attention as a nanostructured material in the form of NC.

8.2.1.2 Applications of NC

Nanocrystalline cellulose, nanofibrillated cellulose, and BNC are the three kinds of cellulose utilised as a nanoscale component to biopolymer films (Phanthong et al., 2018). NCC is mostly used to strengthen polymeric matrices in nanocomposite composites. The use of NCC as reinforcing fillers in poly(styrene-co-butyl acrylate) (poly(S-co-BuA)-based nanocomposites was reported in 1995 (Favier et al., 1995).

Many exhilarating applications of NC have been explored, particularly in the fabrication of polymer nanocomposites. NC has several uses in healthcare sector, petroleum and gas, packaging, paper and board, composites, printed and flexible electronics, textiles, filtration, rheology modifiers, 3D printing, aerogels, coating films, etc. Agricultural wastes are attractive sources for NC production in recent years. The research regarding NC is not only about its extraction from biomass, but also the new applications in numerous fields (Phanthong et al., 2018). Materials made from NC are carbon-neutral, long-lasting, recyclable, and non-toxic. As a result, they have the potential to be truly green nanomaterials with a wide range of beneficial and surprising features. Heavy metal removal, dye removal, carbon sequestration, and antibacterial activity have mostly been proven in NC-based nanocomposites, making them ideal for environmental remediation. Food packaging (Li et al., 2013), cellular orientation (Mallkpour & Naghdi, 2018), and electronics have all been proposed as applications for CNC films. Ultrathin, aligned CNC films have recently been proven to have a significant piezoelectric response and so have potential in flexible electronic devices (Csoka et al., 2012). Because of its good biocompatibility and low toxicity, as well as its unique geometry, surface chemistry, rheology, crystallinity, and self-assembly behaviour, NC is a promising biomaterial for medical applications. NC-based materials for drug delivery, like membranes, tablet coatings, and composite-biopolymer delivery systems, are another hot topic. These materials can be pre-loaded with the drug of choice and have a regulated release profile. Furthermore, surface modification of CNCs has been employed to construct novel carriers (Dash & Ragauskas, 2012). One especially versatile alteration of CNCs uses an aromatic linker to enable both binding and selective release of amine-containing medicines. NC fluorescence labelling with a variety of fluorophores is gaining popularity in bioimaging, targeting, and sensing applications (Dong & Roman, 2007). Since the porosity network provides a physical barrier against external infections while enabling release of pre-loaded antimicrobial drugs, NC BC wound dressings can also help with antimicrobial treatment (Abitbol et al., 2016). NC-based materials are carbon-neutral, sustainable, recyclable, and non-toxic. As a result, they have the potential to be truly green nanomaterials with a wide range of beneficial and surprising features.

8.2.2 NANOSTARCH

The versatile nature of starch is a promising advantage for manufacturing environmentally friendly materials. The nutritious content, easy storage and transportation, adaptability to diverse growing conditions, etc., made starch a staple ingredient of the human diet for a long time. Starch is a renewable natural resource, extracted from different parts of the plants, like leaves, seeds, fruits, tubers, and roots. A variety of vegetables like potato, peas, maize, tapioca, corn, cassava, avocado, etc., have enormous source of starch. Starch can either be used as "native starch", or it undergoes chemical modifications to obtain specific properties, and is called "modified starch".

As a biodegradable thermoplastic polymer and as biodegradable particulate filler, starch has received enormous attention of researchers during the past decades. The superiority of starch fillers are their availability, biodegradability, renewability, high specific strength, non-abrasive character which lets easier processing even at high filling levels, and a relatively reactive surface that can be modified by adding reactive groups compatible with the polymeric matrix (Lamanna et al., 2013).

Starch is a biopolymer composed of linear amylase with α-(1–4)-linked d-glucose units and branched amylopectin with α-(1–4)-d-glucose backbones, and about 5% α-(1–6)-d-glucose units co-exist. Starches obtained from different botanical origins vary in their morphology and functional properties. The content, morphology, and functional properties vary depending on the botanical source. Molecular weight of the components determines the functionality and applicability. The properties of starch are determined by the amylase/amylopectin ratio as well as the non-starch contents like lipids, proteins, and phosphates. Starch granules (Figure 8.5) exist in different shapes and sizes as spheres, polygon, ellipsoids, platelets, irregular tubules,

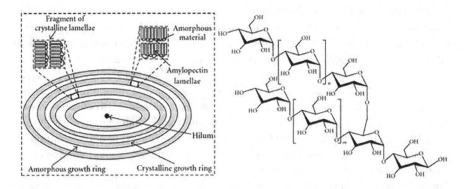

FIGURE 8.5 Schematic illustration of starch granule structure and chemical structure of amylopectin (Lin et al., 2011).

etc., with diameters ranging from about 0.1 to 200 μm depending on their botanical origin (Spinozzi et al., 2020).

Nanostarches (particle sizes ranging from 1 to 1000 nm) have recently sparked a flurry of research and public interest due to their unusual surface area and reaction activity. Starch nanoparticles (SNPs) and starch nanocrystals (SNCs) are the most common forms of nanostarch. SNPs have a higher surface area and a modified amorphous/crystallinity ratio than native starch.

Various approaches have been developed to create nanosized particles from the starch biopolymer throughout the last few decades. "Top-down" (e.g., acid hydrolysis, enzyme hydrolysis, or physical treatments) and "bottom-up" techniques are used to make nanostarches (e.g., self- assembly and nanoprecipitation) (Dukare et al., 2021). The use of SNCs as reinforcing fillers in a matrix was first investigated using the analogy of cellulose whiskers. Sulphuric acid hydrolysis followed by ultrasonic treatment is the traditional method of nanostarch synthesis (Rajisha et al., 2014). Researchers have recently attempted to prepare nanostarch from native starch using acid hydrolysis (Lunardi et al., 2021), homogenisation (Apostolidis & Mandala, 2020), rapid ultrasonication (Chang et al., 2019), microemulsion (Qi & Tester, 2016), ionic gelation (Liu et al., 2020), rapid antisolvent nanoprecipitation (Dong et al., 2021), alkali freezing and cross-linking (Xiao et al., 2020), acid hydrolysis and ultrasound technique (Shabana et al., 2019), and enzyme-based approach (Cuthbert et al., 2017).

8.2.2.1 Preparation of Nanostarch

SNCs were made by hydrolysing waxy maize starch and amaranth starch with H_2SO_4 (3.16 M) at 40°C for 3, 5, and 10 days, according to Sanchez de la Concha et al. (2018). The yield and particle size ranged from 3.6% to 26.0% and 322 to 577 nm, respectively, with lower values indicating longer hydrolysis times. The H_2SO_4 concentration of 3.16 M, the reaction temperature of 40°C, and the reaction duration of 5 days serve as the foundation for SNCs industrial production. Furthermore, waxy maize starch was found to be a suitable and widely available raw material for making SNCs. Because the crystalline regions of starch are mostly made up of amylopectin, the materials used may influence the yield of SNCs. In acid hydrolysis process, a reactor submerged in an oil bath was filled with 117.5 g of wheat or waxy maize starch and 800 ml of sulphuric acid 3.16 mol/l, which was heated to 40°C with a stirring speed of 100 rpm. The reaction was halted after 7 days, and the suspension was rinsed with distilled water and centrifuged for 10 minutes at 5000 rpm (2236 g) until neutralisation was obtained. For 1 week, the final suspension was refrigerated at 6°C (Angellier et al., 2004). The acid hydrolysis process has a number of drawbacks, including long preparation times, pollution, and low yields (15%), all of which make industrial-scale SNC manufacture difficult. Chemical treatment changes the surface features of SNCs, which improves interfacial interactions with the polymer matrix and so affects the performance of nanomaterials (Lin et al., 2011). Strong interfacial adhesion aids stress transfer and acts as a reinforcement for nanocrystals. Furthermore, functional modification of SNCs, such as grafting fluorescent molecules, could broaden their application in biological materials.

8.2.2.1.1   Preparation of SNPs by Enzymatic Treatment with Pullulanase

A 5% w/v starch suspension in a sodium acetate buffer (pH 4.5) was made and cooked at 80°C–90°C with steady stirring within 15 minutes. The suspension was then transferred to an autoclave and incubated at 121°C for 60 minutes, where the starch gelatinisation process occurred. The suspension was chilled after gelatinisation, and 1-ml pullulanase (1000 ASPU/g dry starch) was added at 55°C. The reaction was stopped after 24 hours, and the suspension was placed in a boiling water bath for 30 minutes to deactivate the enzyme, followed by 2 hours of natural cooling to laboratory temperature. The samples and EM3 were then kept at 4°C in the fridge (Chena Aldao et al., 2018). The production of nanoparticles was mostly attributable to the recrystallisation of linear glucans following native starch enzymolysis. The linear glucans were vulnerable to retrogradation. As a result, the SNP was formed by crystallisation in two steps: first, linking linear glucans into double helices and forming clusters with hydrogen bonds, and second, rearrangement of clustering into the SNP (Sun et al., 2014). SNCs' main drawback is their hydrophilicity, which limits their employment as a reinforcing phase in hydrophilic polymers. To increase the affinity between the filler and hydrophobic polymer matrices, many ways of surface chemical modification of SNCs have been intensively researched.

8.2.2.2   Application of Nanostarch

SNCs have been incorporated into natural rubber latex for the manufacture of nanocomposites using an aqueous suspension of waxy maize SNCs as the reinforcing phase (Angellier et al., 2005). Nanostarch has received a lot of attention as a food-grade Pickering emulsion stabiliser because of its biodegradability, non-toxicity, small size, and large specific surface area. Nanostarch has been used as a filler in biocompatible films since it is affordable and abundant in the environment. New nanotechnology trends and views have paved the way for the use of starch nanofillers as active agents in the packaging sector. This method is applicable to the development of biodegradable packaging materials based on biopolymers and nanofillers. Synthetic and biodegradable polyester matrices, such as poly(lactic acid) and polycaprolactone, have also employed chemically modified SNCs as reinforcing fillers. The structure of SNCs is less robust than that of cellulose nanocrystals, especially during functional modification, which hinders the development of SNCs for application in biomaterials. Nanofillers, which can serve as both a reinforcing and an active element, are currently being used as a unique ingredient to improve the properties of biopolymer films. The most commonly used polysaccharides in the production of biopolymer films are starch and CS.

8.2.3   NANOCHITIN

After cellulose, chitin is the second most abundant natural organic resource. It is usually found in the exoskeletons of mollusks, arthropods, and in some fungi (Mincea et al., 2012). Chitin is a high-molecular-weight straight-chain polysaccharide linked by [N-acetyl-2-amido-2-deoxy-D-glucose] via the β-(1,4) glycosidic bond (Fernandes et al., 2011). The structure of chitin is similar to that of cellulose in which the hydroxyl at the C2

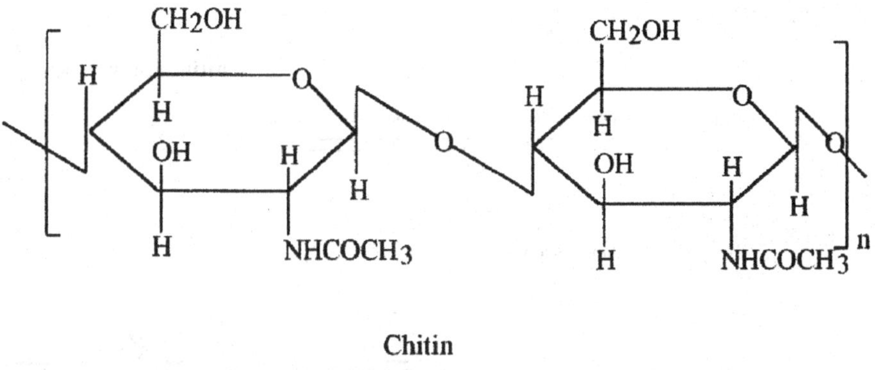

Chitin

FIGURE 8.6   Structure of chitin (Ravi Kumar, 2000).

position has been replaced with an acetamido group (Figure 8.6). It is a nitrogen-containing polysaccharide, which is hard, white, inelastic, semi-transparent substance with low chemical reactivity, and insoluble in majority of the organic solvents. The high percentage of nitrogen in chitin makes it a better chelating agent. Chitin has the characteristics of biodegradability, biocompatibility, non-toxicity, and antibacterial activity (Barikani et al., 2014). Nanochitin is a high-potential nanomaterial that not only maintains the properties of the original chitin, but also has a high aspect ratio, large surface area, low density, and large number of functional groups, which contributes to surface functionalisation. Due to its linear structure with two hydroxyl groups and one acetamide group, natural chitin is highly crystalline with strong hydrogen bonds (Azuma et al., 2014). The shellfish (shrimp or crab shells) industry produces large amounts of waste from shellfish shells that contain about 30% chitin, results in relatively low price of chitin. In living organisms, chitin exists as a semi-crystalline biopolymer which creates microfibrillar arrangements. These microfibrils consist of alternate crystalline and amorphous domains. It has a highly ordered crystal structure comprising three crystal forms: α-chitin, β-chitin, and γ-chitin. These crystal forms of chitin differ in packing arrangements (Rinaudo, 2006). Chitin can be synthesised from different types of cells. Therefore, various chitin derivatives can be obtained using different species of crustaceans and different production methods. The extraction and purification of chitin involves demineralisation, deproteinisation, and decolourisation (Soon et al., 2018).

8.2.3.1   Preparation of Nanochitin

The purification of chitin is the first step in the process of the preparation of nanochitin (elKnidri et al., 2018). The purification involves the treatment with alkali (NaOH, KOH) or enzyme to remove protein, bleaching with CH_3CH_2OH or $NaClO_3$ to remove pigments, and inorganic minerals are eliminated by the treatment with acid (HCl). Generally, the preparation of nanochitin involves two routes: top-down and bottom-up. The isolation of nanochitin from the crustacean shell follows top-down method. This method comprises the bulk separation of nanochitin from natural chitin source. It involves the purification of chitin followed by microfibrillation. The synthesis of nanochitin from crustacean

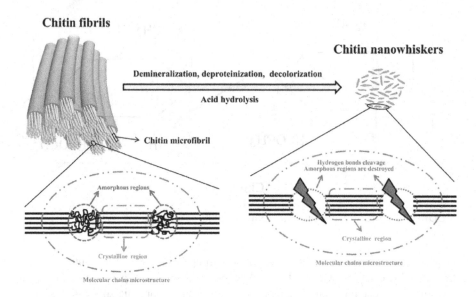

FIGURE 8.7   Mechanism of acid hydrolysis of chitin (X. Yang et al., 2020).

shells by processes like acid hydrolysis, partial deacetylation, mechanical treatment, and TEMPO oxidation are top-down methods. Gopi et al. reported the preparation of chitin nanowhiskers by the acid hydrolysis of native chitin by stirring with hydrochloric acid for a specific time. The reaction is stopped by the addition of distilled water, and the resultant suspension is centrifuged followed by ultrasonication. The morphology of chitin nanowhiskers thus obtained depends on the time of acid hydrolysis and the concentration of the acid (Figure 8.7). The acids have a tendency to attack amorphous regions first in the acid hydrolysis method. Dufresne et al. extracted chitin nanowhiskers which is 100–600 nm in length and 4–40 nm in width from crab shells (Gopi et al., 2016).

The mechanical treatment involves the use of methods like ultrasonication, grinding, and high-pressure homogenisation which are used individually or simultaneously under neutral or acidic conditions to obtain nanochitin. The morphology of nanochitin differs based on the chitin source, the time, and the intensity of the ultrasound used. Ultrasound can even make chitin powder and chitin gels to nanosized dimensions. Mechanical grinding can be used for the separation of chitin nanofibres. Ifuku et al. used grinding method for the separation of chitin nanofibres from mushroom cell walls (Ifuku et al., 2015). Wu et al. reported the preparation of chitin nanofibres by using high-pressure homogenisation method (Wu et al., 2014).

TEMPO-mediated oxidation (TEMPO-mediated oxidation) is an important method used for the isolation of chitin nanowhiskers. In this method, the purified chitin undergoes oxidation in the presence of compounds like NaBr and NaClO with pH of 10. The water-soluble polyuronic acid and water-insoluble whiskers are the major products remaining after oxidation. These whiskers will then undergo ultrasonication treatment to obtain chitin nanowhiskers (Fan et al., 2008).

In the bottom-up method, chitin solution is prepared by dissolving the crustacean shell in limited amount of solvent, and nanochitin is synthesised from this solution by electrospinning, dissolution-regeneration, and self-assembly method. The morphology of the chitin nanofibres formed by electrospinning method varies with the concentration of the chitin solution. Process conditions, solution properties, and environmental parameters are the factors affecting electrospinning method. In this method, the molecular weight of chitin is reduced by ultrasonication or microwave irradiation and then it is dissolved in highly toxic solvents like 1,1,1,3,3,3-hexafluoro-2-propanol. 1-ethyl-3-methylimidazolium acetate is an ionic liquid which is used as an alternative for volatile organic solvents, can dissolve crustacean shells, and then spin nanofibres from the extract solution (Singh, 2019).

The dissolution-regeneration method is a process in which chitin is first dissolved in a solution, and then the chitin molecules are regenerated into chitin nanofibres using a precipitating agent. Wet spinning is an example of dissolution-regeneration method (elSeoud et al., 2021).

In self-assembly method, chitin is dissolved in certain solvents, and chitin nanofibres are regenerated from this solution. The solvents used in self-assembly method include water-based solvents, toxic solvents, and green solvents, which helps in the cleavage of hydrogen bonds (Kadokawa, 2015). Duan et al. prepared chitin nanofibre microspheres in urea–NaOH aqueous solution by thermally induced self-assembly (Duan et al., 2018).

8.2.3.2 Applications of Nanochitin

Nanochitin is a material which is used for various purposes. It can be utilised in water purification, paper finishing, textile industry, heavy metal chelating agents, and biomedical applications. The biomedical applications consist of tissue engineering, drug delivery, wound dressings, and antibacterial coatings (Jayakumar et al., 2010). Due to the presence of hydroxyl, carboxyl, and amino groups, surface modifications can be easily done in nanochitin which makes them highly suitable for drug delivery applications.

Nanochitin has been used in tissue engineering and surgical sutures due to the presence of chitinase and similar proteins in the human body. It has the ability to stimulate immune cells and exhibit significant anti-cancer activity. The nanofibrils and nanocrystals of chitin are used in skin protective formulations due to its anti-ageing properties, moisturising effect, and UV protection ability. Torres-Rendon et al. used sacrificial templating of nanochitin and NC hydrogels for the manufacture of bioactive gyroid scaffolds for biomimetic tissue engineering (Torres-Rendon et al., 2015). Nanochitin is an excellent raw material for the fabrication of rapidly degradable and inexpensive bioplastics. Chitin nanomaterials can be used as reinforcing agents to enhance the mechanical properties of starch-based food films. Due to the presence of abundant hydroxyl and acetamide functional groups and the large surface-to-volume ratio, nanowhiskers act as an excellent adsorbent. Nanochitin can be used as reinforcing fillers in biopolymer-based nanocomposites to improve its properties (Joseph et al., 2020).

8.2.4 NANOCHITOSAN

CS is a natural, cationic amino polysaccharide which is the *N*-deacetylated derivative of chitin. The degree of deacetylation ranges from 70% to 95%. It is obtained from the crustacean shells. CS (Figure 8.8) is a copolymer of β-(1→4)-*N*-acetyl-D-glucosamine

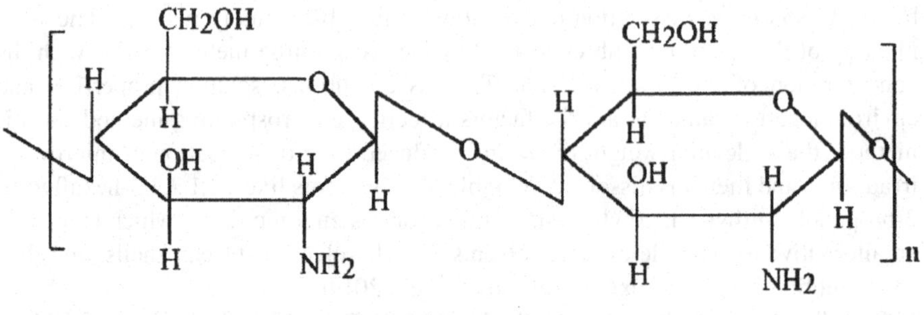

Chitosan

FIGURE 8.8 Structure of CS (Ravi Kumar, 2000).

and β-(1→4)-D-glucosamine (Riva et al., 2011). It is highly crystalline, hydrophilic, and soluble in wide range of organic solvents. The solubilisation is due to the protonation of -NH_2 functional group on the C-2 position of the repeated D-glucosamine unit. Thus, in acidic medium, this polysaccharide gets converted to polyelectrolyte. Since it is soluble in aqueous solutions, it is extensively used in various applications such as fibres, films, gel, or solutions. Nanochitosan is a natural, environment friendly material with excellent physicochemical properties and has received great attention due to their nanosize, large surface area, high aspect ratio, biocompatibility, and good biodegradability (Thomas et al., 2019). Compared to chitin, CS is much easier to process, but due to their hydrophilic character, the stability of CS materials is generally lower. The superior characteristics of CS include anticoagulant properties, water solubility, fluidity, high water-reducing ratio, and biological properties like controlled drug delivery, gelation, and targeting the colon.

8.2.4.1   Preparation of Nanochitosan

Different strategies are used for the preparation of CS nanoparticles. It also involves bottom-up and top-down approaches. The major challenge for the synthesis of CS nanoparticles was that the method should be of low cost, simple, and high yield. Obtaining fibres from CS was relatively easy because it dissolves in dilute acids like acetic acid. The methods such as precipitation or coagulation, microemulsion, ionic cross-linking, covalent cross-linking, and polyelectrolyte complexation can be used for the preparation of nanochitosan. These techniques follow a bottom-up approach in which nanoparticles are formed from individual macromolecules via self-assembly. The first method for the preparation of nanochitosan was reported by Ohya et al. in 1994 (Grenha, 2012). Berthold et.al used sodium sulphate as the precipitating agent for the synthesis of CS nanoparticles. Initially, Tween 80 is added to the CS acetic acid solution, and then sodium sulphate solution is added to this solution with constant stirring followed by ultrasonication (Biró et al., 2008).

Tang et al. used 2.0% acetic acid to dissolve CS followed by the dropwise addition of sodium tripolyphosphate (TPP) with constant stirring (Tang et al., 2007). Bodmeier et al. used the ionic cross-linking method to prepare CS nanoparticles by adding TPP cations to CS solution. The intermolecular cross-linking reactions between the TPP cations and the free amino groups of CS form a bead gel of CS (Huang et al., 2009).

Moura et al. prepared nanochitosan by dissolving CS and maleic acid in distilled water and kept for 30 minutes under magnetic stirring and heated for 3 hours at a constant temperature of 70°C. They added $K_2S_2O_8$ to this solution and cooled in ice bath to obtain a milky emulsion, centrifuged, and then freeze-dried using freeze-dryer (Zareie, 2013). In emulsion cross-linking method, aqueous solution of CS is emulsified in the oil phase, and a surfactant is used to stabilise the aqueous droplets. CS nanoparticles are precipitated by adding glutaraldehyde as the cross-linking agent (Agnihotri et al., 2004).

In top-down approach, the bulk material is subjected to mechanical grinding to form nanosized particles. Beads mill is an instrument which is used to synthesise nanoparticles via top-down approach. Rochima et al. reported the preparation of CS nanoparticles by bead-milling process and proposed the mechanism of surface modification of the CS nanoparticles thus produced (Rochima et al., 2017).

8.2.4.2 Applications of Nanochitosan

Nanochitosan is used to prepare fibres, films, and hydrogels (Afshar et al., 2020). Due to its biocompatibility, it is mostly used in biomedical domain. In artificial organs, it is used as a controlled release drug carrier for gene transfer (H.-C. Yang et al., 2010). Nanochitosan has been used to enhance the strength, antibacterial effects, and washability of the textiles. In addition, it is used in the fabrication of sensors, packaging materials, and tissue engineering. Nanochitosan polymeric membranes are used for the purification of contaminated water (Divya & Jisha, 2018).

Because of the presence of cationic amino group, it is reported that its antimicrobial property is high at low pH. It is used in dentistry, orthopaedics, and ophthalmology due to its adhesive nature (Sivanesan et al., 2021).

8.2.5 ALGINATE NANOPARTICLES

Alginate is a naturally occurring polysaccharide generated from brown algal cell walls that has unique properties that make it ideal for a variety of biological and pharmacological applications. Alginate (Alg) is a naturally occurring anionic polysaccharide of linear copolymers made up of blocks of β-D-mannuronate (M units) and α-L-guluronate (G units) residues linked by β-glycosidic bonds (i.e., β—1,4 linkages) (Figure 8.9) (Dodero et al., 2021). Alginate extraction is a multi-step process that typically involves the use of aqueous acid and alkali solution treatments, salt-induced precipitation, and subsequent purification (Rinaudo, 2008). One of the most truly amazing properties of alginate, which, by the way, allows for its widespread industrial use, is its ability to form gels through the coordination of divalent (e.g., Ca^{2+}, Sr^{2+}, Ba^{2+}, etc.) or trivalent cations (e.g., Al^{3+}, Fe^{3+}, etc.), as well as in acidic environment owing to polymer chain shrinkage.

Alginate-based nanomaterials hold extreme promises owing to their superior biocompatibility, mucoadhesive properties, non-toxicity, hydrophilicity, bioavailability, relatively low cost, and possible mass production; hence, they can be employed in several application fields with particular emphasis on the biomedical and

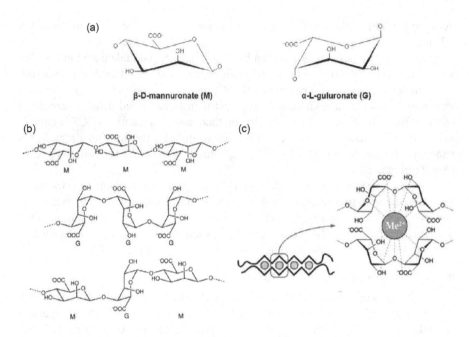

FIGURE 8.9 (a) Alginate monomer structural units, (b) alginate block distribution and chain conformation, and (c) alginate egg-box gelation model with divalent cations (Dodero et al., 2021).

pharmaceutical industries where their unique biological properties play a fundamental role. Antimicrobial products, medication and gene delivery, tissue engineering, cancer therapy, wound dressing, and biosensors are just a few of the many applications for alginate-based nanomaterials (Rinaudo, 2008). In this regard, alginate is used to make nanostructured scaffolds, membranes, gels, beads, and sponge morphologies, as well as NPs, nanocapsules, NFs, and nanocoating. NPs (Lopes et al., 2017; Nayak & Hasnain, 2020; Zhu et al., 2021) are by far the most studied alginate-based nanostructures in pharmaceutical applications for the delivery of drugs and other chemicals (e.g., enzymes, factors, vitamins, etc.) at the nanoscale level directly to the site of interest. To date, there are numerous methods for making alginate-based NPs. Ion-induced alginate gelation, covalent cross-linking methodology (Zhu et al., 2021), electro-spraying technique, emulsification technique, and other procedures are common in this regard. A polyelectrolyte complexation phenomenon (Nayak & Hasnain, 2020) can also be used to make alginate-based NPs.

Alg-NPs containing curcumin and resveratrol were studied for the treatment of prostate cancer and shown to have a significant cytotoxicity effect on tumour cells without causing hemolysis, making them safe for intravenous administration (Saralkar & Dash, 2017). Similarly, when compared to equivalent free drug formulations and pH-responsive delivery for the targeted treatment of breast cancer, oxidised alginate NPs conjugated with doxorubicin and containing curcumin showed decreased toxicity towards health cells (Gao et al., 2017). Alginate–CS NPs loaded with a short peptide generated from a neural growth factor have also showed potential for the treatment of brain degenerative diseases, such as Alzheimer's disease, by promoting the differentiation of stem cells into adult neurons (Lauzon et al., 2018). Because of their great biocompatibility, alginate-based NFs incorporating curcumin and cross-linked with trifluoroacetic acid have been proposed for a variety of applications, including tissue engineering. Alginate NFs (H. Lu et al., 2021), on the other hand, have been examined as a viable platform for immobilising lipase, an enzyme with significant potential in bone regeneration, with the ability to preserve its activity for up to 14 days (İspirliDoğaç et al., 2017). By comparing their physical–chemical and biological qualities to commercially available goods, alginate NFs were combined with natural spider silk and evaluated as wound dressing materials (İspirliDoğaç et al., 2017). It's no surprise that alginate-based nanofibrous scaffolds have showed significant promise in replacing and/or mending both soft and hard tissues, such as bones, cartilage, nerves, and skin, due to their adaptability.

Despite the many benefits of alginate NPs and alginate electrospun NFs, the fact that they are naturally derived means they have an intrinsically variable structure that can affect the material's physical–chemical properties, which can be further influenced by the extraction and processing procedures. Furthermore, despite pure alginate's overall biocompatibility, the inclusion of contaminants in industrially generated products could cause unforeseen negative effects in the human body. Another significant barrier to large-scale usage of alginate-based nanostructures is the significantly lower production rate compared to synthetic polymers.

8.3 CONCLUSION

There is a growing demand for products manufactured from non-petroleum-based resources that are renewable and sustainable. In relation to their inherent features like biocompatibility and biodegradability in the human body for some of them, an important development of polysaccharide applications may be projected for the next few years; they are also renewable and have fascinating physical properties (film-forming, gelling, and thickening properties). Furthermore, they can be processed into a variety of forms, including beads, films, capsules, and fibres. CS is a highly fascinating film-forming base due to its biological (antimicrobial and antioxidant activity) and physical (thermal or mechanical) qualities.

The potential of cellulosic nanoparticles or NC for unique functional nanomaterials has been established.

8.3.1 Nanomaterials: Future

The European Commission considers nanotechnology to be a key enabling technology. Because of the intriguing features of nanoparticles, a growing number of novel materials are being developed, and assessing their dangers necessitates a unique approach to each nanomaterial. Nanomaterials are subject to the same strong regulatory framework that governs the safe use of all chemicals and combinations in the EU, namely the REACH and CLP guidelines. This means that the dangerous qualities of nanoforms of substances must be evaluated, and their safe usage must be guaranteed. The presence of nanoparticles in foods and cosmetics must be disclosed on the label in the ingredient list. Due to various concerns regarding the safety of employing nanomaterials, further research is needed to determine whether and particular nanoparticles can be a viable replacement to existing materials for use in a variety of applications. Still, more study with a focus on regulatory issues is required, particularly in regards to the application of the nanomaterials definition.

REFERENCES

Abitbol, T., Rivkin, A., Cao, Y., Nevo, Y., Abraham, E., Ben-Shalom, T., ... & Shoseyov, O. (2016). Nanocellulose, a tiny fiber with huge applications. *Current Opinion in Biotechnology*, 39, 76–88.

Afshar, M., Dini, G., Vaezifar, S., Mehdikhani, M., & Movahedi, B. (2020). Preparation and characterization of sodium alginate/polyvinyl alcohol hydrogel containing drug-loaded chitosan nanoparticles as a drug delivery system. *Journal of Drug Delivery Science and Technology*, 56, 101530. https://doi.org/10.1016/j.jddst.2020.101530

Agnihotri, S. A., Mallikarjuna, N. N., &Aminabhavi, T. M. (2004). Recent advances on chitosan-based micro- and nanoparticles in drug delivery. *Journal of Controlled Release*, 100(1), 5–28. https://doi.org/10.1016/j.jconrel.2004.08.010

Angellier, H., Choisnard, L., Molina-Boisseau, S., Ozil, P., & Dufresne, A. (2004). Optimization of the preparation of aqueous suspensions of waxy maize starch nanocrystals using a response surface methodology. *Biomacromolecules*, 5(4), 1545–1551. https://doi.org/10.1021/BM049914U

Angellier, H., Molina-Boisseau, S., & Dufresne, A. (2005). Mechanical properties of waxy maize starch nanocrystal reinforced natural rubber. *Macromolecules,* 38(22), 9161–9170.

Apostolidis, E., & Mandala, I. (2020). Modification of resistant starch nanoparticles using high-pressure homogenization treatment. *Food Hydrocolloids*, 103, 105677. https://doi.org/10.1016/J.FOODHYD.2020.105677

AppliedChem | Special Issue : Polysaccharides from Biomass Conversion. (n.d.). A Special Issue of AppliedChem (ISSN 2673–9623). Retrieved February 17, 2022, from https://www.mdpi.com/journal/appliedchem/special_issues/Polysaccharides_Biomass_Conversion

Azuma, K., Ifuku, S., Osaki, T., Okamoto, Y., & Minami, S. (2014). Preparation and biomedical applications of chitin and chitosan nanofibers. *Journal of Biomedical Nanotechnology*, 10(10), 2891–2920. https://doi.org/10.1166/jbn.2014.1882

Barikani, M., Oliaei, E., Seddiqi, H., & Honarkar, H. (2014). Preparation and application of chitin and its derivatives: a review. *Iranian Polymer Journal*, 23(4), 307–326. https://doi.org/10.1007/s13726-014-0225-z

Biró, E., Németh, Á. Sz., Sisak, C., Feczkó, T., & Gyenis, J. (2008). Preparation of chitosan particles suitable for enzyme immobilization. *Journal of Biochemical and Biophysical Methods*, 70(6), 1240–1246. https://doi.org/10.1016/j.jprot.2007.11.005

Boldizar, A., Klason, C., Kubát, J., Näslund, P., &Sáha, P. (2006). Prehydrolyzed cellulose as reinforcing filler for thermoplastics. *International Journal of Polymeric Materials and Polymeric Biomaterials*, 11(4), 229–262. https://doi.org/10.1080/00914038708078665

Chang, R., Ji, N., Li, M., Qiu, L., Sun, C., Bian, X., Qiu, H., Xiong, L., & Sun, Q. (2019). Green preparation and characterization of starch nanoparticles using a vacuum cold plasma process combined with ultrasonication treatment. *Ultrasonics Sonochemistry*, 58, 104660. https://doi.org/10.1016/J.ULTSONCH.2019.104660

Chena Aldao, D., Šárka, E., Ulbrich, P., &Menšíková, E. (2018). Starch Nanoparticles-Two Ways of their Preparation. *Czech J. Food Sci*, 36(2), 0–6. https://doi.org/10.17221/371/2017-CJFS

Csoka, L., Hoeger, I. C., Rojas, O. J., Peszlen, I., Pawlak, J. J., & Peralta, P. N. (2012). Piezoelectric effect of cellulose nanocrystals thin films. *ACS Macro Letters*, 1(7), 867–870.

Cuthbert, W. O., Ray, S. S., & Emmambux, N. M. (2017). Isolation and characterisation of nanoparticles from tef and maize starch modified with stearic acid. *Carbohydrate Polymers, 168,* 86–93. https://doi.org/10.1016/J.CARBPOL.2017.03.067

Dantas de Oliveira, A., & Augusto Gonçalves Beatrice, C. (2019). Polymer nanocomposites with different types of nanofiller. *Nanocomposites - Recent Evolutions.* https://doi.org/10.5772/intechopen.81329

Dash, R., & Ragauskas, A. J. (2012). Synthesis of a novel cellulose nanowhisker-based drug delivery system. *RSC Advances, 2*(8), 3403–3409.

Divya, K., & Jisha, M. S. (2018). Chitosan nanoparticles preparation and applications. *Environmental Chemistry Letters, 16*(1), 101–112. https://doi.org/10.1007/s10311-017-0670-y

Dodero, A., Alberti, S., Gaggero, G., Ferretti, M., Botter, R., Vicini, S., & Castellano, M. (2021). *An Up-to-Date Review on Alginate Nanoparticles and Nanofibers for Biomedical and Pharmaceutical Applications.* https://doi.org/10.1002/admi.202100809

Dong, H., Zhang, Q., Gao, J., Chen, L., & Vasanthan, T. (2021). *Comparison of morphology and rheology of starch nanoparticles prepared from pulse and cereal starches by rapid antisolvent nanoprecipitation. Food Hydrocolloids.*

Dong, S., & Roman, M. (2007). Fluorescently labeled cellulose nanocrystals for bioimaging applications. *Journal of the American Chemical Society, 129*(45), 13810–13811.

Dong, X. M., Revol, J. F., & Gray, D. G. (1998). Effect of microcrystallite preparation conditions on the formation of colloid crystals of cellulose. *Cellulose 1998 5:1, 5*(1), 19–32. https://doi.org/10.1023/A:1009260511939

Duan, B., Huang, Y., Lu, A., & Zhang, L. (2018). Recent advances in chitin based materials constructed via physical methods. *Progress in Polymer Science, 82,* 1–33. https://doi.org/10.1016/j.progpolymsci.2018.04.001

Dufresne, A. (2013). Nanocellulose: a new ageless bionanomaterial. *Materials Today, 16*(6), 220–227. https://doi.org/10.1016/J.MATTOD.2013.06.004

Dukare, A. S., Arputharaj, A., Bharimalla, A. K., Saxena, S., & Vigneshwaran, N. (2021). Nanostarch production by enzymatic hydrolysis of cereal and tuber starches. *Carbohydrate Polymer Technologies and Applications, 2,* 100121. https://doi.org/10.1016/J.CARPTA.2021.100121

Echeverria, C., Almeida, P. L., Feio, G., Figueirinhas, J. L., & Godinho, M. H. (2015). A cellulosic liquid crystal pool for cellulose nanocrystals: structure and molecular dynamics at high shear rates. *European Polymer Journal, 72,* 72–81. https://doi.org/10.1016/J.EURPOLYMJ.2015.09.006

elKnidri, H., Belaabed, R., Addaou, A., Laajeb, A., & Lahsini, A. (2018). Extraction, chemical modification and characterization of chitin and chitosan. *International Journal of Biological Macromolecules, 120,* 1181–1189. https://doi.org/10.1016/j.ijbiomac.2018.08.139

elSeoud, O. A., Jedvert, K., Kostag, M., & Possidonio, S. (2021). Cellulose, chitin and silk: the cornerstones of green composites. *Emergent Materials.* https://doi.org/10.1007/s42247-021-00308-0

Fan, Y., Saito, T., & Isogai, A. (2008). Chitin nanocrystals prepared by TEMPO-mediated oxidation of α-chitin. *Biomacromolecules, 9*(1), 192–198. https://doi.org/10.1021/bm700966g

Favier, V., Canova, G. R., Cavaillé, J. Y., Chanzy, H., Dufresne, A., & Gauthier, C. (1995). Nanocomposite materials from latex and cellulose whiskers. *Polymers for Advanced Technologies, 6*(5), 351–355. https://doi.org/10.1002/PAT.1995.220060514

Fernandes, S. C., Freire, C. S., Silvestre, A. J., Pascoal Neto, C., & Gandini, A. (2011). Novel materials based on chitosan and cellulose. *Polymer International, 60*(6), 875–882. https://doi.org/10.1002/pi.3024

Fukuzumi, H., Saito, T., Iwata, T., Kumamoto, Y., & Isogai, A. (2008). Transparent and high gas barrier films of cellulose nanofibers prepared by TEMPO-mediated oxidation. *Biomacromolecules, 10*(1), 162–165. https://doi.org/10.1021/BM801065U

Gao, C., Tang, F., Gong, G., Zhang, J., Hoi, M. P. M., Lee, S. M. Y., & Wang, R. (2017). pH-Responsive prodrug nanoparticles based on a sodium alginate derivative for selective co-release of doxorubicin and curcumin into tumor cells. *Nanoscale, 9*(34), 12533–12542. https://doi.org/10.1039/C7NR03611F

Gopi, S., Pius, A., & Thomas, S. (2016). Enhanced adsorption of crystal violet by synthesized and characterized chitin nano whiskers from shrimp shell. *Journal of Water Process Engineering, 14*, 1–8. https://doi.org/10.1016/j.jwpe.2016.07.010

Grenha, A. (2012). Chitosan nanoparticles: a survey of preparation methods. *Journal of Drug Targeting, 20*(4), 291–300. https://doi.org/10.3109/1061186X.2011.654121

Henriksson, M., Henriksson, G., Berglund, L. A., & Lindström, T. (2007). An environmentally friendly method for enzyme-assisted preparation of microfibrillated cellulose (MFC) nanofibers. *European Polymer Journal, 43*(8), 3434–3441. https://doi.org/10.1016/J.EURPOLYMJ.2007.05.038

Huang, K.-S., Sheu, Y.-R., & Chao, I.-C. (2009). Preparation and properties of nanochitosan. *Polymer-Plastics Technology and Engineering, 48*(12), 1239–1243. https://doi.org/10.1080/03602550903159069

Ifuku, S., Hori, T., Izawa, H., Morimoto, M., & Saimoto, H. (2015). Preparation of zwitterionically charged nanocrystals by surface TEMPO-mediated oxidation and partial deacetylation of α-chitin. *Carbohydrate Polymers, 122*, 1–4. https://doi.org/10.1016/j.carbpol.2014.12.060

İspirliDoğaç, Y., Deveci, İ., Mercimek, B., & Teke, M. (2017). A comparative study for lipase immobilization onto alginate based composite electrospun nanofibers with effective and enhanced stability. *International Journal of Biological Macromolecules, 96*, 302–311. https://doi.org/10.1016/J.IJBIOMAC.2016.11.120

Jamróz, E., Kulawik, P., & Kopel, P. (2019). The effect of nanofillers on the functional properties of biopolymer-based films: a review. *Polymers, 11*(4). https://doi.org/10.3390/POLYM11040675

Jayakumar, R., Prabaharan, M., Nair, S. V., & Tamura, H. (2010). Novel chitin and chitosan nanofibers in biomedical applications. *Biotechnology Advances, 28*(1), 142–150. https://doi.org/10.1016/j.biotechadv.2009.11.001

Joseph, B., Mavelil Sam, R., Balakrishnan, P., J. Maria, H., Gopi, S., Volova, T., C. M. Fernandes, S., & Thomas, S. (2020). Extraction of nanochitin from marine resources and fabrication of polymer nanocomposites: recent advances. *Polymers, 12*(8), 1664. https://doi.org/10.3390/polym12081664

Kadokawa, J. (2015). Fabrication of nanostructured and microstructured chitin materials through gelation with suitable dispersion media. *RSC Advances, 5*(17), 12736–12746. https://doi.org/10.1039/C4RA15319G

Kargarzadeh, H., Ioelovich, M., Ahmad, I., Thomas, S., & Dufresne, A. (2017). Methods for extraction of nanocellulose from various sources. *Handbook of Nanocellulose and Cellulose Nanocomposites*, 1–49. https://doi.org/10.1002/9783527689972.CH1

Lamanna, M., Morales, N. J., Lis, N., & Goyanes, S. (2013). Development and characterization of starch nanoparticles by gamma radiation: potential application as starch matrix filler. *Carbohydrate Polymers, 97*(1), 90–97. https://doi.org/10.1016/j.carbpol.2013.04.081

Lauzon, M. A., Marcos, B., & Faucheux, N. (2018). Characterization of alginate/chitosan-based nanoparticles and mathematical modeling of their SpBMP-9 release inducing neuronal differentiation of human SH-SY5Y cells. *Carbohydrate Polymers, 181*, 801–811. https://doi.org/10.1016/J.CARBPOL.2017.11.075

Li, J., Cha, R., Mou, K., Zhao, X., Long, K., Luo, H., Zhou, F., & Jiang, X. (2018). Nanocellulose-based antibacterial materials. *Advanced Healthcare Materials, 7*(20), 1800334. https://doi.org/10.1002/ADHM.201800334

Lin, N., Huang, J., Chang, P. R., Anderson, D. P., & Yu, J. (2011). Preparation, modification, and application of starch nanocrystals in nanomaterials: a review. *Journal of Nanomaterials*. https://doi.org/10.1155/2011/573687

Liu, Q., Cai, W., Zhen, T., Ji, N., Dai, L., Xiong, L., & Sun, Q. (2020). Preparation of debranched starch nanoparticles by ionic gelation for encapsulation of epigallocatechin gallate. *International Journal of Biological Macromolecules*, *161*, 481–491. https://doi.org/10.1016/J.IJBIOMAC.2020.06.070

Loow, Y. L., New, E. K., Yang, G. H., Ang, L. Y., Foo, L. Y. W., & Wu, T. Y. (2017). Potential use of deep eutectic solvents to facilitate lignocellulosic biomass utilization and conversion. *Cellulose*, *24*(9), 3591–3618. https://doi.org/10.1007/S10570-017-1358-Y

Lopes, M., Abrahim, B., Veiga, F., Seiça, R., Cabral, L. M., Arnaud, P., Andrade, J. C., & Ribeiro, A. J. (2017). Preparation methods and applications behind alginate-based particles. *Expert Opinion on Drug Delivery*, *14*(6), 769–782. https://doi.org/10.1080/17425247.2016.1214564

Lu, H., Butler, J. A., Britten, N. S., Venkatraman, P. D., & Rahatekar, S. S. (2021). Natural antimicrobial nano composite fibres manufactured from a combination of alginate and oregano essential oil. *Nanomaterials*, *11*(8). https://doi.org/10.3390/NANO11082062

Lu, P., & Hsieh, Y. Lo. (2010). Preparation and properties of cellulose nanocrystals: rods, spheres, and network. *Carbohydrate Polymers*, *82*(2), 329–336. https://doi.org/10.1016/J.CARBPOL.2010.04.073

Lunardi, V. B., Soetaredjo, F. E., Putro, J. N., Santoso, S. P., Yuliana, M., Sunarso, J., Ju, Y. H., & Ismadji, S. (2021). Nanocelluloses: sources, pretreatment, isolations, modification, and its application as the drug carriers. *Polymers*, *13*(13). https://doi.org/10.3390/POLYM13132052

Microfibrillated cellulose: morphology and accessibility (Conference) | OSTI.GOV. (n.d.). Retrieved January 26, 2022, from https://www.osti.gov/biblio/5039044

Mincea, M., Negrulescu, A., & Ostafe, V. (2012). Preparation, modification, and applications of chitin nanowhiskers: a review. *Reviews on Advanced Materials Science*, *30*(3), 225–242.

Nanocellulose: From Nature to High Performance Tailored Materials - Alain Dufresne - Google Books. (n.d.). Retrieved January 26, 2022, from https://books.google.co.in/books?hl=en&lr=&id=vElADwAAQBAJ&oi=fnd&pg=PR7&ots=E8sugBcvvt&sig=u4nqrFzKl7zA2Ep43pz0gHXx5AE&redir_esc=y#v=onepage&q&f=false

Nayak, A. K., & Hasnain, M. S. (Eds.). (2020). *Alginates in drug delivery*. Academic Press.

Ouros, A. C. D., Souza, M. O. D., & Pastore, H. O. (2014). Metallocene supported on inorganic solid supports: an unfinished history. *Journal of the Brazilian Chemical Society*, *25*, 2164–2185.

Pääkko, M., Ankerfors, M., Kosonen, H., Nykänen, A., Ahola, S., Österberg, M., Ruokolainen, J., Laine, J., Larsson, P. T., Ikkala, O., & Lindström, T. (2007). Enzymatic hydrolysis combined with mechanical shearing and high-pressure homogenization for nanoscale cellulose fibrils and strong gels. *Biomacromolecules*, *8*(6), 1934–1941. https://doi.org/10.1021/BM061215P

Peng, B. L., Dhar, N., Liu, H. L., & Tam, K. C. (2011). Chemistry and applications of nanocrystalline cellulose and its derivatives: A nanotechnology perspective. *Canadian Journal of Chemical Engineering*, *89*(5), 1191–1206. https://doi.org/10.1002/CJCE.20554

Phanthong, P., Reubroycharoen, P., Hao, X., Xu, G., Abudula, A., & Guan, G. (2018). Nanocellulose: Extraction and application. *Carbon Resources Conversion*, *1*(1), 32–43. https://doi.org/10.1016/J.CRCON.2018.05.004

Qi, X., & Tester, R. F. (2016). Effect of native starch granule size on susceptibility to amylase hydrolysis. *Starch - Stärke*, *68*(9–10), 807–810. https://doi.org/10.1002/STAR.201500360

Rajisha, K. R., Maria, H. J., Pothan, L. A., Ahmad, Z., & Thomas, S. (2014). Preparation and characterization of potato starch nanocrystal reinforced natural rubber nanocomposites. *International Journal of Biological Macromolecules*, *67*, 147–153. https://doi.org/10.1016/J.IJBIOMAC.2014.03.013

Ravi Kumar, M. N. V. (2000). A review of chitin and chitosan applications. *Reactive and Functional Polymers, 46*(1), 1–27. https://doi.org/10.1016/S1381-5148(00)00038-9

Rinaudo, M. (2006). Chitin and chitosan: properties and applications. *Progress in Polymer Science (Oxford)*, 31(7), 603–632. https://doi.org/10.1016/j.progpolymsci. 2006.06.001

Rinaudo, M. (2008). Main properties and current applications of some polysaccharides as biomaterials. *Polymer International, 57*(3), 397–430. https://doi.org/10.1002/PI.2378

Riva, R., Ragelle, H., des Rieux, A., Duhem, N., Jérôme, C., & Préat, V. (2011). *Chitosan and Chitosan Derivatives in Drug Delivery and Tissue Engineering* (pp. 19–44). https://doi. org/10.1007/12_2011_137

Rochima, E., Azhary, S. Y., Pratama, R. I., Panatarani, C., & Joni, I. M. (2017). Preparation and Characterization of Nano Chitosan from Crab Shell Waste by Beads-Milling Method. *IOP Conference Series: Materials Science and Engineering, 193*, 012043. https://doi. org/10.1088/1757-899X/193/1/012043

Maiti, S., Jayaramudu, J., Das, K., Reddy, S.M., Sadiku, R., Ray, S. S., & Liu, D. (2013). Preparation and characterization of nano-cellulose with new shape from different precursor. *Carbohydrate Polymers, 98*, 562–567.

Sanchez de la Concha, B. B., Agama-Acevedo, E., Nuñez-Santiago, M. C., Bello-Perez, L. A., Garcia, H. S., & Alvarez-Ramirez, J. (2018). Acid hydrolysis of waxy starches with different granule size for nanocrystal production. *Journal of Cereal Science, 79*, 193–200. https://doi.org/10.1016/J.JCS.2017.10.018

Saralkar, P., & Dash, A. K. (2017). Alginate nanoparticles containing curcumin and resveratrol: preparation, characterization, and in vitro evaluation against DU145 prostate cancer cell line. *AAPS PharmSciTech, 18*(7), 2814–2823. https://doi.org/10.1208/S12249-017-0772-7

Shabana, S., Prasansha, R., Kalinina, I., Potoroko, I., Bagale, U., & Shirish, S. H. (2019). Ultrasound assisted acid hydrolyzed structure modification and loading of antioxidants on potato starch nanoparticles. *Ultrasonics Sonochemistry, 51*, 444–450. https://doi. org/10.1016/J.ULTSONCH.2018.07.023

Singh, S. K. (2019). Solubility of lignin and chitin in ionic liquids and their biomedical applications. *International Journal of Biological Macromolecules, 132*, 265–277. https://doi. org/10.1016/j.ijbiomac.2019.03.182

Sivanesan, I., Muthu, M., Gopal, J., Hasan, N., Kashif Ali, S., Shin, J., & Oh, J.-W. (2021). Nanochitosan: commemorating the metamorphosis of an ExoSkeletal waste to a versatile nutraceutical. *Nanomaterials, 11*(3), 821. https://doi.org/10.3390/nano11030821

Soon, C. Y., Tee, Y. B., Tan, C. H., Rosnita, A. T., & Khalina, A. (2018). Extraction and physicochemical characterization of chitin and chitosan from Zophobas morio larvae in varying sodium hydroxide concentration. *International Journal of Biological Macromolecules, 108*, 135–142. https://doi.org/10.1016/j.ijbiomac.2017.11.138

Spinozzi, F., Ferrero, C., & Perez, S. (2020). The architecture of starch blocklets follows phyllotaxic rules. *Scientific Reports*, 1–16. https://doi.org/10.1038/s41598-020-72218-w

Sun, Q., Li, G., Dai, L., Ji, N., & Xiong, L. (2014). Green preparation and characterisation of waxy maize starch nanoparticles through enzymolysis and recrystallisation. *Food Chemistry, 162*, 223–228. https://doi.org/10.1016/J.FOODCHEM.2014.04.068

Tang, Z.-X., Qian, J.-Q., & Shi, L.-E. (2007). Preparation of chitosan nanoparticles as carrier for immobilized enzyme. *Applied Biochemistry and Biotechnology, 136*(1), 77–96. https://doi.org/10.1007/BF02685940

Thomas, M. S., Koshy, R. R., Mary, S. K., Thomas, S., & A. Pothan, L. (2019). *Starch, Chitin and Chitosan Based Composites and Nanocomposites*. Springer International Publishing. https://doi.org/10.1007/978-3-030-03158-9

Torres-Rendon, J. G., Femmer, T., de Laporte, L., Tigges, T., Rahimi, K., Gremse, F., Zafarnia, S., Lederle, W., Ifuku, S., Wessling, M., Hardy, J. G., & Walther, A. (2015). Bioactive gyroid scaffolds formed by sacrificial templating of nanocellulose and nanochitin hydrogels as instructive platforms for biomimetic tissue engineering. *Advanced Materials*, 27(19), 2989–2995. https://doi.org/10.1002/adma.201405873

Vaishnav, D. G. (2016). *Chapter - Introduction to Nanomaterials. December 2011.*

Wang, N., Ding, E., & Cheng, R. (2007). Thermal degradation behaviors of spherical cellulose nanocrystals with sulfate groups. *Polymer*, 48(12), 3486–3493. https://doi.org/10.1016/J.POLYMER.2007.03.062

Wiegand, C., Moritz, S., Hessler, N., Kralisch, D., Wesarg, F., Müller, F. A., Fischer, D., & Hipler, U. C. (2015). Antimicrobial functionalization of bacterial nanocellulose by loading with polihexanide and povidone-iodine. *Journal of Materials Science. Materials in Medicine*, 26(10). https://doi.org/10.1007/S10856-015-5571-7

Wu, J., Zhang, K., Girouard, N., & Meredith, J. C. (2014). Facile route to produce chitin nanofibers as precursors for flexible and transparent gas barrier materials. *Biomacromolecules*, 15(12), 4614–4620. https://doi.org/10.1021/bm501416q

Wypych, G. (2016). *Handbook of Fillers: Fourth Edition*, 1–922. ChemTec Publishing. https://doi.org/10.1016/C2015-0-01953-0

Xiao, H., Yang, F., Lin, Q., Zhang, Q., Zhang, L., Sun, S., Han, W., & Liu, G. Q. (2020). Preparation and characterization of broken-rice starch nanoparticles with different sizes. *International Journal of Biological Macromolecules*, 160, 437–445. https://doi.org/10.1016/J.IJBIOMAC.2020.05.182

Yang, H.-C., Wang, W.-H., Huang, K.-S., & Hon, M.-H. (2010). Preparation and application of nanochitosan to finishing treatment with anti-microbial and anti-shrinking properties. *Carbohydrate Polymers*, 79(1), 176–179. https://doi.org/10.1016/j.carbpol.2009.07.045

Yang, X., Liu, J., Pei, Y., Zheng, X., & Tang, K. (2020). Recent progress in preparation and application of nano-chitin materials. *Energy & Environmental Materials*, 3(4), 492–515. https://doi.org/10.1002/eem2.12079

Zareie, C. (2013). Preparation of nanochitosan as an effective sorbent for the removal of copper ions from aqueous solutions. *International Journal of Engineering*, 26(8 (B)). https://doi.org/10.5829/idosi.ije.2013.26.08b.04

Zhu, F. D., Hu, Y. J., Yu, L., Zhou, X. G., Wu, J. M., Tang, Y., Qin, D. L., Fan, Q. Z., & Wu, A. G. (2021). Nanoparticles: a hope for the treatment of inflammation in CNS. *Frontiers in Pharmacology*, 12, 1114. https://doi.org/10.3389/FPHAR.2021.683935/BIBTEX

9 Formulation of Polymer Nano-Composite NPK Fertilizer [Cellulose-*Graft*-Poly (Acrylamide)/ Nano-Hydroxyapatite/ Soluble Fertilizer] and Evaluation of Its Nutrient Release

Kiplangat Rop, George N. Karuku,
and Damaris Mbui
University of Nairobi

CONTENTS

9.1   Introduction 258
9.2   Materials and Methods 259
    9.2.1   Synthesis of Hydroxyapatite (HA) Nano-Particles...........................259
    9.2.2   Formulation of Cellulose-g-Poly(Acrylamide)/Nano-HA/Soluble
        Fertilizer Composite 260
    9.2.3   Chemical Characterization of Nano-HA and Nano-Composite
        Fertilizer 260
    9.2.4   Evaluating Nutrient Release from Nano-Composite Fertilizer by
        Soil Incubation Experiment..260
        9.2.4.1   Study Site 260
        9.2.4.2   Soil Sampling and Preparation for Incubation Experiment 261
        9.2.4.3   Soil Characterization Before the Onset of Incubation
               Experiment...261
        9.2.4.4   Fertilizer Nano-Composite Samples and Laboratory
               Incubation Experiment...261
        9.4.4.5   Estimation of Nitrogen Mineralization Potential (N_0).......262
    9.2.5   Statistical Data Analysis...263

DOI: 10.1201/9781003279372-9

9.3   Results and Discussion 263
          9.3.1   FTIR Spectroscopic Analysis...263
          9.3.2   Transmission Electron Microscopy 266
          9.3.3   Chemical Characteristics of Soil at the Onset of Incubation
                      Experiment..267
          9.3.4   Nitrogen Mineralization 267
          9.3.5   Available Phosphorous...273
          9.3.6   Exchangeable Potassium..274
9.4   Conclusion 275
References..276

9.1   INTRODUCTION

Soil fertility decline contributes to low-crop yields due to the lack of nutrient resources, imbalanced nutrient mining, reduced fallow periods, fewer rotations and soil erosion, among other limiting factors (Kimetu et al., 2007; Mucheru-Muna et al., 2014). To increase and sustain crop yields, farmers apply chemical fertilizers such as diammonium phosphate (DAP), triple superphosphate, nitrogen, phosphorous and potassium (NPK), monoammonium phosphate, single super phosphate (SSP), calcium ammonium nitrate (CAN), urea (Mathenge, 2009; Shoals, 2012; Oseko and Dienya, 2015) and to some extent, organic manure, to supply the most-limiting nutrients (NPK). However, the use of considerable amounts of chemical fertilizers in the sub-humid zones results in low nutrient use efficiency (NUE) due to leaching. Split application by top dressing is known to improve NUE, but small-scale poor-resource farmers consider it a luxury or apply below recommended rates (Mucheru-Muna et al., 2014) leading to poor crop performance. On the other hand, exclusive use of organic manure is limited by bulkiness, low nutrient quality, low nutrient mineralization (Makokha et al., 2001; Gaskell and Smith, 2007; Leye and Omotayo, 2014) and extra labour that adds to the production cost.

   Of the amount of fertilizers applied to the farms, only a small percentage is utilized by plants while the rest is eventually washed into water bodies (Tolescu et al., 2009) through leaching and surface runoff or lost by volatilization under reduced conditions. About 40%–70% N and 80%–90% P of fertilizers applied in the farms are lost to the environment resulting not only in economic and resource losses but also environmental pollution (Guo et al., 2005; Naderi and Danesh, 2013; Giroto et al., 2017). Efforts have been made to minimize these challenges by developing new-generation fertilizers, the so-called "smart" fertilizers. Among them are slow-release fertilizers (SRF) which contain at least one nutrient that either delays its availability and utilization processes or is available for plant uptake for a longer period than a standard fertilizer considered "quickly available" (Zeroual and Kossir, 2012; Chen et al., 2013). The availability of nutrients is prolonged by either slowing the release or altering reactions leading to losses (Olson-Rutz et al., 2011). The enhancement of NUE implies more efficient food production and reduced cost for environmental protection (Naderi and Danesh, 2013; Yuejin et al., 2013). Further, SRF can be applied as a pre-plant application, and the need for split application is eliminated, reducing production costs (Chen et al., 2013).

In the face of resource scarcity and ever-increasing population, development in agriculture can be achieved exclusively through the effective use of modern technologies. To date, intensive research is directed towards integrating nano-technologies into fertilizer development or formulation. Due to the high surface-area-to-volume ratio, nano-fertilizers are expected to be more effective than polymer-coated conventional SRF as they enhance NUE, reduce toxicity and minimize the potential negative effects associated with an excess application such as groundwater pollution (DeRosa et al., 2010; Naderi and Danesh, 2013; Monreal et al., 2015; Montalvo et al., 2015). Hydroxyapatite (HA) nano-particles are rated as one of the prominent candidates for potential agricultural nutrient sources (Kottegoda et al., 2011). However, much of the available data on HA is mainly focused on biomedical applications (Mateus et al., 2007; Pang et al., 2010; Pataquiva-Mateus et al., 2013) while the agricultural application is lacking.

More recently, there has been an increasing interest in the use of polymer hydrogels (PHGs) in agricultural production. PHGs are macromolecular networks with the ability to swell or shrink in the presence or absence of water due to hydrophilic groups such as $-SO_3H$, $-CONH-$, $-CONH_2$, and $-OH$ and a slightly cross-linked structure which resists dissolution (Sannino et al., 2009; Qiu and Hu, 2013; Laftah and Hashim, 2014; Ahmed, 2015). Polyacrylamide (PAM) is used as a chemical intermediate in the production of PHGs with high absorption capacity (super-absorbents) such as disposable diapers and medical and agricultural products, among others (Charoenpanich, 2013; Laftah and Hashim, 2014). High-molecular-weight PAM is added into the soil through irrigation water as anti-erosion additive (Charoenpanich, 2013) and it has been reported to be degraded by native soil bacterial species such as *Bacillus, Pseudomonas and Rhodococcus,* among others and also fungi (*Aspergillus*) which are capable of accessing N through amidase activity (Kay-Shoemake et al., 1998; Yu et al., 2015). Extracellular amidase enzyme catalyzes the hydrolysis of the C-N bond of the amides, resulting in the generation of NH_3, NH_4^+ and carboxylic acid groups ($-COOH$). The production of NH_3 under moist conditions contributes to mineral N in the soil, whereas carboxylic acid is further degraded by microorganisms as the source of carbon (energy) to CO_2 and H_2O, thus being environment-friendly. PAM-treated agricultural soil has been experimentally demonstrated by Kay-Shoemake et al. (1998) to exhibit higher bacterial counts and high inorganic N concentration and amidase activity, hence considered healthier soil than the untreated ones. In this study, cellulose-grafted PAM polymer hydrogel was utilized in the formulation of slow-release nano-composite fertilizer, and the release of nutrients was assessed using laboratory incubation experiment.

9.2 MATERIALS AND METHODS

9.2.1 SYNTHESIS OF HYDROXYAPATITE (HA) NANO-PARTICLES

The methodology used in the synthesis was adopted from Kottegoda et al. (2011) with some modifications namely introducing a surfactant. $Ca(OH)_2$ 7.716 g was weighed into a beaker and 0.22 mM TX-100 (non-ionic surfactant) solution was added to make a total volume of 100 mL and the mixture was stirred for 30 minutes with a motorized stirrer. A 100 mL solution of 0.6 M H_3PO_4 was added into the suspension of $Ca(OH)_2$ drop-wise (15 mL/min) from the burette while stirring vigorously at 1000 rpm. After the reaction, the dispersion was stirred for 10 minutes and then allowed to age for 2 hours. It was then oven-dried at 105°C to constant weight and then pulverized into a fine powder. The surfactant was removed by washing the powder with methanol.

9.2.2   Formulation of Cellulose-g-Poly(Acrylamide)/ Nano-HA/Soluble Fertilizer Composite

Thirty millilitres (pre-determined volume containing 0.8 g dry weight) of cellulose fibres, 1.0 g nano-HA and varied amounts of soluble NPK fertilizer blend $(NH_4)_2HPO_4$, urea and K_2SO_4 weight ratio 3:5:2, respectively, were transferred into a 3-necked flask. The flask was fitted with a reflux condenser and nitrogen line and then placed in a thermostated water bath equipped with a magnetic stirrer. Nitrogen gas was bubbled through the mixture for 10 minutes, as the temperature was gradually raised to 70°C. Ammonium persulphate 0.1 g was added to the mixture and stirred for 30 minutes to generate radicals. Acrylic acid (AA) 2.7 mL partially neutralized with NH_3 to a 70% degree of neutralization was added to 0.25 g of *N,N*-methylene– *bis*-acrylamide, stirred to dissolve and then introduced into the reaction mixture. The total volume of the reaction mixture was adjusted to 40 mL. The mixture was stirred for an additional 1 minute after which the reaction was allowed to proceed for 2 hours. The reaction product was then cooled to room temperature, removed from the flask and then cut into regular pieces. NH_3 solution (1:1) was added drop-wise to adjust the pH to 8. The nano-composite fertilizer was then oven-dried at 60°C to constant weight and then pulverized to pass through a 1 mm sieve.

9.2.3   Chemical Characterization of Nano-HA and Nano-Composite Fertilizer

Nano-HA and cellulose-g-poly(acrylamide)/nano-HA/soluble fertilizer composite were characterized using Fourier-transform infrared (FTIR) spectrophotometer, Shimadzu IRAffinity-1S, by scanning the samples between 4000 and 400 cm⁻¹. The morphology was assessed by transmission electron microscopy (HRTEM, Tecnai G2 F20 X-TWIN MAT).

9.2.4   Evaluating Nutrient Release from Nano-Composite Fertilizer by Soil Incubation Experiment

9.2.4.1   Study Site

The study was carried out at the Faculty of Agriculture, Upper Kabete Campus, University of Nairobi, located in Kiambu County, coordinates 1° 15′ S and 36° 44′ E and an altitude of 1940 m above sea level. The site is representative of large areas of central highlands of Kenya in terms of soils and climate, and the geology is composed of the Nairobi Trachyte of the Tertiary age. The soils are very deep (>180 m), well-drained, dark-red to dark-reddish-brown, friable clays with moderate-to-high inherent fertility (Gachene, 1989; Kimetu et al., 2007; Karuku et al., 2012; Mucheru-Muna et al., 2014) and are classified as humic Nitisols (WRB, 2015). The climate is characterized as semi-humid according to the Kenya Soil Survey agro-climatic zonation (Sombroek et al., 1982). The annual average rainfall to annual potential evaporation (r/Eo) is 58% (Karuku et al., 2012, 2014a, 2014b). The average monthly maximum and minimum temperatures are 23.8°C and 12.6°C, respectively. The area experiences bimodal rainfall distribution (long rains from mid-March to May and short rains from October to December) and the average annual precipitation is 1000mm (Jaetzold et al., 2006). Crops grown in the area include kale (*Brassica oleracea*), tomatoes (*Lycopersicon esculentum*), cabbage (*Brassica oleracea*), carrots (*Daucus carota*), onions (*Allium fistulosum*), beans (*Phaseolus vulgaris*), maize (*Zea mays*) and coffee (*Coffea arabica*).

9.2.4.2 Soil Sampling and Preparation for Incubation Experiment

The surface litter that included leaves, sticks, stumps and other materials was removed gently to expose the surface soil. Soil samples at a depth of 0–20cm were randomly collected in a transect at selected points in a field measuring 50 × 50m using a 600cm^3 soil auger and bulked to make a composite sample. The composite sample was air-dried in the laboratory and crushed to pass through a 4mm sieve to remove large pieces of surface materials. A portion of the composite sample was set aside for chemical characterization and the other portion for incubation experiments.

9.2.4.3 Soil Characterization Before the Onset of Incubation Experiment

A sub-soil sample was air-dried in the laboratory and crushed to pass through a <1mm sieve. Total N was determined by the micro-Kjeldahl method (Bremner, 1996) and available P by the Mehlich-1 method (Mehlich, 1953). Exchangeable cations (Ca, K and Mg) were extracted with 1M NH_4OAc; Ca and Mg contents were then determined in the leachate by atomic absorption spectroscopy and K by flame photometry. Organic carbon (OC) was determined using the wet oxidation method (Nelson and Sommers, 1996) and the organic matter (OM) was calculated by multiplying the % of OC by 1.724. Soil texture was determined by the hydrometer method (Glendon and Doni, 2002). Soil pH-H_2O was determined with a pH metre (with glass electrodes) at a 1:2.5 soil-to-water (salt) ratio while the electrical conductivity was measured at a 1:2.5 (soil-to-water ratio) extract using a conductivity metre. Cation exchange capacity (CEC) was determined by leaching the soil with NH_4OAc at pH 7, and the NH_4^+ concentration in the leachate was determined through steam distillation by the micro-Kjeldahl method (Bremner, 1996).

9.2.4.4 Fertilizer Nano-Composite Samples and Laboratory Incubation Experiment

Table 9.1 shows the composition of the nano-composite fertilizer and the amount of NPK added to the soil for incubation experiments. T1 is cellulose-g-poly(acrylamide) polymer hydrogel (PHG) and T2-T6 was formulated to contain the cellulose-g-poly

TABLE 9.1

The Composition of the Formulated Nano-Composite Fertilizer and the Amounts of N, P and K, in the Treatments

Treatment	Composition of Nano-Composite Fertilizer (% w/w)				Soil Treatment (mg/kg)		
	N-P$_2$O$_5$-K$_2$O	PHG	SF	Nano-HA	N	P$_2$O$_5$	K$_2$O
T1	14: 0: 0	100	0	0	50	0	0
T2	13.8: 18.8: 4.6	68.8	16.0	16.0	50	68.1	16.6
T3	15.4: 20: 4.8	62.7	22.3	15.0	50	65.0	15.6
T4	16.8: 21.5: 6.2	57.7	28.5	14.3	50	64.0	18.4
T5	20: 24: 9.7	44.5	44.4	11.1	50	60.0	24.2
T6	21.3: 25.2: 11.6	25.0	66.0	9.0	50	58.5	27.2
T7	22: 24: 11	0	100	0	50	54.5	24.9

PHG, polymer hydrogel; SF, soluble fertilizer; nano-HA. nano-hydroxyapatite.

(acrylamide)/nano-HA/soluble fertilizer. The amount of soluble fertilizer (SF) in the composites was increased from T2 to T6 while the content of PHG and nano-HA decreased, and T7 represented a conventional fertilizer (CF).

The formulated nano-composite fertilizer (<1 mm) was added to the soil at the rate of 50 mgN/kg, thoroughly mixed and then placed in plastic incubation bags. This corresponded to 100 kgN/ha, recommended for N application for maize in Kiambu County (planting, 250 kg/ha NPK 23:23:0; top dressing with CAN, 125 kg/ha). The amount of the fertilizer added to 1 kg of soil includes T1–397 mg, T2–362 mg, T3–325 mg, T4–298 mg, T5–250 mg, T6–235 mg and T7–227 mg. The treatments were replicated three times, with untreated soil serving as the control. Distilled water was added to field capacity (30% w/w) and the bags were sealed and incubated in the dark at 20°C for 16 weeks. The amount of mineral N (NH$_4$-N and NO$_3$-N), P and K was followed biweekly since the onset of incubation. The soil was kept moist at field capacity throughout the incubation period by adding distilled water where the feel method was used to establish the necessity. Aerobic conditions were maintained by opening plastic bags periodically to allow aeration. Each of the samples was divided into two portions at the time of sampling. For one portion, available N (NH$_4$-N and NO$_3$-N) was extracted and quantified. The other portion was air-dried before analysing for total N, available P and K.

9.4.4.5 Estimation of Nitrogen Mineralization Potential (N$_0$)

Potentially mineralizable N (N_0), which is the fractional quantity N susceptible to mineralization, was estimated using single first-order kinetics employed by Stanford et al. (1974) and later adopted by Karuku (1989), and Karuku and Mochoge (2018), Equation 9.1.

$$\frac{dN}{dt} = -kN \qquad (9.1)$$

The integration gives Equation 9.2,

$$\text{Log } (N_0 - N_t) = \log N_0 - kt/2.303 \tag{9.2}$$

where N_t is the cumulative N mineralized at time t (days), N_0 is the amount of potentially mineralizable N and k is the first-order rate constant (day^{-1}). Stanford et al. (1974) and Karuku and Mochoge (2018) found the rate constant k to be reasonably equal for a large number of soils, and a period of 2 weeks short-term pre-incubation was sufficient to estimate the mineralization potential (N_0) using simplified Equation 9.3.

$$N_0 = 9.77 \ N_t \tag{9.3}$$

where N_0 is nitrogen mineralization potential and N_t is nitrogen mineralized in 2 weeks.

9.2.5 Statistical Data Analysis

The data were subjected to ANOVA using IBM SPSS Statistics version 20. A Tukey honest significant difference post hoc test was used to compare and assess the significance of the mean values at a probability level of $P \le 0.05$.

9.3 RESULTS AND DISCUSSION

9.3.1 FTIR Spectroscopic Analysis

The Fourier-transform infrared (FTIR) spectrum of HA nano-particles is shown in Figure 9.1. The absorption bands at 1419 and 875 cm^{-1} correspond to CO_3^{2-} ions, attributed to the physical interaction of HA with CO_2 during the synthesis at ambient conditions (Iyyappan and Wilson, 2013). The spectrum observed in the study is similar to that of Costescu et al. (2010), who reported decreased intensity of the peaks related to CO_3^{2-} at high calcination temperatures of 600 and 1000°C. The broad and weak bands at 3600–3000 cm^{-1} and 1635 cm^{-1} are attributed to H-O-H of lattice water (Costescu et al., 2010) and also stretching vibrations of –OH ions in the nano-HA lattice (Cisneros-Pineda et al., 2014; Gayathri et al., 2018). The characteristic bands for PO_4^{3-} group appear at 1022 and 964 cm^{-1} due to stretching vibrations and those at 601 and 563 cm^{-1} corresponding to bending vibrations. The band characteristics of C-H stretch at 2928 and 2856 cm^{-1} due to –CH$_3$ and –CH$_2$ respectively were found to be absent in the FTIR spectrum. This confirms a complete removal of T-X 100 upon washing HA nano-particles with methanol.

FTIR spectrum of nano-composite fertilizer is shown in Figure 9.2. The broad and strong bands at 2500–3500 cm^{-1} can be assigned to the O–H stretch due to carboxylic acid (AA) and alcoholic group (cellulose), and also, N–H for the amide group (acrylamide). The bands at 3182 and 1543 cm^{-1} are assigned to the N–H stretching vibration for primary amides (Bundela and Bajpai, 2008).

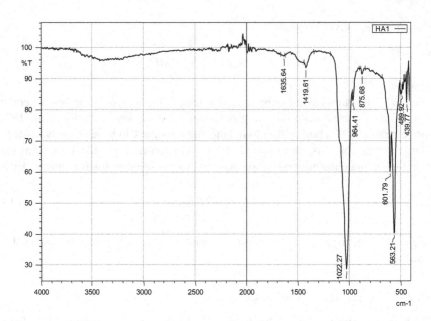

FIGURE 9.1 FTIR spectrum of hydroxyapatite nano-particles.

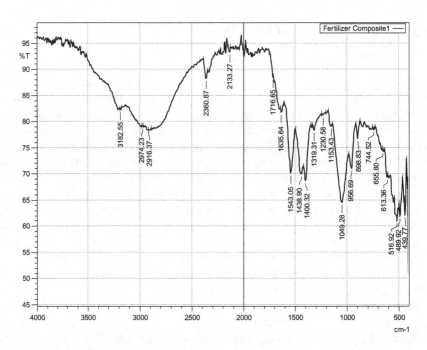

FIGURE 9.2 FTIR spectrum of cellulose-g-poly(acrylamide)/nano-HA/soluble fertilizer composite.

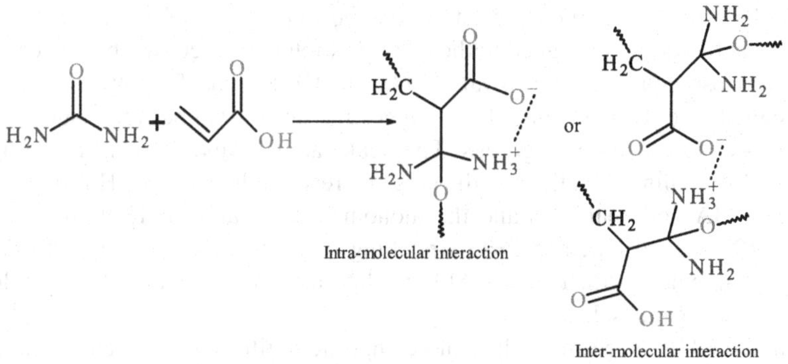

SCHEME 9.1 Radical polymerization between urea and acrylic acid.

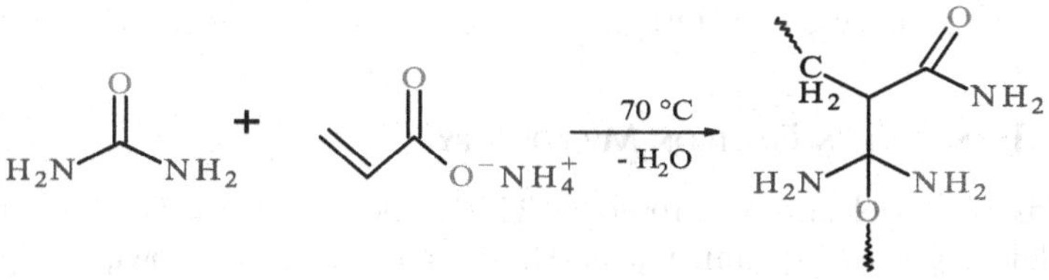

SCHEME 9.2 Radical polymerization between urea and ammonium acrylate.

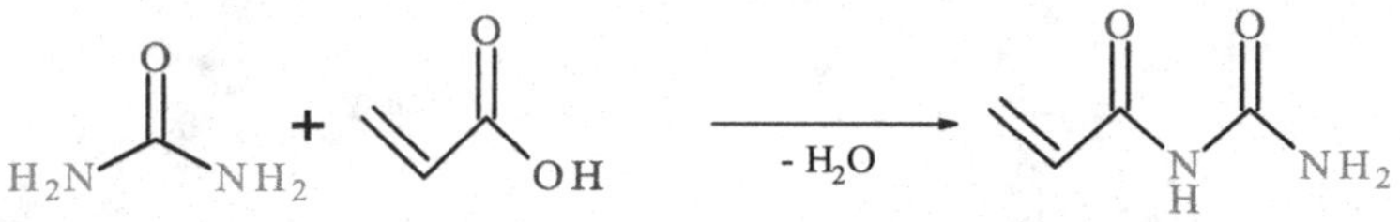

SCHEME 9.3 Condensation reaction between urea and acrylic acid.

Eritsyan et al. (2006) and Fernandes et al. (2015) proposed a radical polymerization reaction mechanism between AA and urea via the carbonyl carbon as shown in Scheme 9.1. According to these authors, the moderately strong band at 1635 cm^{-1} (Figure 9.2) is assigned to the adsorption of NH_3^+ and COO^- groups as a result of intra- and intermolecular interactions between –COOH and –NH_2 which lead to the formation of a salt.

Since AA was partially neutralized with NH_3, ammonium acrylate could react the same way according to Scheme 9.2. Alongside radical polymerization, a condensation reaction between urea and AA may also occur (Fernandes et al., 2015), yielding a branched copolymer according to Scheme 9.3.

The strong band at 1438–1400 cm^{-1} (Figure 9.2) corresponds to the O–H bending vibration for carboxylic acid, revealing incomplete neutralization of AA. Spectral bands at 1153 to 1049 cm^{-1} are assigned to C–O–C bridging resulting from the reaction between ammonium acrylate (monomer) and the –OH group of cellulose. The band at 898 cm^{-1} is assigned to the C–O–C stretch of glycosidic bonds for amorphous cellulose (Synytsya and Novak, 2014). The FTIR peaks from 1049 to 920 cm^{-1} are assigned to P–O–C (Figure 4.13), suggesting an overlap between bands attributed to C–O–C and P–O–C groups. The peaks at 1049 and 956 cm^{-1} are assigned to P–O–C stretching vibrations, indicating the reaction between –OH groups at the surface of HA nano-particles and the monomer. The bands at 1319 and 516 cm^{-1} (Figure 9.2) are attributed to P=O stretching vibrations for the PO_4^{3-} ion due to $(NH_4)_2HPO_4$, whereas the bands at 613 and 439 cm^{-1} correspond to SO_4^{2-} ion due to K_2SO_4, i.e. inorganic salts.

From the FTIR spectrum of the nano-composite fertilizer, there is an indication of the existence of chemical interactions between (i) the monomer, cellulose and nano-HA and (ii) monomer and urea molecules towards the formation of 3-D network structure. Additionally, due to the large surface area of HA nano-particles, Kottegoda et al. (2011, 2017) associated the formation of urea-HA nano-hybrid (molar ratio, 6:1) to the existence of H-bonds between the –OH group on the surface of HA and –NH$_2$ group of urea.

9.3.2 Transmission Electron Microscopy

The transmission electron microscopy (TEM) micrographs of HA nano-particles and cellulose-g-poly(acrylamide)/nano-HA/SF composite are shown in Figure 9.3.

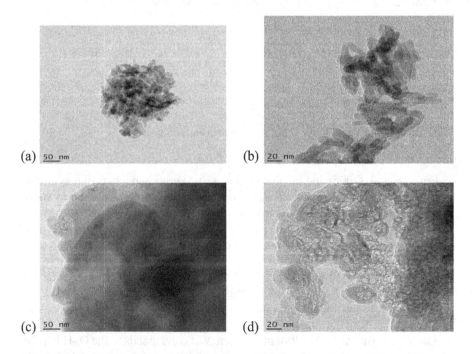

FIGURE 9.3 TEM micrographs of (a,b) HA nano-particles, (c,d) cellulose-g-poly(acrylamide)/nano-HA/soluble fertilizer) composite, at 50 and 20 nm scales.

The images of nano-HA displayed rod-shaped nano-particle agglomerates with particle size <50 nm (Figure 9.3a, b). This is consistent with the observation made by Iyyappan and Wilson (2013) who attributed the formation of rod-shaped nano-HA to a hydrophobic ring complex formed through the ion-dipole interaction between the Ca^{2+} and polyoxyethylene group of TX-100. The agglomeration of nanoparticles observed in Figure 9.3a, b was attributed to the high specific surface energy that led to aggregation (Pramanik et al., 2009; Ragu and Sakthivel 2014). TEM images of the fertilizer composite (Figure 9.3c, d) showed the dispersion of HA nano-particles and the salt crystals. Similarly, Ragu and Sakthivel (2014) observed diminished agglomeration of nano-HA in polymethyl methacrylate/nano-HA composite at about 20–50 nm. This was attributed to the reduction in surface energy of nano-HA by the polymer through its pendent PO_4^{3-} group.

9.3.3 CHEMICAL CHARACTERISTICS OF SOIL AT THE ONSET OF INCUBATION EXPERIMENT

Table 9.2 shows the salient characteristics of the soil used in the study. The soils at the site were acidic with low available P content. The soil acidity could be attributed to the humid conditions in the central highlands of Kenya which lead to the leaching of Ca, Mg and K and other basic cations. Low amounts of available P could be attributed to the soil acidity which renders P unavailable through fixation by Al, Fe and Mn and also continuous removal by crops.

9.3.4 NITROGEN MINERALIZATION

The mineral N (NH_4-N + NO_3-N) content during 16 weeks of incubation is shown in Table 9.3. The results show low mineral N content in the initial stages of incubation with some decrease in the 4th week followed by a significant increase through the 8th week and a peak at the 12th week and then a decline in the 16th week. Low

TABLE 9.2
Some Salient Chemical Characteristics of Soil Used in Incubation Experiment

Parameters	Units	Values
pH (soil:H_2O, 1:2.5)	–	5.25
pH ($CaCl_2$:1:2.5)	–	4.50
Electrical conductivity	ds/m	0.26
Cation exchange capacity	Cmol(+) kg^{-1}	15.62
Total N	%	0.29
Available P	ppm	8.50
Exchangeable K	Cmol(+) kg^{-1}	1.10
Ca	Cmol(+) kg^{-1}	8.51
Mg	Cmol(+) kg^{-1}	4.26

TABLE 9.3

Mineral N (NO₃-N + NH₄-N) Content (ppm) During 16 Weeks of Incubation Period

	Incubation Period (Weeks)					
Treatment	2	4	8	12	16	Cumulative Mineral N at the 16th week
Cntrl	43.6a	24.2a	107.2a	145.8a	85.1a	405.8a
T1	50.8ab	33.0b	152.9b	176.6ab	117.9b	531.1b
T2	59.0abc	44.1cd	176.4c	200.1bc	128.4cd	608.0cd
T3	55.0abc	41.5c	189.5cd	192.2bc	119.3b	597.4c
T4	55.3abc	46.0cd	205.1de	266.1e	139.1e	711.7e
T5	63.5bc	42.9c	190.1cde	210.6cd	124.7bcd	631.8cd
T6	58.3abc	48.8cd	211.7e	242.1de	131.2de	692.1e
T7	73.2c	51.7d	184.9cd	207.5bc	129.6d	646.9d

Notes: Different letters in the same column are significantly different ($p \leq 0.05$ level). Cntrl = no treatment, T1 = 14: 0: 0, T2 = 13.8: 18.8: 4.6, T3 = 15.4: 20: 4.8, T4 = 16.8: 21.5: 6.2, T5 = 20: 24: 9.7, T6 = 21.3: 25: 11.2, T7 = 24: 22: 11.

mineralization of N in the initial stages of incubation reflects the lag phase (Deenik and Yost, 2008) associated with the immobilization of nutrients by microorganisms to nourish and diversify their species and biomass (Karuku and Mochoge, 2016, Karuku and Mochoge, 2018; Tambone and Adani, 2017). The inorganic nutrients, sufficient water, carbon sources and trace elements are essential for the maintenance and population growth of microorganisms. The incubation period between the 4th and 12th week relates to the microbial exponential growth phase where microbes proliferated and were able to act on the substrate. After they are satiated, the remaining portion of the substrate is mineralized into the soil. The decline in mineral N content after the 12th week of incubation may be attributed to the depletion of the mineralizable substrate and the possibility of immobilization by microbes as their population had proliferated during the long incubation period (Karuku, 2019). This mineralization pattern may favour annual crops such as maize since the uptake of N is slow at establishment and moderate at development and reproductive growth stages and declines at maturity as the plant reaches senescence.

The release of N from the SRF composites T2 to T6 [cellulose-g-poly(acrylamide)/nano-HA/SF composite] is thought to have occurred in two phases: (i) diffusion of urea-N and NH₄-N and (ii) hydrolysis of amide-N according to Liu et al. (2007). The highest content of mineral N in the first 4 weeks was observed in CF (T7) and was attributed to the urea-N hydrolysis and release of NH₄-N in DAP. However, no significant difference was observed between T7 and T2 to T6 in the initial period of incubation. This non-significant difference may be attributed to the small particle sizes of SRF composites (<1 mm) that enabled rapid diffusion of SF into the soil. Thus, the rate of diffusion of soluble N may be decreased by increasing the particle size of the fertilizer composite. Significant higher mineral N content was observed in T6 in the 8th and 12th weeks and T4 in the 12th and 16th weeks, relative to T7. The lower values observed in T7 (relative to T4 and T6) during this period (8th–16th week) may be attributed to early exposure of NH_4-N to exchangeable sites in the soil and the possibility of immobilization through fixation. The higher mineral N content in T4 and T6 (than T7) in the later stages of incubation (8th–12th week) was attributed to its reduced interaction with soil particles in the early stages of incubation and release of hydrolyzed amide-N. NH_4^+ fixation occurs in clay soils due to the formation of NH·O bond in the hexagonal holes and the balancing of positive charge deficiency which arises from the isomorphous substitution of Si^{4+} and Al^{3+} ions (Chen, 1997). No significant difference in mineral N content was observed in the 12th week between T7 and T1, T2 and T3, indicating considerable mineralization of N in all SRF treatments. The observed increase in mineral N contents between the 8th and 16th week in SRF treatments (T1 to T6) may be attributed to the release through hydrolysis of amide-N.

Cumulative mineral N generally increased from T1 to T6 due to the increased proportion of SF and decreased content of nano-HA in the fertilizer formulation (Table 3.1). N mineralization is a biological process; hence, its release depends on the chemical constituent of the fertilizer such as N content and C:N ratio (Masunga et al., 2016) as well as lignin and lingocellulose ratio, among other factors (Karuku et al., 2014b; Karuku, 2019). The C:N ratio influences the mineralization rate, as microorganisms have to immobilize N to break carbon bonds/chains in the organic material for their energy requirement (Dong et al., 2012; Karuku and Mochoge, 2016; Tambone and Adani, 2017). The significant difference observed in T1 mineral N content compared to all other SRF treatments, from the 4th to the 16th week, was related to the carbon content in the formulation. The slow mineralization of N in T1 (100% of PHG which is composed of carbonaceous material of acrylic and cellulose chains) may have been due to its higher carbon content, compared to other SRF treatments (Table 9.1). The proportion of carbon decreased on incorporating SF and nano-HA into PHG.

Nevertheless, significantly higher mineral N content in T1 compared with the control indicates substantial hydrolysis of the amide-N within the incubation period. Cellulose chains being part of the polymer composite provide easily degradable C to the microorganisms, enhancing the breakdown of the copolymer, hence N release. The addition of organic material to PAM-amended soil has been reported to effect degradation through increased microbial activity. Higher amounts of soil aggregating fungi were reported by TonThat et al. (2008) in macro-aggregates generated from PAM-wheat residue-amended soil compared to the control. Award et al. (2012) also reported a stimulating effect of synthesized PAM biopolymer (BP) and biochar (BC) on the decomposition of soil OM and maize residue. These workers observed higher enzymatic activities in both BP- and BC-amended soil compared to the control and, fungi were implicated for contributing largely to plant residue decomposition. Watson et al. (2016) observed stimulation of nitrification and C mineralization in maize straw-amended soil conditioned with PAM, a phenomenon attributed to improved microbial conditions and partial utilization of PAM as a substrate.

The addition of mineral N from an inorganic source to organic fertilizer enhances the decomposition of organic material (Abbasi and Khaliq 2016, Karuku, 2019;

Karuku and Mochoge, 2018). Further, cultures of bacteria derived from agricultural soils have been reported to utilize PAM as an N source (Kay-Shoemake et al., 1998). The bacterial strain (*Pseudomonas putida* H147) studied by Yu et al. (2015) showed a 31.1% of degradation efficiency of PAM in 7 days and exceeded 45% under optimum culture conditions. Degraded PAM showed low-molecular-weight oligomer derivatives while acrylamide monomers did not accumulate. The observed increase in the content of mineral N in SRF treatments T2 to T6 could be attributed to the increased amount of soluble N from DAP and easily hydrolyzable N from urea. The microbes were provided with an easy source of N, and to acquire energy (carbon), the polymer had to be degraded. The polymer medium supplemented with mineral N, liquid paraffin and sucrose has been shown to contribute to PAM degradation and microbial biomass compared to the control (Yu et al., 2015). The significantly (p ≤0.05) lower content of mineral N observed in SRF treatment T5 relative to T4 and T6 could be attributed to the experimental errors.

Table 9.4 shows ammonium-N and nitrate-N contents during different incubation stages. The NH₄-N content in T7 was significantly higher (p ≤0.05) than T1, T2, T4

TABLE 9.4
NH₄-N and NO₃-N Content (ppm) During 16 Weeks of Incubation Period

	Incubation Period (Weeks)					
Treatment	2	4	8	12	16	Cumulative Mineral N after 16 Weeks
NH₄-N						
Cntrl	29.4 a	17.1 a	72.5 a	87.0 a	53.3 a	259.0 a
T1	34.0 ab	24.5 b	83.0 ab	99.6 ab	69.3 b	310.3 b
T2	38.8 c	26.8 b	97.4 bc	116.8 bc	78.9 cd	358.7 cd
T3	34.0 ab	25.0 b	95.4 bc	114.5 bc	73.6 bc	342.5 bc
T4	35.1 ab	29.3 b	109.8 c	131.7 c	86.4 d	392.3 bc
T5	35.8 ab	23.5 ab	86.29 ab	103.5 ab	78.9 cd	328.0 bc
T6	38.8 c	28.3 b	113.7 c	136.4 c	78.9 cd	396.1 d
T7	38.8 c	27.9 b	89.5 ab	107.4 ab	73.6 bc	342.6 c
NO₃-N						
Cntrl	14.2 a	7.2 a	34.6 a	58.8 a	31.7 a	146.5 a
T1	16.8 ab	8.7 b	69.9 b	77.0 ab	48.5 b	220.8 b
T2	20.2 abc	17.2 cd	79.1 c	83.3 bc	49.5 cd	249.2 c
T3	20.9 abc	16.5 c	94.1 cd	77.7 bc	45.7 bc	254.9 c
T4	20.2 abc	16.7 c	95.4 de	134.4 e	52.7 e	319.3 d
T5	27.6 bc	19.5 c	103.9 cde	107.1 cd	45.7 bcd	303.8 d
T6	19.4 bc	20.5 cd	98.1 e	105.7 de	52.3 de	295.9 d
T7	34.3 c	23.8 d	95.4 cd	100.1 bc	56.0 d	309.7 d

Notes: Different letters in the same column are significantly different (*p* ≤ 0.05 level). Cntrl = no treatment, T1 = 14: 0: 0, T2 = 13.8: 18.8: 4.6, T3 = 15.4: 20: 4.8, T4 = 16.8: 21.5: 6.2, T5 = 20: 24: 9.7, T6 = 21.3: 25: 11.2, T7 = 24: 22: 11.

and T5 in the 2nd week, whereas for NO_3-N, T7 showed a significantly higher value than T1, T3, T4 and T5 in the 4th week of incubation. No significant difference in both NH_4-N and NO_3-N contents was observed between T7 and SRF T2, T3 and T5 in the 8th, 12th and 16th weeks, but significantly higher values were observed in T4 and T6 compared to T7. Significantly higher N content in T7 at early stages of incubation (2nd and 4th weeks) reveals the availability of N which crops may not fully utilize as it may be lost through leaching or fixation by clay minerals, while in SRFs it may be preserved for future use, hence better synchronization. From the 8th to the 16th week, T7 and most of the SRFs showed nearly equal N content, implying that the plant can utilize the N released by SRF more efficiently than T7 which might get depleted sooner due to the susceptibility to early losses.

NH$_4$-N content was higher than NO_3-N throughout the incubation. Subsequently, the cumulative NH_4-N content at the end of the incubation period recorded higher values than NO_3-N content and generally increased from T1 to T6. No significant difference was observed in cumulative NH_4-N content between T7 and SRF T2, T3, T4 and T5, and also, in the cumulative NO_3-N between T7 and SRF T4, T5 and T6. The higher NH_4-N content compared to NO_3-N may be attributed to the acidity of the soil which could have inhibited the growth and activities of nitrifying bacteria. The pH of the soil during the incubation period showed some increase in the 12th and 16th weeks (Table 9.5), particularly in SRF T2 to T6, though not significant among the treatments. The lowest pH value was 5.15 recorded in the 2nd week, while the highest was 5.97 recorded in the 16th week. Nitrification (biological oxidation of NH_4^+ to NO_3^-) occurs in the soil at pH values ranging from 5.5 to 10.0, is optimal at a pH value of about 8.5, and is inhibited at a pH less than 5 (Sahrawat, 2008).

Similar to the current study, Omar and Ismail (1999) observed a higher content of NH_4-N than NO_3-N in the soil treated with a mixture of urea and $CaCl_2$ or K_2SO_4. The population of bacteria and fungi decreased in both urea and $CaCl_2$ or K_2SO_4

TABLE 9.5

Soil pH During the Incubation Period

Treatment	Incubation Period (Weeks)				
	2	4	8	12	16
Cntrl	5.34 a	5.60 a	5.53 a	5.33 a	5.42 a
T1	5.31 a	5.54 a	5.43 a	5.66 b	5.71 b
T2	5.28 a	5.23 a	5.56 a	5.74 b	5.76 bc
T3	5.21 a	5.69 a	5.30 a	5.63 ab	5.67 b
T4	5.33 a	5.46 a	5.36 a	5.68 b	5.83 bc
T5	5.15 a	5.47 a	5.44 a	5.59 ab	5.71 bc
T6	5.28 a	5.80 a	5.34 a	5.47 ab	5.92 c
T7	5.24 a	5.65 a	5.23 a	5.52 ab	5.68 b

Notes: Different letters in the same column are significantly different ($p \leq 0.05$ level).

Cntrl = no treatment, T1 = 14: 0: 0, T2 = 13.8: 18.8: 4.6, T3 = 15.4: 20: 4.8, T4 = 16.8: 21.5: 6.2, T5 = 20: 24: 9.7, T6 = 21.3: 25: 11.2, T7 = 24: 22: 11.

treatments. The soil pH increased in urea amendment but was decreased in inorganic salt amendments to values lower than that of the control. The toxic effect of urea and inorganic salts was reduced when they were applied as a mixture. Giroto et al. (2017) observed higher pH values (6.3–7.9) after 42 days of aerobic incubation of soil amended with urea/HA and thermoplastic starch urea/HA compared to that of untreated soil, whereas the pH of the soil amended with HA and SSP remained close to the pH of the control (≈5). The increase in pH in nano-composite amendments was attributed to the high hydrolysis of urea in the soil with low CEC and low buffering capacity. The existence of more mineral N in the form of NH_4-N is beneficial because it is not susceptible to leaching losses.

Table 9.6 shows N mineralization potential (N_0), mineralization rate constant (K), coefficient of determination (R^2) and time taken for 50% of potentially mineralizable N ($t_{1/2}$), to be mineralized. T7 had the highest potentially mineralizable nitrogen (N_0) compared to all other treatments though not significantly different from T2, T3, T4, T5 and T6. The low N mineralization observed in SRF treatments may be attributed to a slow release in nutrients in the initial stages of incubation. N_0 related well to the observed cumulative mineral N at the 16th week of incubation. However, no significant difference was observed for the cumulative mineral N between T7, T2 and T5 at the 16th week of incubation, implying that the incorporation of SF into the polymer composite enhances the mineralization of N. Thus, the significantly higher ($p \leq 0.05$) mineral N values observed in T4 and T6, relative to T7, may be attributed to improved mineralization, leading to the release of higher amounts of mineral N in the later stages of incubation.

The coefficient of determination, R^2, ranged from 0.742 to 0.917, indicating a good fit of the experimental data to the single-order kinetic model. The mineralization

---

TABLE 9.6
Nitrogen Mineralization Potential (N_o), Mineralization Rate Constant (K), Half-Life ($t_{1/2}$) and Cumulative Mineral Nitrogen

Treatment	N_o (ppm)	R^2	K (Week^{-1})	$t_{1/2}$ (Weeks)	Observed Cumulative Mineral N at 16 Weeks (ppm)
Cntrl	425 a	0.903	0.052	13.3	405 a
T1	495 ab	0.917	0.051	13.6	531 b
T2	576 abc	0.829	0.059	11.7	608 cd
T3	536 abc	0.716	0.051	13.6	597 c
T4	539 abc	0.865	0.056	12.4	712 e
T5	619 bc	0.742	0.060	11.6	632 cd
T6	569 abc	0.910	0.053	13.1	692 e
T7	714 c	0.831	0.057	12.2	647 d

N_o = nitrogen mineralization potential, K = mineralization rate constant and $t_{1/2}$ = time taken for 50% of potentially mineralizable nitrogen to be mineralized. Cntrl = no treatment, T1= 14: 0: 0, T2 = 13.8: 18.8: 4.6, T3 = 15.4: 20: 4.8, T4 = 16.8: 21.5: 6.2, T5 = 20: 24: 9.7, T6 = 21.3: 25: 11.2, T7 = 24: 22: 11.

rate constant ranged from 0.051 to 0.056 week^{-1} which resulted in the half-life time ($t_{1/2}$) ranging from 11.6 to 13.6 weeks, suggesting that mineralization of most of the N would occur within the growing period of most annual crops of about 20 weeks. The $t_{1/2}$ values obtained in the experiment were similar to the average value of 12.8 weeks reported by Stanford and Smith (1972) on evaluating $t_{1/2}$ of 39 soil types in the United States. Since there was less variation in the $t_{1/2}$ among the treatments, the advantage of SRF over T7 could be attributed to the slow initial N mineralization, leading to the release of significantly (p ≤ 0.5) higher amounts in the later stages of incubation. The incubation experiment was, however, carried out at optimal conditions of moisture, temperature and aeration for the growth and activity of soil microbes, and hence, the N mineralization rate might be lower/higher in the field than in the laboratory due to varying conditions that could affect the performance of microorganisms.

9.3.5 AVAILABLE PHOSPHOROUS

Available P at different incubation times is shown in Table 9.7. The lowest P values were recorded in the 4th week and highest in the 8th week and remained nearly constant in the 12th and 16th weeks of incubation. The decline in P content between the 2nd and 4th week could be attributed to microbial immobilization and adsorption of soluble P into the soil. The increased P availability after 4 weeks in all SRF treatments may be attributed to its release through microbial solubilization of nano-HA and degradation of the copolymer. Insoluble phosphates such as apatite have been shown to be solubilized by native soil microorganisms. Phosphate-solubilizing bacteria (*Pseudomonas, Enterobacter and Anthrobacter*) and fungi (*Aspergillus and Penicillium*) present in the soil and the rhizosphere have been reported to hydrolyze the insoluble P by secreting low-molecular-mass organic acids, chelate mineral ions or lower the pH (Khan et al., 2014; Alori et al., 2017). Besides organic acids, mineral

TABLE 9.7
Available P Content (ppm) at Different Incubation Times (Weeks)

Treatment	Incubation Period (Weeks)				
	2	4	8	12	16
Cntrl	21.0 a	10.4 a	25.3 a	23.9 a	26.1 a
T1	24.1 ab	12.9 abc	25.9 a	26.3 a	28.5 a
T2	24.1 ab	12.5 ab	39.3 b	37.8 b	41.1 b
T3	22.3 ab	13.6 abc	46.5 bc	46.6 c	46.2 bc
T4	26.1 ab	14.4 bc	55.2 cd	54.3 d	51.6 cd
T5	25.6 ab	16.3 cd	63.5 d	66.2 e	66.2 e
T6	27.4 ab	16.3 cd	76.3 e	80.4 f	76.6 f
T7	27.8 b	19.6 d	54.2 cd	53.7 d	54.3 d

Notes: Different letters in the same column are significantly different (*p* ≤ 0.05 level).
Cntrl = no treatment, T1= 14: 0: 0, T2 = 13.8: 18.8: 4.6, T3 = 15.4: 20: 4.8, T4 = 16.8: 21.5: 6.2,
     T5 = 20: 24: 9.7, T6 = 21.3: 25: 11.2, T7 = 24: 22: 11.

acids such as HCl, HNO_3 and H_2SO_4 produced by chemoautotrophs and H^+ pump, for instance in *Penicillium rugulosum,* have been reported to solubilize P (Khan et al., 2014). Soil fungi such as mycorrhizae have been shown to be better solubilizers of P than bacteria as they traverse longer distances within the soil and also produce and secrete more acids such as gluconic, citric, lactic, 2-ketogluconic, oxalic, tartaric and acetic acids (Alori et al., 2017). Additionally, the assimilation of NH_4^+ within microbial cells releases H^+ that solubilizes P without the production of organic acids. Acidification of microbial cells and their surroundings release P through the substitution of H^+ for Ca^{2+} ions. The release of Ca^{2+} ions into the soil could be the reason for the observed increase in soil pH towards the end of the incubation period (Table 9.5). Ca^{2+} ions are bases and have the effect of neutralizing soil acidity (Mucheru-Muna et al., 2014).

No significant difference was observed in the 2nd week between T7 and SRF T1 to T6, while in the 4th week T1 to T4 recorded a significantly lower P content compared to T7. From the 8th to 16th week, the highest P value was observed in T6 which was also significantly different from all the treatments. The observation could be attributed to the solubilization of nano-HA and release of soluble P which was physically protected by the composite from adsorption into the soil in the initial stages of incubation.

Fertilizer nano-composites were quantified to deliver a specific amount of N (50 mgN/kg of soil) into the soil regardless of NPK formulae; hence, the amount of P in the amendments varied as: T2 = 68.1 mg/kg, T3 = 65 mg/kg, T4 = 64 mg/kg, T5 = 60 mgkg^{-1}, T6 = 58.5 mg/kg and T7 = 54.5 mg/kg (Table 9.1). The available P increased significantly ($p \leq 0.05$) from T2 to T6, an observation attributed to the increased content of soluble P and decreased content of HA. No significant difference was observed between T7 and T4, except in the 4th week, a fact attributed to the balance between the amount of P in the treatment (T7 < T4) and availability in the soil. Nitisols are strong P-sorping soils (WRB, 2015), and hence, the lower content of P observed in T7, relative to T5 and T6, could be attributed to the soil retention capacity which increases with contact time (Naima et al., 2015; Giroto et al., 2017). The slow microbial solubilization of nano-HA and encapsulation by the copolymer composite could have reduced the available P time of contact in T5 and T6. No significant difference was observed between the control and T1 since they did not contain P in the shipments. Due to the varied content of P in the amendments (Table 9.1), it was not possible to ascertain the optimum amount to be incorporated into the nano-composite fertilizer.

9.3.6 EXCHANGEABLE POTASSIUM

Exchangeable K at different incubation times is shown in Table 9.8. The K content showed less variation during the incubation period suggesting a short release time. The small particle size of the SRF nano-composite and high water solubility of the K_2SO_4 salt could have enabled faster diffusion of K into the soil. No significant difference ($p \geq 0.05$) was observed between CF T7 and SRF T1 to T6 in the first 4 weeks, but a significant difference was observed between the same from the 8th to the 16th week. The control did not differ significantly from T1 throughout the

TABLE 9.8
Exchangeable K Content (Cmol/kg) at Different Incubation Times (Weeks)

Treatment	Incubation Period (Weeks)				
	2	4	8	12	16
Cntrl	1.63 a	1.70 a	1.87 a	1.58 a	1.70 a
T1	1.75 ab	1.85 ab	1.90 ab	1.85 a	1.80 a
T2	1.87 ab	1.83 ab	1.98 bc	2.05 b	2.08 b
T3	2.00 ab	1.92 ab	1.90 c	2.06 b	2.18 c
T4	2.05 ab	2.05 bc	1.95 d	2.00 c	1.95 d
T5	2.10 ab	2.10 bc	2.05 d	2.10 d	2.37 e
T6	2.17 ab	2.13 bc	2.25 e	2.17 e	2.27 e
T7	2.10 b	2.07 bc	2.30 d	2.37 c	2.28 d

Notes: Different letters in the same column are significantly different ($p \leq 0.05$ level).
Legend: Cntrl = No treatment, T1= 14: 0: 0, T2 = 13.8: 18.8: 4.6, T3 = 15.4: 20: 4.8, T4 = 16.8: 21.5: 6.2, T5 = 20: 24: 9.7, T6 = 21.3: 25: 11.2, T7 = 24: 22: 11

incubation period and this was expected as K was not included in the formulation. The contents of K in the treatments were different (Table 9.1), and hence, just like in P, it was not possible to ascertain the optimum amount to be incorporated into the nano-composite fertilizer.

9.4 CONCLUSION

A slow-release nano-composite fertilizer was formulated and characterized. FTIR spectroscopy revealed the existence of chemical interaction between nano-HA, urea molecules and the copolymer structure besides physical entrapment within the copolymer. The incubation experiment revealed low mineral N content in the first 4 weeks and a peak in the 12th week corresponding to the most active and nutrient demand stages of development and reproduction of most crops, hence proper synchronization of SRF. The highest mineral N content was observed in CF in the first 4 weeks, whereas between the 8th and 16th week, both CF and SRF showed a similar mineral N content with some SRFs, for instance T4 and T6, releasing significantly higher amounts. The single first-order kinetic model predicted well N mineralization, and the half-life time ($t_{1/2}$) showed less variation among the treatments. Low contents of P were observed in the first 4 weeks, increased to a maximum in the 8th week and remained constant thereafter. Availability of P increased significantly in SRF with increased content of soluble P and decreased content of nano-HA. SRF T5 and T6 could provide synchronized release of N and P, although the release of K was almost immediate. The SRF nano-composite would be more suitable for use in annual crops and should be applied during planting so as to match nutrient release with the crop uptake. The release of nutrients from nano-composite fertilizer was, however, based on a laboratory incubation experiment, and the evaluation should be done under field conditions before they can be recommended with confidence.

REFERENCES

Abbasi, M.K. and Khaliq, A. (2016), Nitrogen mineralization of a loam soil supplemented with organic–inorganic amendments under laboratory incubation, *Frontiers in Plant Science*, **7**: 1038.

Ahmed, E.M. (2015), Hydrogel: Preparation, characterization and applications: A review, *Journal of Advanced Research,* **6**(2): 105–121.

Alori, E.T., Glick, B.R. and Babalola, O.O. (2017), Microbial phosphorus solubilization and its potential for use in sustainable agriculture, *Frontiers in Microbiology*, **8**: 971.

Award, Y.M., Blagodatskaya, E., Ok, Y.S. and Kuzyakov, Y. (2012), Effect of polyacrylamide, biopolymer and biochar on decomposition of soil organic matter and plant residues as determined by ^{14}C and enzyme activities, *European Journal of Soil Biology*, **48**: 1–10.

Bremner, J.M. (1996), Total nitrogen, in: Sparks, D.L. (Eds), *Methods of Soil Analysis, Part 3: Chemical Methods,* Soil Science Society of America and American Society of Agronomy, Madison, Wisconsin, pp. 1085–1086.

Bundela, H. and Bajpai, A.K. (2008), Designing of hydroxyapatite-gelatin based porous matrix as bone substitute: Correlation with biocompatibility aspects, *Polymer Letters*, **2**(3): 201–213.

Charoenpanich, J. (2013), Removal of acrylamide by microorganisms, in: Yogesh, P. and Prakash R. (Eds), *Applied Bioremediation - Active and Passive Approaches*, IntechOpen Limited, London, UK, pp. 101–121.

Chen, C., Gao, Z., Qiu, X. and Hu, S. (2013), Enhancement of the controlled-release properties of chitosan membranes by cross-linking with suberoyl chloride, *Molecules,* **18**: 7239–7252.

Chen, J.F. (1997), Adsorption and diffusion of ammonium in soils, in: Zhu, Z., Wen, Q. and Freney, J.R. (Eds), *Nitrogen in Soils of China, Development in Plant Soil Sciences.* Springer, Dordrecht, The Netherlands, p. 74.

Cisneros-Pineda, O.G., Kao, W.H., Loría-Bastarrachea, M.I., Veranes-Pantoja, Y. Cauich-Rodríguez, J.V. and Cervantes-Uc, J.M. (2014), Towards optimization of the silanization process of hydroxyapatite for its use in bone cement formulations, *Materials Science and Engineering,* C, **40**: 157–163.

Costescu, A., Pasuk, I., Ungureanu, F., Dinischiotu, A., Costache, M., Huneau, F., Galaup, S., Le Coustumer, P. and Predoi, D. (2010), Physico-chemical properties of nano-sized hexagonal hydroxyapatite powder synthesized by sol-gel, *Digest Journal of Nanomaterials and Biostructures,* **5**(4): 989–1000.

Deenik, J.L. and Yost, R.S. (2008), Nitrogen mineralization potential and nutrient availability from five organic materials in an atoll soil from the Marshall Islands, *Soil Science,* **173**(1): 54–68.

DeRosa, C., Monreal, C., Schnitzer, M., Walsh, R. and Sultan, Y. (2010), Nanotechnology in fertilizers, *Nature Nanotechnology,* **5**(2): 91–94.

Dong, W., Zhang, X., Wang, H., Dai, X., Sun, X., Qiu, W. and Yang, F. (2012), Effect of different fertilizer application on the soil fertility of paddy soils, in red soil region of Southern China, *PLoS ONE,* **7**(9): e44504.

Eritsyan, M.L., Gyurdzhyan, L.A., Melkonyan, L.T. and Akopyan, G.V. (2006), Copolymers of acrylic acid with urea, *Russian Journal of Applied Chemistry,* **79**(10): 1666–1668.

Fernandes, B.S., Pinto, J.C, Cabral-Albuquerque, E.M. and Fialho, R.L. (2015), Free-radical polymerization of urea, acrylic acid and glycerol in aqueous solutions, *Polymer Engineering and Science,* **55**(6):1219–1229.

Gachene, C.K.K. (1989), Soils of the erosion research farm, Kabete Campus, *Department of Agricultural Engineering,* University of Nairobi.

Gaskell, M. and Smith, R., (2007), Nitrogen sources for organic vegetable crops, *Hortechnology,* **17**(4): 431–441.

Gayathri, B., Muthukumarasamy, N., Velauthapillai, D., Santhosh, S.B. and Asokan, V. (2018), Magnesium incorporated hydroxyapatite nanoparticles: Preparation, characterization, antibacterial and larvicidal activity, *Arabian Journal of Chemistry,* **11**: 645–654.

Giroto, A.S., Guimarães, G.F., Foschini, M. and Ribeiro, C. (2017), Role of slow-release nano-composite fertilizers on nitrogen and phosphate availability in soil, *Scientific Reports,* **7**: 46032.

Glendon, W.G. and Doni, O.R. (2002), Particle-size analysis, in: Dane, J.H. and Topp, G.C. (Eds), *Methods of Soil Analysis. Part 4: Physical Methods,* Soil Science Society of America, Inc., Madison, WI, pp. 255–293.

Guezennec, A.G., Michael, C., Bru, K., Touze, S., Desroche, N., Mnif, I. and Motelica-Heino, M. (2015), Transfer and degradation of polyacrylamide-based flocculants in hydrosystems: A review, *Environmental Science and Pollution Research,* **22**(9): 6390–6406.

Guo, M., Liu, M., Hu, Z., Zang, F. and Wu, L. (2005), Preparation and properties of a slow release NP compound fertilizer with superabsorbent and moisture preservation, *Journal of Applied Polymer Science,* **96**: 2132–2138.

Iyyappan, E. and Wilson, P. (2013), Synthesis of nanoscale hydroxyapatite particles using Triton X-100 as an organic modifier, *Ceramics International,* **39**: 771–777.

Jaetzold, R., Schmidt, H., Hornetz, B. and Shisanya, C. (2006), Farm management handbook of Kenya Vol. II., Natural conditions and farm management information (2nd Ed.) Part A: West Kenya, Subpart A1: Western Province, *Ministry of Agriculture Kenya in Cooperation with the German Agency for Technical Cooperation (GTZ),* Nairobi, Kenya.

Karuku, G.N. (2019), Effect of lime, N and P salts on nitrogen mineralization, nitrification process and priming effect in three soil types, andosols, luvisols and ferralsols, *Journal Agriculture and Sustainability,* **12**(1): 74–106.

Karuku, G.N. and Mochoge, B.O. (2016), Nitrogen forms in three Kenyan soils nitisols, luvisols and ferralsols, *International Journal for Innovation Education and Research,* **4**(10): 17–30.

Karuku, G.N. and Mochoge, B.O. (2018), Nitrogen mineralization potential (N_o) in three Kenyan soils, nitisols, ferralsols and luvisols, *Journal of Agricultural Science,* **10**(4): 60–78.

Karuku, G.N., Gachene, C.K., Karanja, N., Cornelis W. and Verplancke, H. (2014a), Use of CROPWAT model to predict water use in irrigated tomato (*Lycopersicon esculentum*) production at Kabete, Kenya, *East African Agriculture and Forestry Journal,* **80**(3): 175–183.

Karuku, G.N., Gachene, C.K., Karanja, N., Cornelis, W. and Verplancke, H. (2014b), Effect of different cover crop residue management practices on soil moisture content under a tomato crop (*Lycopersicon esculentum*), *Tropical Subtroptropical Agroecosystems,* **17**(3): 509–523.

Karuku, G.N., Gachene, C.K., Karanja, N., Cornelis, W., Verplancke, H. and Kirochi, G. (2012), Soil hydraulic properties of a nitisol in Kabete, Kenya, *Tropical and Subtropical Agroecosystems,* **15**: 595–609.

Kay-Shoemake, J.L., Wartwood, M., Lentz, R. and Sojka, R. (1998), Polyacrylamide as an organic nitrogen source for soil micro-organisms with potential effects on inorganic soil nitrogen in agricultural soil, *Soil Biology and Biochemistry,* **30**(8–9): 1045–1052.

Khan, M.S., Zaidi, A. and Ahmad, E. (2014), Mechanism of phosphate solubilization and physiological functions of phosphate-solubilizing microorganisms, in: Khan, M.S., Zaidi, A. and Musarrat, J. (Eds), *Phosphate Solubilizing Microorganisms,* Springer, Cham, pp. 31–62.

Kimetu, J.M., Mugendi, D.N., Bationo, A., Palm, C.A., Mutuo, P.K., Kihara, J., Nandwa, S. and Giller, K. (2007), Partial balance of nitrogen in a maize cropping system in humic nitisol of Central Kenya, in: Bationo, A., Waswa, B., Kihara, J. and Kimetu, J. (Eds), *Advances in Integrated Soil Fertility Management in Sub-Saharan Africa: Challenges and Opportunities,* Springer, Dordrecht, The Netherlands, pp. 521–530.

Kottegoda, N., Munaweera, I., Madusanka, N. and Karunaratne, V. (2011), A green slow-release fertilizer composition based on urea-modified hydroxyapatite nanoparticles encapsulated wood, *Current Science*, 73–78.

Kottegoda, N., Sandaruwan, C., Piryrashana, G., Siriwardhana, A., Rathnayake, A.U., Arachchige, D.M.B., Kumarasinghe, A.R., Dahanayake D., Karunaratne, V. and Amaratunga, G.A.J. (2017), Urea-hydroxyapatite nanohybrid for slow release of nitrogen, *ACS Nano,* **11**(2): 1214–1221.

Laftah, W.A. and Hashim, S. (2014), Synthesis, optimization, characterization and potential agricultural application of polymer hydrogel composites based on cotton microfiber, *Chemical Papers,* **68**(6): 798–808.

Leye, A.S. and Omotayo, A.E. (2014), Mineralization rates of soil forms of nitrogen, phosphorus, and potassium as affected by organomineral fertilizer in sandy loam, *Advances in Agriculture,* **2014**: 1–5, Article ID 149209.

Liu, M., Liang, R, Zhan, F., Liu, Z. and Niu, A. (2007), Preparation of superabsorbent slow release nitrogen fertilizer by inverse suspension polymerization, *Polymer International,* **56**: 729–737.

Makokha, S., Kimani, S., Mwangi, W., Verkuij, H. and Musembi, F. (2001), Determinants of fertilizer and manure use in maize production in Kiambu district, Kenya, *Mexico, D.F.: International Maize and Wheat Improvement Center, Kenya Agricultural Research Institute.*

Masunga, R.H, Uzokwe, V.N., Mlay, P.D., Odeh, I., Singh, A., Buchan, D. and De Neve, S. (2016), Nitrogen mineralization dynamics of different valuable organic amendments commonly used in agriculture, *Applied Soil Ecology,* **101**: 185–193.

Mateus, P.A., Barrias, C.C., Ribeiro, C., Ferraz, M.P. and Monteiro, F.J. (2007), Comparative study of nano-hydroxyapatite microspheres for medical applications, *Journal of Biomedical Materials Research,* **86**A: 483–493.

Mathenge M.K. (2009), Fertilizer types, availability and consumption in Kenya, Tegemeo Institute, Egerton University; *Paper presented at the 6th National Fertilizer Conference in Kenya, on "Towards Increased Use of Fertilizer and Improved Seed for Food Security and Economic Growth,"* KARI headquarters, Nairobi, Kenya, August 20–21, 2009.

Mehlich, A. (1953), Determination of P, Ca, Mg, K, Na and NH_4, *North Carolina Soil Testing Laboratory,* Raleigh, University of North Carolina.

Monreal, C.M., DeRosa, M., Mallubhotla, S.C., Bindraban, P.S. and Dimkpa, C. (2015), Nanotechnologies for increasing the crop use efficiency of fertilizer-micronutrients, *Biology and Fertility of Soils,* **52**: 423.

Montalvo, D., McLaughlin, M.J., and Degryse, F. (2015), Efficacy of hydroxyapatite nanoparticles as phosphorus fertilizers in andisols and oxisols, *Soil Science Society of America Journal,* **79**: 551–558.

Mucheru-Muna, M., Mugendi, D., Pypers, P., Mugwe, J., Kungu, J., Vanlauwe, B. and Merckx, R. (2014), Enhancing maize productivity and profitability using organic inputs and mineral fertilizer in Central Kenya small-hold farms, *Experimental Agriculture,* **50**(2): 250–269.

Naderi, M. and Danesh, S. (2013), Nano-fertilizers and their roles in sustainable agriculture, *International Journal of Agriculture and Crop Sciences,* **5**(19): 2229–2232.

Naima, B., Leila, H. and Adil, M. (2015), Effect of incubation period of phosphorous fertilizer on some properties of sandy soil with low calcareous content, Southern Algeria, *Asian Journal of Agricultural Research,* **9**(3): 123–131.

Nelson, E.W. and Sommers, L.E. (1996), Total carbon, organic carbon and organic matter, in: D.L. Sparks (Ed.), *Methods of Soil Analysis, Part 3: Chemical Methods*, Soil Science Society of America, Inc., Madison, WI, pp. 961–1010.

Olson-Rutz, K., Jones, C. and Dinkins, C.P. (2011), Enhanced efficiency fertilizers, *Montana State University Extension*, http://landresources.montana.edu/soilfertility/documents/PDF/pub/EEFEB0188.pdf, Available 25th August 2019.

Omar, S.A. and Ismail, M.A. (1999), Microbial populations, ammonification and nitrification in soil treated with urea and inorganic salts, *Folia Microbiologica,* **44**(2): 205–212.

Oseko, E. and Dienya, T. (2015), Fertilizer consumption and fertilizer use by crop (FUBC) in Kenya, 2010–2013, https://africafertilizer.org/wp-content/uploads/2017/05/FUBC-Kenya-final-report-2015.pdf, Available 25th August 2019.

Pang, X., Zeng H., Liu, J., Wei, S. and Zheng, Y. (2010), The properties of nanohydroxyapatite materials and its biological effects, *Material Sciences and Applications,* **1**: 81–90.

Pataquiva-Mateus, A.Y., Ferraz, M.P. and Monteiro, F.J. (2013), Nanoparticles of hydroxyapatite: Preparation, characterization and cellular approach: An overview, *Revista Mutis,* **3**(2): 43–57.

Pramanik, N., Mishra, D., Banerjee, I., Maiti, T.K., Bhargava, P. and Pramanik, P. (2009), Chemical synthesis, characterization, and biocompatibility study of hydroxyapatite/chitosan phosphate nanocomposite for bone tissue engineering applications, *International Journal of Biomaterials,* **2009:** Article ID 512417, 8 pgs.

Qiu, X. and Hu, S. (2013), "Smart" materials based on cellulose: A review of the preparations, properties, and applications, *Materials,* **6**: 738–781.

Ragu, A. and Sakthivel, P. (2014), Synthesis and characterization of nano hydroxyapatite with polymethyl methacrylate nanocomposites for bone tissue regeneration, *International Journal of Science and Research,* **3**(11): 2282–2285.

Sahrawat, K.L. (2008), Factors affecting nitrification in soils, *Communication in Soil Science and Plant Analysis,* **39**(9–10): 1436–1446.

Sannino, A., Demitri, C. and Madaghiele, M. (2009), Biodegradable cellulose-based hydrogels: Design and applications: A review, *Materials,* **2**: 353–373.

Shoals, M. (2012), Kenya Fertilizer Assessment, The African Fertilizer and Agribusiness Partnership, IFDC. https://africafertilizer.org/wp-content/uploads/2017/04/Kenya-Fertilizer-Market-Assessment-IFDC-AFAP-June-2012.pdf, Available 25th August 2019.

Sombroek, W.G., Braun, H.M. and Van Der Pouw, B.J. (1982), The exploratory soil map and agro climatic zone map of Kenya, 1980 scale 1: 1,000,000. Exploratory Soil Survey Report No. E1, *Kenya Soil Survey,* Nairobi, Kenya

Stanford, G., Carter S.J. and Smith, S.J. (1974), Estimates of potentially mineralizable soil nitrogen based on short term incubations, *Soil Science Society of America Proceedings,* **38**(1): 99–102.

Stanford, G. and Smith, S.J. (1972), Nitrogen mineralization potentials of soils, *Soil Science Society of America Journal,* **36**(3): 465–472.

Synytsya, A. and Novak, M. (2014), Structural analysis of glucans: A review, *Annals of Translational Medicine,* **2**(2): 17.

Tambone, F. and Adani, F. (2017), Mineralization from digestate in comparison to sewage sludge, compost and urea in a laboratory incubation soil experiment, *Journal of Plant Nutrition and Soil Science,* **180**: 355–365.

Tolescu, C., Neamtu, C., Raceanu, G., Popescu, M. and Iovu, H. (2009), Polymeric microstructures for slow release of fertilizers, *Materiale Plastice,* **46**(4): 387–392.

TonThat, T.C., Busscher, W.J., Novak, J.M., Gaskin, J.F. and Kim, Y. (2008), Effects of polyacrylamide and organic matter on microbes associated to soil aggregation of Norfolk loamy sand, *Applied Soil Ecology,* **40**: 240–249.

Watson, C., Singh, Y., Iqbal, T., Knoblauch, C., Simon, P. and Wichern, F. (2016), Short-term effects of polyacrylamide and dicyandiamide on C and N mineralization in a sandy loam soil, *Soil Use and Management,* **32**: 127–136.

WRB (World Reference Base) for Soil Resource (2015), International soil classification for naming soils and creating legends for maps, *World Soil Resources Report,* **106**.

Yu, F., Fu, R., Xie, Y. and Chen W. (2015), Isolation of polyacrylamide-degrading bacteria from dewatered sludge, *International Journal of Environmental Research and Public Health,* **12**: 4214–4230.

Yuejin, W., Xiaoyu, N., Zhengyan, W., Lin, W., Guannan, Q. and Lixiang, Y. (2013), A novel slow-release urea fertilizer: Physical and chemical analysis of its structure and study of its release mechanism, *Biosystems Engineering,* **1159**: 274–282.

Zeroual, Y. and Kossir, A. (2012), Smart fertilizers for sustainable agriculture: The state of the art and the recent developments, *25th AAF International Fertilizer Technology Conference and Exhibition,* Morocco.

10 Natural Nanofibers Classification, Their Properties, and Applications

P. A. Nizam and Sabu Thomas
Mahatma Gandhi University

CONTENTS

10.1 Introduction...281
10.2 Classification of Natural Fibers 282
10.3 An Overview of Animal Fibers 282
10.4 Plant Fibers 284
    10.4.1 Hemp Fiber...285
    10.4.2 Kenaf Fiber...287
    10.4.3 Flax Fiber 288
    10.4.4 Sisal Fibers 289
    10.4.5 Banana and Pineapple Fibers 291
    10.4.6 Wood Fibers 292
    10.4.7 Cellulose and Their Related Fibers 293
10.5 Conclusion 294
References...294

10.1 INTRODUCTION

The exponential growth of the world's population necessitates increased efforts to address their social and economic requirements. The rising environmental consciousness and regulatory demands, the production, usage, and disposal of non-biodegradable wastes are being scrutinized more closely. At present, synthetic materials, which are polluting the environment and endangering individuals, are a big worry. These materials are employed in almost all applications, which eventually fills the lands or oceans causing an imbalance with non-biodegradable materials in the environment.

There has been an increase in ecological consciousness over the last several decades, which has aroused the interest of materials' researchers in the use of natural biofibers created from renewable resources rather than standard fibers (glass, carbon, and aramid) in polymer matrix composites. Extensive study has demonstrated that natural fibers such as banana, sisal, coir, jute, hemp, kenaf, sugarcane bagasse, rice

DOI: 10.1201/9781003279372-10

husk, pineapple leaf, Roselle, and ramie have the potential to be good reinforcing agents for polymer matrix composites. Although natural fibers derived from renewable resources have significant environmental advantages over synthetic fibers, they also have several downsides, including greater moisture absorption, poor thermal stability, and lower impact strength. These properties can be improved by various treatments which enhances their applications in various sectors.

The chapter focuses on several types of natural fibers, their properties, and their use in diverse industries. These materials show promise for a more sustainable future since their flaws may be mitigated via surface modification and use as reinforcements. Proper study of these fibers will undoubtedly lead to the development of novel technologies in the field of natural fiber science.

10.2 CLASSIFICATION OF NATURAL FIBERS

A basic understanding of the classification of natural fibers is necessary to better understand the topic. Natural fibers may be categorized into three types based on their origin: animal fiber, mineral fiber, and plant or vegetable fiber. Proteins make up the majority of animal fibers, the most well-known examples being silk and wool. Fibers derived from various animal sources have varying characteristics. Merino and Cotswold wool are two examples of wool from distinct animal species. The texture of Cotswold is coarse, whereas the texture of the former is soft. It should also be mentioned that the consistency of natural fibers varies, while synthetic fibers are known to be more consistent. Animal fibers have been used to make soft and warm jackets, wraps, blazers, ponchos, shawls, coats, and various garments and accessories. Carpets, coverings, and rugs are often constructed of rougher animal fibers.

Asbestos refers to a group of mineral fibers with great flexibility and resilience to heat, corrosion, wear and tear, humidity, and other factors. These fibers are nonflammable and have a low electrical conductivity. Asbestos is easily incorporated into construction materials (glue, plaster, concrete, etc.). As a result, until recently, significant amounts of asbestos were mined. However, health concerns have significantly curtailed the industry's size in recent years. Wherever feasible, asbestos has been substituted by alternative fibers such as glass or Nomex.

Plant fibers are derived from plants. Also known as crop fibers, they are elongated, thick-walled cells with pointed ends that are made of cellulose, hemicellulose, and lignin. Plant fibers come in a variety of varieties such as jute, flax, cotton, and so on. Fibers are used commercially in the textile industry for weaving cloth, as a filtering medium, and for insulation. In this chapter, we will concentrate on plant fibers, which have recently gained prominence. Figure 10.1 depicts an overview of natural fibers, which are divided into three categories: animal, plant, and minerals. Cellulose is present in all fibers and is the basic unit of all plant species.

10.3 AN OVERVIEW OF ANIMAL FIBERS

Animal fibers are often generated from the animals' hair, fur, skin, or secretions. They are frequently woven or knitted (or sometimes felted) after they have been removed to create beautiful animal textiles. Historically, animal fibers were used to

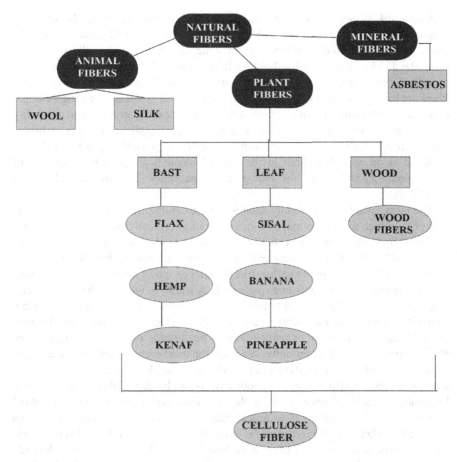

FIGURE 10.1   Classification of natural fibers. Cellulose is a common component of all these fibers.

make soft and warm jackets, wraps, blazers, shawls, ponchos, coats, and other types of apparel and accessories. Carpets, coverings, and rugs are often constructed of rougher animal fibers. The common animal fiber includes silk and wool.

Many people consider silk to be a "natural" protein fiber. The most well-known type of silk is that which is acquired from cocoons grown in captivity by silkworm larvae of the Bombyx mori species. Sericulture is a term used to describe the activity of raising silkworms. It should also be highlighted that degummed fibers derived from the Bombyx mori species have a diameter ranging from 5 to 10 m. Certain varieties of silk are well known for their shimmering look, which is caused by the fiber's triangular prism-like cross-sectional structure. Light incident on the silk fiber is refracted at different angles due to the prism-like fibrous features. The quality of silk fibers is also determined by the health of the silkworm larvae, which is determined by the food and living circumstances supplied to them.

Spider silk is one of the strongest natural fibers known to man. The strongest dragline silk known to man is said to be three times stronger than Kevlar and five times stronger than steel. The elasticity of some varieties of spider silk is also well recognized. The silk made by the ogre-faced spider, for example, is reported to be able to extend to more than five times its length without incurring harm. These fibers are highly biocompatible and biodegradable, also broadcasting a high elongation and contraction. Their properties such as high breaking strength and durability provide opportunities for designing these materials in applications such as bridge cable, aircraft bodies, parachutes, armors, etc. Biodegradable spider silk (SS) is regarded as an appealing fiber in the field of biomedical application because of its sustainability and remarkable biological compatibility (Thirugnanasambantham *et al.*, 2020).

The term "wool" is commonly used to refer to animal fibers obtained from the furs of Caprinae family animals. Although wool is typically produced from sheep fur, wool from other species such as rabbits, goats, and alpacas is not uncommon. The fundamental distinction between sheep's wool and hair is that sheep's wool is known to include overlapping scales (like shingles on a roof). Some varieties of wool contain almost 20 such bends in a single inch. A strand of wool can have a diameter ranging from 17 to 35 μm. Wool fiber's structure and chemical makeup distinguish it from other types of fibers; its enormous diversity, heterogeneity of traits, and benefits are unlike any other natural or artificial fiber. The use of wool in the clothing and textile industry dates back to antiquity. Wool has the most complicated structure of any textile fabric. This complex structure provides good flexibility, moisture absorption, flame resistance, warmth, coolness, odor absorption, biodegradability, recyclability, breathability, and resilience softness, absorbs noise, and is simple to handle. Wool fiber has several restrictions in technical industrial applications owing to cost. This fiber, on the other hand, has received far more attention for technological applications than synthetic fibers. Animal fibers have long been used in a variety of industries. Their exceptional qualities, environmental friendliness, and biodegradability have further increased their usage (Allafi *et al.*, 2020).

10.4   PLANT FIBERS

Plant fibers are among the earliest natural materials known to mankind. Archaeologists have discovered cloth fragments believed to be 10,000 years old in both the Old and New Worlds. Initially, humans collected a variety of wild plant species like leaves or barks for their varied fibers. However, the discovery of cotton and hemp altered the course of history. When manufacturing increased substantially, fibers derived from cultivated plants made a variety of tasks easier. As a result, plant fibers became important in human life in a very short period (Kilinç, Durmuşkahya, and Seydibeyoğlu, 2017). Initially, these materials were restricted to domestic use, but as technology advanced, they were introduced to the industry as well. Cotton, sisal, coir, jute, hemp, flax, banana, bamboo, and other natural fibers are commonly utilized in polymer composites. They are a hybrid material composed of a polymer reinforced by fibers, combining the fiber's excellent mechanical and physical performance with the polymer's appearance, bonding, and physical qualities. The case has changed where fibers are being used as a matrix these days. A comprehensive outlook of different plant fibers is discussed (Ekundayo, 2019).

10.4.1 HEMP FIBER

Hemp fibers are regarded as one of the most powerful members of the bast fibers family and are produced from the Cannabis plant species. Although the hemp plant is native to India and Persia, it has been planted in practically every temperate and tropical country. Russia is the world's largest hemp fiber producer, accounting for around 33% of yearly global output. Other nations that produce significant amounts of hemp fiber include France, Germany, Italy, Yugoslavia, Japan, China, Peru, etc. Hemp is a tall, agricultural plant that can be harvested within 2–3 months of sowing. These fibers, found in the bast, have a rigidity equivalent to glass fibers. The cross section of hemp fiber is uneven and does not remain consistent along its length. The phloem comprises the major bast fibers, which are made up of roughly 70%–74% cellulose, 15%–20% t hemicellulose, 3.5%–5.7% lignin, 0.8% pectin, and 1.2%–6.2% wax. Secondary bast fibers emerge from the cambium and are found in the phloem. With several lumens on sides, these hemp fibers can be seen as a composite of a multi-celled structure. Figure 10.2 depicts a typical fundamental structure of hemp fiber. The cell wall of hemp fiber is multi-layered, with the primary cell wall (the initial layer deposited during cell formation) and the secondary wall (S), which is composed of three layers (S1–3). Lignin holds the primary fibers together in the middle lamella (the concentration of lignin is about 90%). The S2 layer, on the other hand,

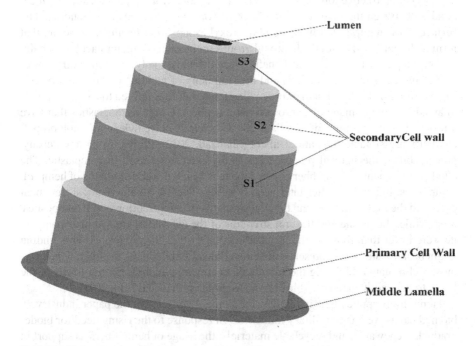

FIGURE 10.2 Fundamental structure of hemp fiber.

has the highest percentage of cellulose, at over 50%. S2 is also the thickest layer, and because of its greater cellulose concentration, it regulates the fiber characteristics (Manaia, Manaia, and Rodriges, 2019).

Hemp fibers have gained widespread recognition as reinforcements in composite materials because of their biodegradability and low density as compared to artificial fibers. These materials also exhibit mechanical, thermal, and acoustic characteristics. Surface functionalization of hemp fibers is critical for expanding its uses. Various researches are being carried out on these fibers to better understand their properties, interaction, etc., to enhance their use in various domains. As lignocellulose is susceptible to moisture, it can cause structural deterioration as well as dimension fluctuation, which affects the interaction between hemp and the matrixes. As a result, their incorporation into a polymer or mineral matrix involves the removal of interface incompatibilities by the application of fiber chemical pretreatments. Some treatments like one with NaOH where treatment with 6% NaOH cleans the fibers by eliminating amorphous substances and increases the crystallinity index of the fiber bundles. The ethylenediaminetetraacetic acid (EDTA) treatment separates the fibers and the complex calcium ions associated with the pectin (Le Troedec *et al.*, 2008). The tensile modulus of silane-treated fiber composites is higher than that of untreated and alkali-treated fiber composites, and the characteristics were noticeably superior to those produced from alkali treatment. A silane content of 1% is optimal for improving mechanical characteristics (Sepe *et al.*, 2018).

The mechanical properties of the thermoset composites are governed by the cumulative mechanical properties of both the matrix and the fibers. Mechanical anchoring, physical attractive forces (van der Waals force and hydrogen bond), and chemical bonding between the matrix and the fiber all contribute to interfacial bonding. The surface of the natural fiber has several hydroxyl groups in its chemical structure that form hydrogen bonds with the hydroxyl groups in the matrix's major backbone chain.

Thermoplastics have several benefits over thermoset polymers, including lower processing costs, design freedom, and the simplicity of molding complicated pieces. Their main drawback is that their processing temperature is limited to less than 230°C to avoid thermal damage of reinforced natural fibers. Only thermoplastics that have a processing temperature of less than 230°C, such as polyethylene and polypropylene (PP), are suitable for natural fibers (Shahzad, 2012). PP due to its low density, processability, mechanical property, etc. is the extensively used thermoplastic. The alkaline treatment of hemp fiber boosted the tensile and flexural strengths of hemp/PP composites, indicating better interfacial bonding. The 4% alkali chemical treatment produced the highest tensile and flexural strengths in hemp-reinforced PP composites. Meanwhile, the tensile and flexural strengths of the 6% NaOH-treated fiber composite were lower than those of the 2% and 4% NaOH-treated composites. This finding might be attributed to the presence of hemicellulose and lignin on the fiber, which was mostly eliminated following the 6% alkali treatment, which permitted fibrils in the fibers to be readily taken out (fibrillation) (Suardana, Piao and Lim, 2011).

Hemp finds applications in textiles, for clothes bags, etc., in the paper industry, in biomedical as well as in military industries. In response to the rising need for biodegradable, renewable, and recyclable materials, the usage of hemp fibers as support in composite materials has lately expanded.

10.4.2   KENAF FIBER

Kenaf fibers are inexpensive and widely available when compared to other fiber families. Kenaf is a dicotyledons plant that belongs to the family of Malvaceae (Hibiscus genus). This 4000-year-old crop was originated in Africa since they were adopted in different parts of the world. This plant requires less water as its lifecycle is 150–180 days. Two types of fibers are collected, which are long fibers from the bast and the small fibers from the center. Kenaf is a well-known cellulose source with both economic and environmental benefits. These robust plants are excellent resistant to climates and insects while using minimum water, fertilizers, and pesticides.

Kenaf has been generally regarded as a promising biological resource and prospective alternative for fossil fuels because of its vast adaptability, high resilience, massive biomass, and high cellulose content. Kenaf is a dicotyledonous plant, which means that the stalk has three layers: an outside cortical layer called phloem, an inner woody core tissue layer called xylem, and a thin central pith layer made up of sponge-like tissue with predominantly non-ferrous cells. These biodegradable fibers have high cellulose content and no silica content which reduces abrasiveness in processing equipment. Kenaf has 40% usable fibers from stock which is double that of other fibers just as hemp, flax, or jute. Kenaf is being used in a variety of novel goods, including paper, absorbents, construction materials, and animal feeds. In addition, there is increased interest in the utilization of kenaf fibers, particularly for composite reinforcement.

Like jute fiber, kenaf fiber also requires surface treatment, due to excess lignocellulose content. This material imparts high hydrophilicity and makes it difficult to reinforce with polymer matrixes. Surface treatments are carried out to improve the interfacial bonding between polymer and fiber using reagents with functional groups that bond with hydroxyl groups of natural fiber. One such alkali treatment where fibers are dipped in NaOH solution for a while. This enhances their roughness and thus mechanical bonding to polymer matrixes. Studies have shown that this treatment shows higher tensile strength for composite when reinforced, which depicts that the surface adhesive properties are improved. This treatment also reduces agglomeration of fiber which enhances a uniform distributed fiber-reinforced composite.

Silane treatments involve the dipping of fibers in a mixture of water/alcohol, where silanol and alcohol are formed by the disintegration of silanes. These silanols interact with cellulose, thus establishing covalent linkages with the cell walls. A stronger bonding between the fiber and matrix is attained by the use of silanes whereby the cross-linking and surface area of fibers are boosted. Silane has several major advantages as a coupling chemical: First, it is inexpensive; second, it has an alkoxysilane group that interacts with hydroxyl groups of fibers at one end, while on the other end, functional groups that can be tailored depending on the matrix to be used. Many other treatments are also studied such as grafting, mercerization, etc.

Kenaf fibers were evaluated as a natural reinforcement with three thermosets, epoxy, vinyl ester, and unsaturated polyester. Plain woven kenaf fibers inherit good mechanical properties to which they were employed for such applications. The tensile, impact, and flexural strength of kenaf/epoxy were superior to others with a linear graph for stress–strain diagram (Salman *et al.*, 2015). In the case of thermosets, polypropylene-reinforced kenaf fibers were studied. Pure PP is poor in tensile and flexural properties. The addition of kenaf fibers in the range of 0–20 enhances their tensile modulus. Furthermore, with more percentage of loading of kenaf, the flexural strength is also improved (Lee, Sapuan and Hassan, 2017). The alkali-treated fiber improves the adhesion between thermoplastic polyurethane and natural rubber, which were shown with an improvement in mechanical strength water absorption, and thickness swelling. The hydrogen bonding due to fiber adhesion has enhanced the water barrier properties of the composite (Azammi *et al.*, 2020). These fibers are used in various domains and applications such as rope, nags and fabrics, fast-food containers, engineering wood as composites, in energy for electricity and bio-fuels, animal bedding, insulated panels, as oil and liquid-absorbent materials, as reinforcing materials, black liquor, bioengineering, and biomedical applications, etc. Furthermore, the potential for kenaf fiber to replace synthetic fibers (glass) in flexural and tensile applications has been carefully assessed. Kenaf fibers due to their impact strength can be employed in a wide range of applications which include structural as well as non-structural. Buildings and construction are concerned with improving the functional qualities of main construction materials such as concrete, steel, wood, and glass.

Kenaf can be used as a substitute for solid woods and plastic composites and can be molded into different forms for applications. This is the first and most reasonably priced plastic timber for usage as a construction material in the housing sector. Furthermore, it is utilized to create a strong, lightweight cement block with excellent insulation and fire resistance. Kenaf core blocks are being utilized to build multi-story and single-family dwellings without the usage of power equipment.

10.4.3 FLAX FIBER

Due to its specific properties, flax fiber (Linum usitatissimum) is regarded as the most essential member of the bast family for composite reinforcing. Bast fibers are extracted from fibrous bundles found within the inner bark of a plant stem. Flax fiber's intrinsic high strength and stiffness, as well as its low elongation to failure, are essential properties that make it appealing for composite research. Flax fibers are not continuous like synthetic fibers, but they do have a hierarchical structure and a structure akin to composites. Their macroscopic qualities are determined by their micro- and nanostructural characteristics. Flax has been utilized as a significant industrial fiber since antiquity. Twisted wild flax strands were used by prehistoric hunters to haft stone weapons, weave baskets, and stitch clothes more than 30,000 years ago.

Flax fibers are found as fiber bundles on the plant stem's outer surface. Their fast-growing nature is very ideal for their use which reaches a height of 100–150 cm in less than 110 days. Cellulose, hemicellulose, lignin, and pectin are the primary components of flax fibers. Oil, wax, and structural water are also found in minor amounts. Cellulosic elements make up both main and secondary cell walls. Cellulose fibrils (diameter 0.1–0.3 lm) are surrounded by concentric lamella, which is made up of around 15% hemicellulose and 2% pectin which contributes to the fibers' thermal breakdown and moisture absorption activity. The location in the stem affects the mechanical characteristics of flax fibers. The flax fibers toward the bottom of the stem have the poorest mechanical characteristics, whereas the fibers in the middle have the best mechanical properties. The investigation revealed that both cellulose and non-cellulosic polymers are abundantly present. Cellulose is the comparable reinforcement material and non-cellulosic materials that facilitate the load transfer from one microfibril to another. They resemble a single cell as that of hemp fiber Figure 10.2.

Most of the fibers possess identical structures and contents. Their moisture susceptibility is one of their main drawbacks. As cellulose and hemicellulose are the main components of plant fibers, hydrogen bondings are predominant in these materials. The hydroxyl (OH)-to-carbon (C) ratio in cellulose and hemicellulose is high. Water molecules may not be able to enter cellulose because it has a very crystalline area. Water molecules, on the other hand, seep into the amorphous portions of cellulose and hemicellulose, breaking intermolecular hydrogen bonds. This permits the cellulose chains' intermolecular distance to rise, causing fiber swelling.

Fundamental flax fibers are processed into mats, roving, textiles, and yarns for application in composites production. To make composites, materials are joined into numerous layers with the matrix material of some sort of resin after selecting a suitable production procedure. The adhesion between these fibers and the hydrophobic matrix material is frequently influenced by their hydrophilic activity. Chemical and physical treatments of the fibers, such as alkali treatment, acetylation, bleaching, isocyanate treatment, peroxide treatment, vinyl grafting, coupling agents, and so on, are employed to solve this issue. Various surface treatments have been briefly explained in the previous sections (Moudood et al., 2019). Polypropylene-based composite with treated and untreated flax fibers was evaluated for their mechanical property and water absorption. The study showed results similar to other fibers where treated fibers enhance the tensile, impact, and water absorption property of the composite (Soleimani et al., 2008).

10.4.4 Sisal Fibers

Sisal fiber is a popular natural fiber that is also quite easy to grow. It is derived from the sisal plant, known commonly as Agave Sisalana. These plants generate rosettes of sword-shaped leaves that are toothed at first but eventually lose their teeth as they mature. Each leaf has several long, straight threads that can be removed by a process called decortication. Decortication is the method by which sisal fiber is created. The leaves of the sisal plant are squeezed and trodden by a spinning wheel installed in this procedure. The kit includes blunt knives, resulting in just fibers remaining. Water washes away the remaining sections of the leaf. Water is also used to clean the decorticated fibers before they are dried in natural heat or by an artificial process using hot air. The quality of fiber is determined by the moisture content, and hence proper drying is required. For higher grades, artificial drying is favored over natural sun drying. After drying, the fibers are untangled and graded using a machine. Retting, followed by scraping, is another method for separating the fiber from leaves. For plant rotting, the retting process employs a mix of bacteria activity and moisture. The procedure removes cellular tissues and sticky substances from the vicinity of bast-fiber bundles, allowing the fiber to split from the stem. For higher grades, artificial drying is favored over natural sun drying. After drying, the fibers are untangled and graded using a machine. Retting, followed by scraping, is another method for separating the fiber from leaves. For plant rotting, the retting process employs a mix of bacteria activity and moisture. The procedure removes cellular tissues and sticky substances from the vicinity of bast-fiber bundles, allowing the fiber to split from the stem.

Sisal is roughly classified into three classes: lower, medium, and high grades. Paper manufacturers employ lower-grade fiber because it contains a high concentration of hemicelluloses and cellulose. Medium-grade fiber is primarily used in the manufacturing of binder twine, ropes, and balers by the cordage industry. These items are typically utilized in agriculture, maritime, and general industrial applications. The carpet business uses the third-quality, a high-grade sisal type, to create strands. For a softer hand, sisal is used alone or in mixes with wool and acrylic in carpets.

In cross section, sisal fiber is made up of hundreds of fiber cells that are smooth, straight, and yellow. Sisal fiber's key characteristic is its strength, which gives it a rough and stiff look. Strength, durability, stretchability, and resistance to deterioration in salt water are just a few of the reasons why sisal is used to make ropes and other related items. The nature of the fiber guarantees that it absorbs dyes quickly and gives the most diverse variety of colored colors of any natural fiber. Because of their great toughness, sisal fiber–polyester composites are more likely to yield high work of fracture than pineapple and banana fiber composites. The fiber is incredibly durable and requires little maintenance due to its low wear and tear. However, sisal fiber is still not employed in the garment sector and is not suitable for moist environments. For fire resistance, sisal leaves are frequently treated with natural borax. Sisal is extensively utilized in the maritime sector for securing small ships, lashing, and cargo handling. Surprisingly, it is also employed as the core fiber of elevator steel wire cables.

Several researchers studied the potential of employing sisal fiber in thermoset polymers because of the low cost of manufacturing and the ability of sisal fibers to be laminated and wrapped. Sisal fiber-based composites have a specific strength equivalent to glass fiber-reinforced polymer composites. Furthermore, unidirectional sisal fiber-reinforced epoxy composites have a tensile modulus of 8.5 GPa. Because of these exceptional properties, sisal fiber-based polymeric composites have been used to construct civil engineering structures, consumer items, and low-cost housing (Joseph et al., 1999). Longitudinally oriented sisal fiber polymer composites outperformed short sisal fiber polymer composites in terms of mechanical characteristics. Fiber length, fiber orientation, fiber dispersion in the matrix, wt percent of fiber, and interfacial strength are all parameters that influence the tensile strength of sisal LDPE composites. The addition of glass fiber to sisal LDPE composites increased their tensile strength. With a crucial fiber length of 6 mm, longitudinally oriented sisal fiber composites have the highest storage modulus. The dielectric constant of sisal LDPE composites was enhanced by using additional sisal fiber in the thermoplastic matrix. Sisal fiber has the potential to be used as a reinforcement in polymer composites. This fiber has prospective uses in the aircraft and automobile industries, in addition to its typical applications (ropes, carpets, mats, and so on) (Naveen et al., 2018).

10.4.5 BANANA AND PINEAPPLE FIBERS

Banana fiber is not a new invention, even though few people are aware of its existence or application. Banana stems have been used to make fiber since the early 13th century in Japan. However, when other fibers such as cotton and silk from China and India grew more popular, the use of banana plants as a source of fiber for textiles diminished. Banana fiber, on the other hand, is making a rebound in the fashion sector. This fiber has recently made a resurgence in a variety of sectors and is used in a variety of items throughout the world, ranging from tea bags to automobile tires to saris and Japanese yen notes. The Banana family (Musa) is a member of the Musaceae family of monocotyledons. The Musaceae plant family is one of the most beneficial in the world. It offers us a wide range of meals as well as industrial raw materials. Musa sapientum provides us with the banana; Musa textiles provide us with the papermaking and cordage fiber abaca or Manila hemp. The Musa textiles plant yields this vegetable leaf fiber.

The most significant Musa species for fiber production is Musa textiles Nee, from which Abaca fiber, commonly known as hemp Manila, is derived. The stalks of abaca are typically slim, and the leaves are narrower, thinner, and tapered than those of a banana plant. Abaca is an extremely robust and lustrous material. Abaca rope is extremely flexible, robust, and resistant to saltwater damage. Along with coir, henequen, and sisal, it is classed as a hard fiber. Abaca has a pleasant natural shine. Its color is determined by the processing circumstances; excellent-grade abaca is off-white; however, some bad-quality fiber is virtually black. Individual fiber cells are cylindrical and have a smooth surface. They can be up to 6 mm (1/4 in) long and have a consistent width. The ends gradually taper to a point. The coarse abaca fibers, which may grow up to 3 m long and have a very high tensile strength, were traditionally used as cordage, particularly for ship rigging. These are mostly pulped processed for vacuum bags, tea bags, banknotes, sausages, cigarette papers, etc. Once a popular rope material, abaca is now showing promise as an energy-saving substitute for glass fibers in vehicles.

Pineapple is a perennial herbaceous plant with a height and breadth of 1–2 m that belongs to the Bromeliaceae family. It is mostly grown in coastal and tropical areas, primarily for its fruits. Pineapple leaf fiber production is abundant for industrial purposes with no further additions. Pineapple fibers (from the leaf PALF) are white, smooth, and glossy like silk, medium in length, and have high tensile strength. It has a softer surface than other natural fibers and absorbs and retains color well. PALF, on the other hand, has high specific strength and stiffness, and it is hydrophilic due to its high cellulose content. The leaves are harvested from the pineapple trunk after which the fibers are scrapped using the technique of decoration. The bundle of cellulose strands is then cleaned and hung to dry at a nearby river. Waxing and whippings are carried out to remove any entanglements or remaining plant tissues. The leaf strands are then one by one knotted to make a continuous filament yarn. This is a time-consuming, labor-intensive operation that frequently involves 30 individuals. Pineapple fiber makes a transparent, rigid, lightweight material that is ideal for usage in subtropical climes for traditional attire and accessories. It is today best known as Piatex®, a stronger weight material used as an alternative to leather in footwear, fashion, etc. This is a cradle-to-cradle material, which means that the entire material stays inside its life cycle—any waste from the manufacturing process may be utilized as fertilizer, and they even transform energy into biomass during the decoration (extraction) process.

Natural fibers with reinforcement composites play an important role in biocomposite and material science. PALF is a good replacement for synthetic fibers because of its low cost and renewable nature. The specific strength of natural fibers contributes to the improvement of the physical and mechanical strength of the polymer matrix without the use of any further processing. PALF is generally recognized in the textile industry and is currently utilized in everyday materials, but we believe that additional research will improve its applicability in the development of many existing goods (Asim *et al.*, 2015).

10.4.6 WOOD FIBERS

Wood fibers are produced up of both live and dead cells in the wood, depending mainly on the age of the tree. Because of the hierarchical structure of wood fibers, this fibrous material has good performance features, such as high strength-to-weight ratio. Wood fibers are extracted from trees via chemical, mechanical, biological, and a variety of combination techniques.

The surface property is one of the most important aspects of wood fibers; it influences the interfacial adherence of resin on the surface of fibers as well as the mechanical properties of fiber-based composites. Fiber shape, chemical makeup, extractive chemicals, and processing conditions all have an impact on this characteristic. Because of the surface's strong polarity, the fibers are less compatible with non-polar resin. As a result, the combination of the inherent polar and hydrophilic properties of wood fibers and the non-polar properties of resins causes difficulties in combining these materials, resulting in inefficient stress transmission of its composites under load. The use of various physical and chemical surface treatment procedures (e.g., coupling agents such as silanes) results in changes in the surface structure of the fibers as well as changes in surface characteristics (Dai and Fan, 2013). As people become more aware of the interconnectedness of global environmental factors, principles such as sustainability, industrial ecology, and ecoefficiency, as well as green chemistry and engineering, are being incorporated into the design of the next generation of materials, products, and processes. Wood fibers are largely employed in the paper and paperboard industries (about 80.5%), accounting for more than 55% of total paper and paperboard output. Composites employ 17.03% of wood fibers, with wood-fiber-based composites accounting for more than 80% of natural fiber-reinforced composites. Wood fibers are the dispersion component in wood-fiber composites. The matrix might be made of an inorganic substance, a natural polymer, or a synthetic resin. The matrix dominates the form, surface appearance, environmental tolerance, and overall durability of the composites, while the fibrous reinforcement bears the majority of the structural load, giving macroscopic stiffness and strength. Wood fiber–inorganic compound composites are one of the composite industry's most successful uses of wood fibers. They have been widely employed in a variety of architectural and agricultural applications as corrugated or flat roofing materials, cladding panels, and water containers. Although large-scale manufacturing of wood nanocellulose is still in the works, a new generation of wood fiber composites will provide a boost in the next decades.

10.4.7 Cellulose and Their Related Fibers

Cellulose is the most prevalent polymer on the globe, accounting for 1.5×10^{12} tons of total yearly biomass production, making it a nearly limitless source of raw material. Not unexpectedly, cellulose-based items are abundant in our society, as seen by the massive global businesses in cellulose derivatives, paper/packaging, textiles, and forest products. Natural cellulose fibers may be obtained from a variety of natural resources, including animal, bird, vegetable, and mineral sources. Natural fibers based on cellulose are derived from diverse sections of vegetables and plants. The physical and mechanical qualities of natural cellulose fibers are determined by the chemical composition, the location in which it is cultivated, the age of the plant, and the extraction procedures (Venkatarajan and Athijayamani, 2020). Cellulose fibers are categorized by their origin and are classified as follows: leaf: abaca, cantala, curaua, date palm, henequen, pineapple, sisal, banana; seed: cotton; bast: flax, hemp, jute, ramie; fruit: coir, kapok, oil palm; grass: alfa, bagasse, bamboo; stalk: straw (cereal). The varieties of bast and leaf (hard fibers) are the most typically employed in composite applications. Plant fibers are made up of cellulose fibers, which are made up of helically wrapped cellulose microfibrils held together by an amorphous lignin matrix. Lignin maintains water in fibers, protects against bacterial attack, and works as a stiffener to offer stem resistance to gravity forces and wind. Hemicellulose, which is abundant in natural fibers, is thought to act as a compatibilizer between cellulose and lignin.

Cellulose nanofibers have a significant potential for usage in a variety of applications, notably as reinforcement in the production of nanocomposites (Rose Joseph et al., 2021). Many investigations on the separation and characterization of cellulose nanofibers from diverse sources have been conducted. Simple mechanical methods or a mix of chemical and mechanical approaches can be used to remove cellulose nanofibers from cell walls. To create composites with improved mechanical characteristics and environmental performance, the hydrophobicity of the cellulose fibers must be increased, as well as the interface between matrix and fibers. The usage of plant cellulose fiber-reinforced composites is less appealing because of weak interfacial adhesion, low melting point, and poor moisture resistance. Pretreatments of cellulose fiber can clean the surface, chemically change the surface, halt moisture absorption, and enhance surface roughness (Kalia et al., 2011).

Cellulose fibers come in a variety of sizes, including microfibrillated and nanofibrillated. Cellulose nanofibers (CNFs) are presently taking over in a variety of applications, including water purification (Nizam et al., 2020), energy (Nizam et al., 2021), medicine, and scaffolding. Their surface chemistry with hydroxy groups, as well as their ease of modification to produce hydrophobicity, increases their use in a variety of applications. Furthermore, the pores in CNF allow other nanoparticles to become trapped in their pores, increasing their characteristics. To avoid aggregation, these holes are frequently used in the synthesis of nanoparticles.

10.5 CONCLUSION

The chapter covers briefly some of the main natural fibers used for various applications. These fibers are currently employed in applications such as carpets, ropes, mattresses, etc, but are promising materials to replace various synthetic materials. Their unique characteristics and composition such as lignin and cellulose are promising for future application.

REFERENCES

Allafi, F. *et al.* (2020) 'Advancements in applications of natural wool fiber: Review', *Journal of Natural Fibers*, 19(1), pp. 1–16. doi: 10.1080/15440478.2020.1745128

Asim, M. *et al.* (2015) 'A review on pineapple leaves fibre and its composites', *International Journal of Polymer Science*. doi: 10.1155/2015/950567

Azammi, A. M. N. *et al.* (2020) 'Physical and damping properties of kenaf fibre filled natural rubber/thermoplastic polyurethane composites', *Defence Technology*, 16(xxxx), pp. 29–34. doi: 10.1016/j.dt.2019.06.004

Dai, D. and Fan, M. (2013) Wood fibres as reinforcements in natural fibre composites: Structure, properties, processing and applications. In *Natural Fibre Composites: Materials, Processes and Applications*. Woodhead Publishing Limited. doi: 10.1533/9780857099228.1.3

Ekundayo, G. (2019) 'Reviewing the development of natural fiber polymer composite: A case study of sisal and jute', *American Journal of Mechanical and Materials Engineering*, 3(1), p. 1. doi: 10.11648/j.ajmme.20190301.11

Joseph, K. *et al.* (1999) 'A review on sisal fiber reinforced polymer composites', *Revista Brasileira de Engenharia Agrícola e Ambiental*, 3(3), pp. 367–379. doi: 10.1590/1807-1929/agriambi.v3n3p367-379

Kalia, S. *et al.* (2011) 'Cellulose-based bio- and nanocomposites: A review', *International Journal of Polymer Science*. doi: 10.1155/2011/837875

Kilinç, A. Ç., Durmuşkahya, C., and Seydibeyoğlu, M. Ö. (2017) 'Natural fibers', *Fiber Technology for Fiber-Reinforced Composites*, pp. 209–235. doi: 10.1016/B978-0-08-101871-2.00010-2

Le Troedec, M. *et al.* (2008) 'Influence of various chemical treatments on the composition and structure of hemp fibres', *Composites Part A: Applied Science and Manufacturing*, 39(3), pp. 514–522. doi: 10.1016/j.compositesa.2007.12.001

Lee, C. H., Sapuan, S. M., and Hassan, M. R. (2017) 'Mechanical and thermal properties of kenaf fiber reinforced polypropylene/magnesium hydroxide composites', *Journal of Engineered Fibers and Fabrics*, 12(2), pp. 50–58. doi: 10.1177/155892501701200206

Manaia, J. P., Manaia, A. T., and Rodriges, L. (2019) 'Industrial hemp fibers: An overview', *Fibers*, 7(12), pp. 1–16. doi: 10.3390/?b7120106

Moudood, A. *et al.* (2019) 'Flax fiber and its composites: An overview of water and moisture absorption impact on their performance', *Journal of Reinforced Plastics and Composites*, 38(7), pp. 323–339. doi: 10.1177/0731684418818893

Naveen, J. *et al.* (2018) Mechanical and physical properties of sisal and hybrid sisal fiber-reinforced polymer composites. In *Mechanical and Physical Testing of Biocomposites, Fibre-Reinforced Composites and Hybrid Composites*. Elsevier Ltd. doi: 10.1016/B978-0-08-102292-4.00021-7

Nizam, P. A. *et al.* (2020) 'Mechanically robust antibacterial nanopapers through mixed dimensional assembly for anionic dye removal', *Journal of Polymers and the Environment*, 28(4), pp. 1279–1291. doi: 10.1007/s10924-020-01681-3

Nizam, P. A. *et al.* (2021) Nanocellulose-based composites. In *Nanocellulose Based Composites for Electronics*. Elsevier Inc. doi: 10.1016/b978-0-12-822350-5.00002-3

Rose Joseph, M. *et al.* (2021) 'Development and characterization of cellulose nanofibre reinforced Acacia nilotica gum nanocomposite', *Industrial Crops and Products*, 161(December 2020), p. 113180. doi: 10.1016/j.indcrop.2020.113180

Salman, S. D. *et al.* (2015) 'Physical, mechanical, and morphological properties of woven kenaf/polymer composites produced using a vacuum infusion technique', *International Journal of Polymer Science*. doi: 10.1155/2015/894565

Sepe, R. *et al.* (2018) 'Influence of chemical treatments on mechanical properties of hemp fiber reinforced composites', *Composites Part B: Engineering*, 133, pp. 210–217. doi: 10.1016/j.compositesb.2017.09.030

Shahzad, A. (2012) 'Hemp fiber and its composites - A review', *Journal of Composite Materials*, 46(8), pp. 973–986. doi: 10.1177/0021998311413623

Soleimani, M. *et al.* (2008) 'The effect of fiber pretreatment and compatibilizer on mechanical and physical properties of flax fiber-polypropylene composites', *Journal of Polymers and the Environment*, 16(1), pp. 74–82. doi: 10.1007/s10924-008-0102-y

Suardana, N. P. G., Piao, Y., and Lim, J. K. (2011) 'Mechanical properties of HEMP fibers and HEMP/PP composites: Effects of chemical surface treatment', *Materials Physics and Mechanics*, 11(1), pp. 1–8.

Thirugnanasambantham, K. G. *et al.* (2020) 'SPIDER silk fiber: A brief review on molecular structure, properties and applications of spider silk', *AIP Conference Proceedings*, 2283(October). doi: 10.1063/5.0024918

Venkatarajan, S. and Athijayamani, A. (2020) 'An overview on natural cellulose fiber reinforced polymer composites', *Materials Today: Proceedings*, 37(Part 2), pp. 3620–3624. doi: 10.1016/j.matpr.2020.09.773

11 Biomedical Applications of Organic Nanofillers in Polymers

Basma M. Eid
National Research Centre

CONTENTS

11.1   Introduction 297
11.2   Common Polymer for Biomedical Applications (Rahmani et al., 2022)....298
11.3   Types of Nanofillers 300
11.4   Cellulose 300
        11.4.1   Cellulose Nanoparticles 302
        11.4.2   Isolation and Preparation of Nanocellulose 302
                  11.4.2.1   Chemical Treatment 302
                  11.4.2.2   Mechanical Process 304
                  11.4.2.3   Chemical Hydrolysis 304
        11.4.3   Bacteria Cellulose 304
        11.4.4   Nanocellulose Filler 305
        11.4.5   Biomedical Applications of Nanocellulose Filler in Polymer 306
11.5   Chitosan 310
        11.5.1   Chitosan Nanoparticles (CsNPs) 311
        11.5.2   Chitosan Nanofibers (CsNFs) 312
        11.5.3   Biomedical Applications of Chitosan Nanoparticle and
                  Chitosan Nanofibers in Polymer 312
11.6   Other Organic Nanofiller 312
11.7   Conclusions 316
References 316

11.1   INTRODUCTION

Biomedical applications include antimicrobial activity, drug delivery, gene delivery, cell imaging, wound healing, tissue engineering and vaccine adjuvants. The materials utilized in biomedical applications should have several significant properties since they have direct contact with the human body. Therefore, biomedical materials should be characterized by biodegradability, biocompatibility and non-toxicity and meet the required specifications for which they will be used (Hule & Pochan, 2007; Rahmani, Maroufkhani, Mohammadzadeh-Komuleh, & Khoubi-Arani, 2022).

DOI: 10.1201/9781003279372-11

Consequently, polymers of both natural and synthetic origin have been widely utilized in various biomedical applications. This is due to their relatively low cost, biodegradability, non-toxic properties, compatibility, facile preparation and fabrication, and they meet the requirements needed for utilization in biomedical applications, which include interaction with the human body. However, their mechanical and physical properties are considered to be their drawbacks.

Incorporating nanofiller into the polymer matrix to form a nanocomposite is carried out to boost the polymer matrix properties. Nowadays, there are wide types of nano-organic/ inorganic filler due to their biological properties that have been selected and investigated for biomedical applications (Hule & Pochan, 2007; Rahmani et al., 2022). The combination of polymer and organic/inorganic nanofiller in the nanocomposite matrix results in a synergetic integration of physical, chemical and biological properties, which have led to obtain novel nanocomposite with outstanding properties, such as excellent strength, dimensional stability, high electrical/ thermal conductivity, tailored permeability, improved biocompatibility and biodegradability, to expand their applications in the biomedical field (Crosby & Lee, 2007; Tripathy, 2017)

Nowadays, various types of nanoparticles are available to upgrade the polymeric matrix nanocomposite. These nanoparticles, with high interfacial area-to-volume ratio, can be embedded into the organic matrices, enhancing the molecular interactions with the polymeric matrix that eventually lead to conferring desirable and novel effects to the nanocomposite in comparison with the traditional one (Koo, 2019; Paul & Robeson, 2008; Schadler, Brinson, & Sawyer, 2007; Schmidt & Malwitz, 2003).

Additionally, nanofillers include two major types: organic, which consist of natural biopolymers (e.g. chitosan and cellulose), and inorganic, which can be subdivided into carbon nanostructure (e.g. fullerenes, graphene and carbon nanotubes) and metal or metal oxide nanoparticles (e.g. AgPs, ZnONPs and TiO_2NPs) (Nasrollahzadeh, Issaabadi, Sajjadi, Sajadi, & Atarod, 2019; Sothornvit, 2019). This chapter will mainly focus on the utilization of organic nanofillers in polymer for biomedical applications.

11.2 COMMON POLYMER FOR BIOMEDICAL APPLICATIONS (RAHMANI ET AL., 2022)

Generally, polymers can be classified into two main groups: (i) natural polymers and (ii) synthetic polymers (Figure 11.1). Natural polymers, which are called biopolymers, can be obtained from plant-based sources such as cellulose and sugarcane or from animal-based sources such as chitin, chitosan and collagen. Natural polymers have been utilized in various biomedical applications, which include wound healing, drug delivery and tissue engineering (Gobi, Ravichandiran, Babu, & Yoo, 2021; Mogoşanu & Grumezescu, 2014), due to their inherent properties such as biocompatibility, non-toxicity, biodegradability and bioactivity, anti-inflammatory and antibacterial properties (Z. Liu, Jiao, Wang, Zhou, & Zhang, 2008; Ribeiro et al., 2019; Singh, Patel, & Singh, 2016). However, these natural polymers have some limitations related to their very low dimensional stability, susceptibility to immunogenic

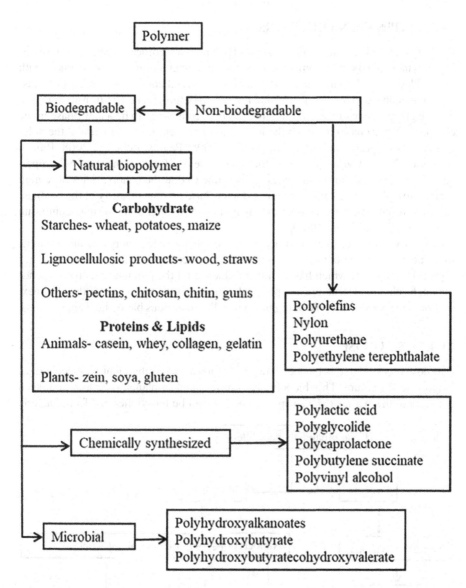

FIGURE 11.1 Classification of polymer.

responses and possibility of pathogen transmission. On the other hand, synthetic polymers, such as poly(ethylene oxide) (PEO), poly (caprolactone) (PCL), polylactic acid (PLA), poly(lactic-co-glycolic acid) (PLGA), poly(vinylpyrrolidone) (PVP) and poly(vinyl alcohol) (PVA), are the most common synthetic polymers used in biomedical applications. The biggest advantages of these polymers are their reproducibility and that they are biologically inert; however, they can cause chronic inflammation (Rahmani et al., 2022).

11.3 TYPES OF NANOFILLERS

Generally, the utilization of nanoparticles (NPs) has a significant impact on developing polymer matrix nanocomposites for the implementation of new products with high added value. Polymer nanocomposite is a material consisted of at least one discontinuous phase (nanofiller) embedded in a matrix (polymer). The transition region between the surface of the NPs and the matrix is considered to be a new phase in the composite (interphase) in which the interaction between the nanofiller and the polymer occurs (Ciprari, Jacob, & Tannenbaum, 2006; Paul & Robeson, 2008; Rallini & Kenny, 2017). The properties of the interphase itself determine the final properties of the obtained nanocomposites. This is due to that the nanoparticles have high surface-to-volume ratio; thus, the formed interphase fills the majority of the composite's volume and can become the fundamental factor in upgrading the nanocomposite properties (Ciprari et al., 2006).

Recently, there are many distinct forms of nanoparticles, inorganic and organic, that can be implanted into organic matrices to create polymer nanocomposites Figure 11.2, each of which has a unique influence on the polymeric matrix depending on the chemical structure and physical dimensions (Ciprari et al., 2006). The next section will focus on the role of organic nanofiller in various biomedical applications.

11.4 CELLULOSE

Cellulose is considered to be one of the most renewable and abundant polysaccharides biopolymer in nature. This biopolymer can be found in plants, such as wood, seed and agriculture wastes (McCarthy, 2003), or it can be biosynthesized from bacteria

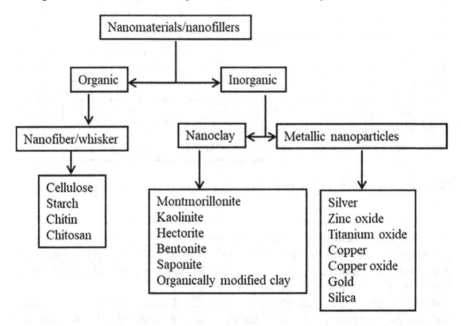

FIGURE 11.2   Classification of nanofiller.

(BC), such as Acetobacter (Gama, Dourado, & Bielecki, 2016; Park, Kim, Kwon, Hong, & Jin, 2009). It can also be found in the cell walls of some fungi algae, such as *Valonia and Microdictyon,* in a highly crystalline form, and very rarely in animal sources, such as in the membrane of marine animal named tunicate (Heinze, 2016).

Cellulose consists of a liner chain of β (1–4) linked D-glucose units with formula $(C_6H_{12}O_6)_n$ where Cellobiose is the repeating unit of cellulose with degree of polymerization ranging from 10,000 to 15,000. Each anhydroglucose unit contains three reactive hydroxyl s (-OH) groups, a primary (OH) at position C6 and two secondary at positions C2 and C3. Cellulose has chains arranged as a basic fibrillar unit (i.e. elementary fibrils), which contain high-ordered crystalline region folded in a longitudinal direction and lower ordered amorphous region with a relatively lower density and more random orientation (Figure 11.3a, b). The degree of crystallinity is ranging

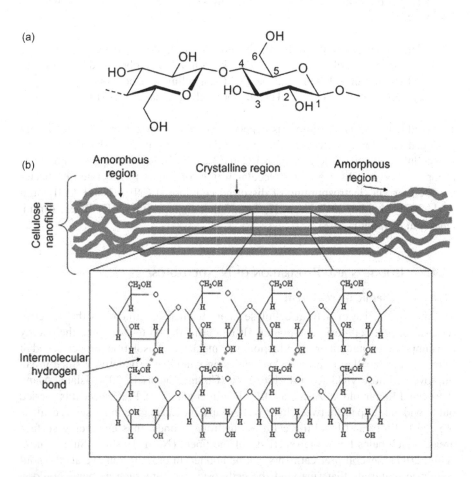

FIGURE 11.3 (a) Chemical structure of cellulose. (b) Crystalline and amorphous distribution in a cellulose nanofibril. (Reproduced from Rallini, M., & Kenny, J. M. (2017). 3-Nanofillers in Polymers. In C. F. Jasso-Gastinel & J. M. Kenny (Eds.), *Modification of Polymer Properties* (pp. 47–86), with permission Elsevier. Copyright 2017.)

from 40% to 70% depending on the cellulose source and production method while it can reach 80% in case of bacteria cellulose (Rebouillat & Pla, 2013a; Schenzel, Fischer, & Brendler, 2005).

11.4.1 CELLULOSE NANOPARTICLES

Nanocellulose can be referred to cellulose molecules with at least on dimension in nanoscale (1–100 nm). The increasing interest in utilizing nanocellulose as a nonofiller in diverse fields especially in biomedical applications related to its biodegradability nature, anisotropic shape, outstanding mechanical properties, biocompatibility and its interesting optical properties (Nabil A Ibrahim, Eid, & Sharaf, 2021; Phanthong et al., 2018).

There are three major types of nanocellulose that can be categorized into the following:

1. Cellulose nanocrystals (CNCs), also known as cellulose whiskers
2. Cellulose nanofibrils (CNFs), also known as nanofibrillated cellulose (NFC), microfibrillated cellulose (MFC) or cellulose nanofibers
3. Bacterial cellulose (BC) and electrospun cellulose nanofibers (ECNFs)

The production of nanocellulose is depending on their types. CNCs and CNFs are produced via a top-to-bottom method where the cellulose fiber can be disintegrated into cellulose nanoparticles, whereas BC and ECNF can be produced via a bottom-to-top method in which a low molecular weight of sugars is generated by bacteria (BC), or via electrospinning of dissolved cellulose (ECNF) (Nabil A. Ibrahim et al., 2021; Mondal, 2017; Nasir, Hashim, Sulaiman, & Asim, 2017; Tayeb, Amini, Ghasemi, & Tajvidi, 2018).

11.4.2 ISOLATION AND PREPARATION OF NANOCELLULOSE

11.4.2.1 Chemical Treatment

The main objective of natural cellulose fiber chemical treatments is the selective removal of non-cellulosic compounds. It is a crucial step to overcome the energy consumption of nanofibers production via mechanical isolation process, and it improves the fibrillation process by increasing nanofiber productivity since it can improve cellulose yield from 43% to 84% (Chinga-Carrasco, 2011; Islam, Alam, Patrucco, Montarsolo, & Zoccola, 2014; Khalil et al., 2014). Pretreatment is carried out to widen the space between hydroxyl groups, increasing the inner surface, altering crystallinity and breaking cellulose hydrogen bonds, thus enhancing surface areas, which helps boost the reactivity of the fibers (Nasir, Hashim, Sulaiman, & Asim, 2017). Several pretreatments can be applied in order to remove all the non-cellulosic materials that surround the cellulosic structure such as hemicellulose, lignin, pectin, wax and fat, then degrading the more accessible cellulose in amorphous region in order to liberate cellulose nanofiber. This step can be carried out by following methods:

11.4.2.1.1   Alkaline-Acid Pretreatment

This step aims to remove lignin, pectin, wax and fats that existed in the outer layer of the fiber cell walls through disrupting the lignin structure to break down the linkages between carbohydrate and lignin. This pretreatment method consists of (Islam et al., 2014; Khalil et al., 2014) the following:

- Soaking in NaOH (12–17 wt%) soaking for 2 hours to increase the cellulose fiber surface area and facilitate the hydrolysis;
- Soaking in HCl (1M) at 60°C–80°C for hemicelluloses solubilization;
- Treating with NaOH (2 wt%) at 60°C–80°C for 2 hours for disrupting the structure of lignin, and breaking the linkages between carbohydrate and lignin.

11.4.2.1.2   Oxidative Pretreatment

Oxidative treatment using TEMPO radicals is used before mechanical treatment to solve the aggregation problem caused by the presence of native cellulose –OH groups. TEMPO-mediated oxidation creates negative charges at surface of the microfibrils by introducing carboxylate and aldehyde functional groups into cellulose under and mild conditions, which in turn led to the repulsion of the nanofibers that made the fibrillation process easier (Johnson, Zink-Sharp, Renneckar, & Glasser, 2009; Saito, Kimura, Nishiyama, & Isogai, 2007; Saito, Nishiyama, Putaux, Vignon, & Isogai, 2006)

11.4.2.1.3   Enzymatic Pretreatment

Enzymatic pretreatment is an eco-friendly, inexpensive and highly specific method. It also increases the solids level and allows a smooth pass during high-pressure homogenization technique (HPH) (Nabil A. Ibrahim et al., 2021; Siddiqui, Mills, Gardner, & Bousfield, 2011). In this method, Laccase enzyme is used to degrade or modify the lignin and hemicellulose contents without attacking cellulose content (Nabil A. Ibrahim et al., 2021; Nasir, Gupta, Beg, Chua, & Asim, 2014; Nasir et al., 2015). However, a set of cellulases enzymes are used to facilitate MFC disintegration (Henriksson & Berglund, 2007; Pääkkö et al., 2007) as follows:

- Cellobiohydrolases: for crystalline cellulose;
- Endoglucanases: for disordered structure of cellulose.

11.4.2.1.4   Ionic Liquids

This method has been tried to produce NFC. ILs are organic salts with interesting properties such as non-flammability, thermal and chemical stability, and infinitely low vapor pressure (Phanthong et al., 2017; Pinkert, Marsh, Pang, & Staiger, 2009). NFCs were produced by treating sugarcane bagasse with 1-butyl-3-methylimidazolium chloride [(Bmim) Cl] as ionic liquid (IL) and followed by HPH technique (J. Li et al., 2012; Phanthong et al., 2017).

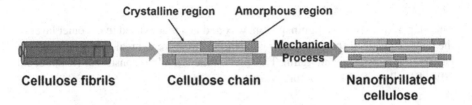

FIGURE 11.4 Schematic of NFC extracted from cellulose chains using mechanical process. (Reproduced from Phanthong, P., Reubroycharoen, P., Hao, X., Xu, G., Abudula, A., & Guan, G. (2018). *Carbon Resources Conversion, 1*(1), 32–43, with permission Elsevier. Copyright 2018.)

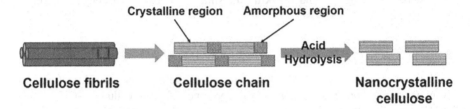

FIGURE 11.5 Schematic of CNC extracted from cellulose chains using acid hydrolyzed. (Reproduced from Phanthong, P., Reubroycharoen, P., Hao, X., Xu, G., Abudula, A., & Guan, G. (2018). *Carbon Resources Conversion, 1*(1), 32–43, with permission Elsevier. Copyright 2018.)

11.4.2.2   Mechanical Process

Mechanical fibrillation can be carried solely without chemical treatment; however, chemo-mechanical treatment can produce finer cellulose having diameter ranging from 5 to 50 nm (Chauhan & Chakrabarti, 2012). There are many mechanical methods that can be utilized to transform cellulosic fiber to nanocellulose including homogenizing (Siró & Plackett, 2010), microfluidization (Bharimalla, Deshmukh, Patil, & Vigneshwaran, 2015; Ferrer, Filpponen, Rodríguez, Laine, & Rojas, 2012; Phanthong et al., 2018), grinding (Panthapulakkal & Sain, 2012), cryocrushing (Chakraborty, Sain, & Kortschot, 2005) and high-intensity ultrasonication (HIUS) (Frone et al., 2011; Johnson et al., 2009).

11.4.2.3   Chemical Hydrolysis

The isolated NFCs from mechanical methods contain both crystalline and amorphous regions within the single cellulose fibril Figure 11.4. To obtain nanocrystalline cellulose (CNCs) with crystallinity more than 90%, the amorphous region should be dissolved by acids Figure 11.5. The acid hydrolysis of cellulose fiber is performed using strong but control acids. The most popular acids for hydrolysis are HCl and H_2SO_4. Nevertheless, phosphoric acid and nitric acid had been also investigated. It is also recommended to apply mechanical agitation such as sonication to avoid agglomeration during the treatment (Bhat, Dasan, Khan, Soleimani, & Usmani, 2017; Phanthong et al., 2018)

11.4.3   Bacteria Cellulose

Bacterial cellulose (BCN) is commonly produced from bacteria (e.g. *Acetobacter xylinum*) as a separate molecule, and no further pretreatment is required to remove non-cellulosic impurities as in the case of CNF and CNC. The biosynthesis of BCN is considered to be a bottom-to-up method. The ribbon-shaped BCN nanofibers are produced when the glucose chains are supplied inside the bacterial body and expelled out through minor pores present on the cell wall. The length of the formed ribbon-like web-shape is ranged between 20 and 100 nm long (S.-P. Lin et al., 2013; Nasir et al., 2017).

11.4.4 NANOCELLULOSE FILLER

To optimize the properties of nanocellulose-based fillers as an effective reinforcement for producing high-strength polymer nanocomposites, nanocellulose-based fillers should be homogenously aggregated in polymer matrices without aggregation; however, the hydrophilic nature of nanocellulose is considered to be the major drawback during the fabrication of polymer nanocomposite using nanocellulose-based fillers. Table 11.1 illustrates the most widely used techniques in order to fabricate nanocellulose-based nanocomposites.

TABLE 11.1
Techniques to Fabricate Cellulose Nanocrystal-Based Bio-nanocomposites

Technique	Description	Advantages & Disadvantages	Ref.
Casting evaporation	NCC is dispersed in a medium that is compatible with the polymer solution Nanocellulose/polymer matrix solution can be obtained via the following techniques: casting on plate followed by evaporation to dry Freeze drying followed by compress molding extruding, followed by molding compression of the mixture	Advantages: • fabrication is easy and simple • preserve the dispersion state of the nanoparticles in the liquid • no need for specific equipment Disadvantages: • the retention of toxic solvent in polymer matrix • the use of solvents can cause denaturation to the proteins and other molecules integrated into the polymer matrix, and • the ultimate product's shape is limited	Bhat et al. (2017); Rebouillat and Pla (2013b)
Electrospinning	• The used polymer is melted out of an orifice (syringe) then a high voltage is used between the orifice and collecting target • Followed by evaporation or solidifying of the polymer and collected as interconnected web of fibers on the collecting target • NCC was covalently bonded to PLA chain via silane-modified PLA in the electrospinning	Fabrication of nanofibers with diameter ranging from 1 μm down to nm	Rahmat, Karrabi, Ghasemi, Zandi, and Azizi (2016)

(Continued)

TABLE 11.1 (*Continued*)

Techniques to Fabricate Cellulose Nanocrystal-Based Bio-nanocomposites

Technique	Description	Advantages & Disadvantages	Ref.
Extrusion	Suspension of NCC can be pumped into the polymer melt during the extrusion process.	This method resulted in improving in mechanical properties and moisture sensitivity.	Bhat et al. (2017)
Impregnation	Film of NCC is immersed in the thermosetting resin under vacuum conditions. The used polymer should be in a solubilizing state to ensure good impregnation.	Advantages: • No specific equipment is required. Disadvantages: • Inefficient techniques from the industrial point of view because the obtained shape of final product is limited.	Bondeson and Oksman (2007); Peng, Dhar, Liu, and Tam (2011)
Layer-by-layer (LBL)	The LBL method is defined as the successive adsorption of oppositely charged components in solution by dipping them alternatively. Cellulose nanocrystal multilayer composites were fabricated with poly (diallyldimethyl-ammonium chloride) via LBL assembly technique.	• Simple technique • Uniform coverage of the produced nanocomposites • Nanoscale thickness can be control • No specific device or components is required	Podsiadlo et al. (2005); Tang, Li, Du, and He (2012)

11.4.5 BIOMEDICAL APPLICATIONS OF NANOCELLULOSE FILLER IN POLYMER

The potential applications of cellulose nanoparticles and their nanocomposites have been extensively studied in multiple fields. Cellulose nanoparticles are highly applicable in biomedical fields, which include drug release, tissue engineering and scaffolds due to their appealing features, such as sustainability, non-toxicity, biocompatibility and excellent mechanical properties as well as low production cost in addition to its high surface area which in turn offers vast possibilities for chemical modifications (N. Lin & Dufresne, 2014). Table 11.2 summarizes the utilization of nanocellulose as filler in biomedical fields.

CNC is considered to be an excellent candidate as biomaterial for drug delivery system since it has reactive functional groups on its surface where drugs, nanoparticles or targeting molecules could be attached (Bober et al., 2014). On the other hand, nanocellulose can be a promising material for tissue engineering as material for cell attachment and proliferation due to their outstanding mechanical properties and biocompatibility. Among nanocellulose types, BNC is considered to be

TABLE 11.2

Some Potential Applications of Nanocellulose Filler in Biomedical Fields

Biomedical Application	Type of Nanocomposite	Main Findings/Remark	References
Drug delivery	PVA/CNC/PLGA-NPs bio-nanocomposites [Polyvinyl alcohol/cellulose nanocrystals/poly(D,L lactide-co-glycolide) (nanoparticles loaded with bovine serum albumin fluorescein isothiocyanate conjugate)	• The incorporation of CNC in the produced bio-nanocomposites increased the elongation properties without adversely affecting other mechanical properties. • Biopolymeric nanoparticles can be successfully delivered to adult bone marrow mesenchymal stem cells using the prepared PVA/CNC/PLGA-NPs bio-nanocomposite films, suggesting a novel tool for drug delivery techniques.	Rescignano et al. (2014)
	CNCs/CS$_{OS}$ (oxidized cellulose nanocrystal/chitosan oligosaccharide)	CNC–CS$_{OS}$ particles were loaded with procaine hydrochloride PrHy at pH 8. The obtained bio-nanocomposite has potential applications as fast response drug carriers in wound dressings and local drug delivery.	Akhlaghi, Berry, and Tam (2013)
	NCC/cyclodextrin/polymer inclusion	The results revealed that the new nanocomposite hydrogels can be a good candidate as a controlled delivery vehicle.	Zhang et al. (2010)
Tissue engineering	Collagen/CNCs composite film	• The results showed that the maximum mechanical properties for the collagen/CNCs composite were reached by adding 7 wt% of CNCs. • CNCs exhibited good biocompatibility and showed no negative effect on the cell morphology, viability, and proliferation and possess. • The used method for incorporating CNCs in collagen was simple and it can be a promising method to reinforce collagen films without compromising biocompatibility. • The obtained composite can be promising in the field of skin tissue engineering.	Li et al. (2014)
	(HA/CNC/SF) scaffold hydroxyapatite/cellulose nanocrystals/silk fibroin	The results showed that the HA/CNC/SF scaffold exhibited better thermostability and mechanical properties compared with SF, CNC/SF, or HA/SF scaffolds as well as outstanding biocompatibility and remarkable osteoconductivity	Chen, Zhou, Chen, and Chen (2016)

(Continued)

TABLE 11.2 (*Continued*)

Some Potential Applications of Nanocellulose Filler in Biomedical Fields

Biomedical Application	Type of Nanocomposite	Main Findings/Remark	References
		The HA/CNC/SF scaffold revealed to be a promising candidate for repairing of bone defects in bone tissue engineering since their efficiency for bone regeneration in the rat calvarial defect model demonstrated that the rat calvarial defect healed with new-formed bone within 12 weeks of implantation and the degradation rate of the scaffold was a good match to the bone regeneration rate.	
	PLA-g-silane/NCC nanocomposites Poly(lactic acid)-g-silan/nanocrystalline cellulose	• The obtained results revealed a noticeable enhancement in tensile strength owing to the incorporation of NCC into the produced nanocomposite. • The biocompatibility of the modified nanocomposite was also evaluated through cytotoxicity test. • The results also confirmed that obtained nanocomposite can be a suitable candidate for tissue engineering applications.	Rahmat et al. (2016)
	3D CNF/HG hydrogel (CNF/hyaluronan-gelatin composite hydrogel)	• This work was aimed to create 3D culture environment for HepaRG liver progenitor cells. • The results demonstrated that 3D NFC/HG hydrogels accelerate the hepatic differentiation of HepaRG liver progenitor cells better than the standard 2D culture environment as it improved cell morphology, expression and localization of hepatic markers, and metabolic activity. • The CNF/HG hydrogel can be favorable material for hepatic cell culture and tissue engineering.	Malinen et al. (2014)
	CNF-based/gelatin/β-tricalcium phosphate	• CNF was incorporated in these scaffolds was to decrease the rate of degradation, controlling the release of osteoinductive biomolecule (simvastatin). • The produced scaffolds improved bone formation and enhanced collagen matrix deposition compared to the control. • The obtained nanocomposite exhibited a good potency for bone tissue engineering.	Sukul, Min, Lee, and Lee (2015)

(Continued)

TABLE 11.2 *(Continued)*

Some Potential Applications of Nanocellulose Filler in Biomedical Fields

Biomedical Application	Type of Nanocomposite	Main Findings/Remark	References
	BC/Gel/Hap nanoscaffolds Bacterial cellulose/ gelatin/ hydroxyapatite	• A regular vertical pore arrays BC was produced using a laser patterning technique. • The cytocompatibility test of the obtained scaffolds using chondrogenic rat cells showed the capability of the scaffold to keep the cell viable via supporting the chondrogenic rat cell attachment and proliferation.	Jing et al. (2013)
	PAM/BC nanocomposite hydrogel Hybrid polyacrylamide/ bacterial cellulose nanofiber	• The obtained hydrogel composite showed an outstanding large elongation at break of 2200%, and a high fracture stress of 1.35 MPa. • The results revealed that the PAM/BC hybrid hydrogel was benign according to the cell viability test and the obtained hydrogel can be a suitable candidate for biomedical applications such as bone and cartilage repair materials.	Yuan et al. (2016)
	PLA/PBS/CNFs fibrous scaffolds poly (lactic acid)/poly (butylene succinate) (PBS)/cellulose nanofibrils	• The incorporation of CNF in PLA/PBS scaffold for reinforcement produced scaffolds with multi-scaled structure and interconnected porous morphology. • The results revealed that the incorporation of CNFs in PLA/PBS/CNFs composite scaffolds showed a better support to the attachment and proliferation of human fibroblast cells than PLA, PBS or their blends. • Conclusively, the obtained composite scaffolds showed a good possibility for vascular tissue engineering application.	Abudula et al. (2019)
Wound healing	PDA-TH/TOCNFs hydrogel (tetracycline hydrochloride loaded polydopamine/ (TEMPO)-oxidized cellulose nanofibrils	The obtained hydrogel of nanocellulose and polydopamine crosslinked physically with Ca2+ exhibited an excellent drug release for wound healing process. The results that the obtained hydrogel showed good antibacterial activity against different types of bacteria	Liu et al. (2018)
	Chitosan/PVP/ Nanocellulose	The fabricated chitosan/PVP/nanocellulose composite dressing in the presence of stearic acid was applied as biological wound healing on albino rats. The obtained results showed that the healing period was shorter than the control wounds.	Poonguzhali, Basha, and Kumari (2017)

(Continued)

TABLE 11.2 (*Continued*)

Some Potential Applications of Nanocellulose Filler in Biomedical Fields

Biomedical Application	Type of Nanocomposite	Main Findings/Remark	References
Antimicrobial	NFC/PPy-Ag (nanofibrillated cellulose/ polypyrrole/AgNPs)	• The addition of NFC to PPy increased the material's capacity to form film. The obtained composites showed good mechanical and electrical properties. • The results revealed that the NFC/PPy-Ag composite films exhibited strong antibacterial activity against G+ve bacteria, e.g. *Staphylococcus aureus*. • The given data demonstrated that the obtained nanocomposite with good electrical conductivity and excellent antimicrobial activity can be a good candidate for various applications in the field of biomedical treatments and diagnostics.	Bober et al. (2014)

bacterial most suitable choice for the medium of cell culture owing to its high porosity, biodegradability, and low toxicity. It can also be an appropriate replacement for traditional skin restoration materials due to its high water holding capacity, high elasticity and comfort, as well as high mechanical strength (Czaja, Young, Kawecki, & Brown, 2007).

11.5 CHITOSAN

Chitosan is one of the outstanding natural polymers with a wide range of applications, and it is one of the most promising biopolymers for the production of advanced materials next to cellulose (Mhd Haniffa, Ching, Abdullah, Poh, & Chuah, 2016; Xu, Huang, Zhu, & Ye, 2015). Chitosan (Cs) is a polyamino-saccharide composed of β-(1–4)-2-amino-2-deoxy-D-glucose units that include active primary amine, primary hydroxyl and secondary hydroxyl groups on its molecular chain Figure 11.6 (N. A. Ibrahim & Eid, 2017). Cs is commonly obtained by deacetylation of chitin, a natural linear copolymer that is vastly found in marine crustaceans (e.g. shrimp, crab), cuticles of insects and micro-organisms cell walls.

Chitosan is also a cationic polymer that exhibits great potential in biomedical applications because of its distinctive properties such as antibacterial, antioxidant, chelating, no toxicity, pH sensitivity, biodegradability and biocompatibility as well as excellent film-forming ability.

For obtaining Cs with lower molecular weight, depolymerization of Cs can be performed by one of the following methods (N. A. Ibrahim & Eid, 2017; Mourya & Inamdar, 2008):

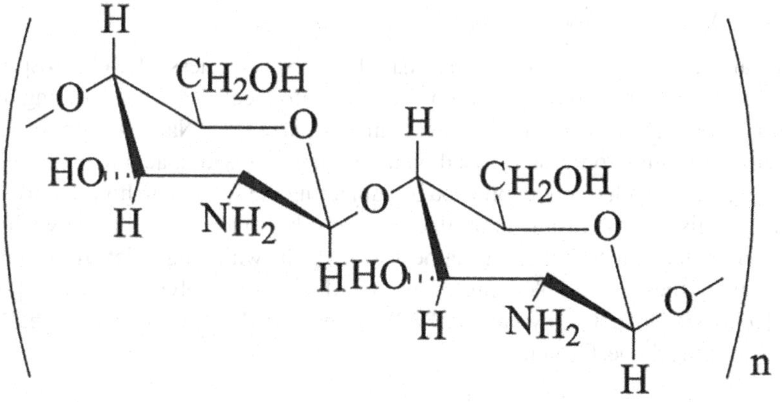

FIGURE 11.6   Chemical structure of chitosan (CS).

- Acid hydrolysis at 80°C for short time using concentrated hydrochloric acid (HCl) or orthophosphoric acid (H_3PO_4) at room temperature for longer time,
- Oxidative degradation via generation of hydroxyl radicals by using concentrated nitrous acid (HNO_2) or hydrogen peroxide (H_2O_2),
- Salt treatment in presence of microwave heating or by
- Enzymatic treatment using specific chitinases or non-specific mixture of enzymes such as cellulases and lipases.

The existence of three free functional groups in chitosan—an amino group and primary and secondary hydroxyl groups at C2, C3 and C6 positions—facilitate chemical modification of the chitosan structure without altering its biophysical properties (Divya & Jisha, 2018; Gao & Wu, 2022). The developed functional properties in the produced chitosan derivatives have been exploited successfully to create a novel product based on chitosan that is highly promising in a vast number of applications, especially in biomedical applications.

11.5.1   CHITOSAN NANOPARTICLES (CSNPS)

There are five simple methods to prepare chitosan nanoparticles namely ionotropic gelation, microemulsion, emulsification solvent diffusion, polyelectrolyte complex and reverse micellar; however, ionotropic gelation and polyelectrolyte complex are the most common and widely used for CsNP preparation (Divya & Jisha, 2018; Gao & Wu, 2022). The ionotropic gelation method is based on electrostatic interaction between the amine group of chitosan and a negatively charged group of polyanion such as tripolyphosphate (Shiraishi, Imai, & Otagiri, 1993). However, polyelectrolyte complex is carried out by self-assembly of the cationic charged polymer and plasmid DNA where the CsNPs is formed after adding DNA solution to dissolved chitosan in acetic acid media using mechanical stirring at room temperature (Erbacher, Zou, Bettinger, Steffan, & Remy, 1998). CsNPs are characterized by chitosan properties and nanoparticle advantages, such as large surface area and interface effect.

11.5.2 Chitosan Nanofibers (CsNFs)

There are many methods to produce nanofiber. Nevertheless, the electrospinning technology has gained much attention because it produces nanofibers ranging in size from micrometers to nanometers (Jayakumar, Prabaharan, Nair, & Tamura, 2010). The electrospinning process is based on dissolving chitosan in acid to protonate and changing it to polyelectrolyte followed by subjecting the solution to high electric field. However, this process results in producing beads due to low charge density which in turn causes inadequate stretch of filaments during the whipping of jet. To prevent the formation of these beads, blending chitosan with synthetic polymers, such as polyvinyl alcohol (PVA), poly ethyl oxide (PVO) and polyethylene terephthalate (PET), has been successfully performed.

11.5.3 Biomedical Applications of Chitosan Nanoparticle and Chitosan Nanofibers in Polymer

It is well known that chitosan, as a biopolymer, has outstanding importance in all biomedical applications due to its attractive inherited properties. Besides, the biological activities of chitosan can be broadened by chemical modification of chitosan through its three active groups which include an amino group and two hydroxyl groups. On the other hand, nanomaterials with small size and large surface area have brought great attention toward biomedical applications as they show high efficiency, better performance better biological distribution and site-specific drug delivery due to their unique physical and chemical properties. Therefore, chitosan-based nanomaterials have shown a great contribution in the field of biomedical applications, such as antibacterial property, drug and gene delivery, cancer and hyperthermia therapy, cell imaging, restorative dentistry, wound healing, tissue engineering and other biomedical fields. Nevertheless, this section focuses on the role of chitosan nanoparticle-polymer composite in biomedical applications. Table 11.3 summarizes some of these activities.

11.6 OTHER ORGANIC NANOFILLER

Several organic nanofillers have been incorporated into polymers for versatile biomedical applications. The following are some of the examples:

- The LNP@Ag–Fe-PAA hydrogel showed remarkable self-healing ability and excellent antibacterial activities against both the G-ve (*E. coli*) and G-ve (*S. aureus*) due to the presence of Ag NPs and lignin in the polymer matrix. The remarkable antibacterial efficacy of LNP@Ag–Fe-PAA hydrogels arise their potential for biomedical applications (Cui et al., 2021).
- Carboxylated lignin nanoparticles (CLNPs) with an increased amount of carboxyl groups as reactive sites were grafted in PEG, polyhistidine (PHIS) and a cell-penetrating peptide by EDC/NHS peptide coupling chemistry. The obtained results revealed that an increased circulation in the bloodstream and also pH-responsive behavior. Additionally, the study showed a high biocompatibility ability release the drug is dependent on the pH media (Figueiredo et al., 2017).

TABLE 11.3
Some Potential Applications of CsNPs/Polymer Composites in Biomedical Fields

Biomedical Application	Type of Nanocomposite	Main Findings/Remark	References
Drug delivery	Fucoidan/CsNPs/MTX) (fucoidan/chitosan nanoparticles/ methotrexate)	• The MTX-loaded fucoidan/chitosan nanocomposite is higher than MTX alone against A549 lung cancer cells for about seven times. • These results revealed that fucoidan/chitosan nanocomposite can be suitable for oral delivery in cancer therapy.	Barbosa, Costa Lima, and Reis (2019)
		• Fucoidan/chitosan nanocomposite was prepared for the treatment of skin inflammatory diseases by topical delivery of methotrexate. • Fucoidan/chitosan nanocomposite show no effect on cell vitality and exhibited lower cytotoxicity than MTX alone. • The composite exerts the anti-inflammatory effect and improves skin permeation therefore it can be a good candidate for methotrexate topical delivery.	Coutinho, Costa Lima, Afonso, and Reis (2020)
Wound dressing	Cs/PVP/ZnO nanofibrous chitosan/polyvinyl alcohol/zinc oxide nanofibrous membrane	• The results revealed that the prepared Cs/PVP/ZnO nanofibrous exhibited higher antibacterial activity, antioxidant and accelerated wound healing much higher than Cs/PVP nanofibrous. • Electrospun scaffolds made of Cs/PVP/ZnO nanofibrous could be suitable for diabetic wound dressings.	Ahmed et al. (2018)
	Cs/SF composite nanofiber (chitosan/Silk fibroin composite nanofiber)	The results showed that the Nanofibrous membranes made of CS/SF composite boosted cell adhesion and proliferation. The antibacterial activity of the membrane increased by increasing the Cs proportion which in turn promotes the prepared Cs/SF nanofibrous to utilize as wound dressings.	Cai et al. (2010)
	PCL/mupirocin+ Cs/LID DLS (polycaprolactone/ mupirocin and chitosan/ lidocaine hydrochloride double-layer nanofibrous scaffolds)	• The DLS scaffold exhibited excellent porosity and swelling behaviors, which are favorable for absorbing extra exudates and keeping moist on the wound surface. Tensile strength and Young's modulus of the prepared DLS nanofibers scaffold had been remarkably increased. • The multifunctional DLS exhibited high antibacterial efficiency against *S. aureus*, *E. coli*, and *P. aeruginosa* with no toxicity to fibroblasts.	Li, Wang, Yang, Liu, and Zhang (2018)

(Continued)

TABLE 11.3 (*Continued*)
Some Potential Applications of CsNPs/Polymer Composites in Biomedical Fields

Biomedical Application	Type of Nanocomposite	Main Findings/Remark	References
	Cs/PVA/GO NF Chitosan/poly (vinyl alcohol)/graphene oxide nanofibrous membrane	• Cs/PVA/GO NF membrane loaded with ciprofloxacin antibiotic was successfully fabricated. • The fabricated NF membranes remarkably enhanced the antibacterial efficiency against *E. coli*, *S. aureus* and *B. subtilis* after the addition of ciprofloxacin. • NF membranes loaded with Cipro showed excellent cytocompatibility with Melanoma cells. • The existence of GO slightly enhanced the rate of drug release.	Yang, Zhang, and Zhang (2019)
Tissue engineering	PLGA–chitosan/PVA composite NF Poly lactide-co-glycolide/ chitosan/poly (vinyl alcohol) nanofibrous membrane	The composite membrane exhibited moderate tensile strength. Cell culture test demonstrated that the PLGA–chitosan/PVA membrane was able to elevate fibroblast attachment and proliferation. Conclusively, PLGA–chitosan/PVA composite nanofiber could be suitable for skin reconstruction	Duan et al. (2006)
	PCL/NFC composite Polycaprolactone/ nanofibrillated chitosan composite	• NFC was used as dispersing phase in the PCL matrix to fabricate electrospun nanocomposite fibrous scaffolds. • PCL/NC scaffolds exhibited significant improvement in enhancement in mechanical properties in terms of tensile strength and Young's modulus compared to individual PCL scaffold. • The wettability of scaffolds was improved after adding NC. • Incorporated NFC to the composite enhanced proliferation and adhesion of cells. • Conclusively, PCL/NC can be proposed as a suitable scaffold for tissue engineering applications.	Fadaie, Mirzaei, Geramizadeh, and Asvar (2018)

(*Continued*)

TABLE 11.3 (*Continued*)

Some Potential Applications of CsNPs/Polymer Composites in Biomedical Fields

Biomedical Application	Type of Nanocomposite	Main Findings/Remark	References
	K/PVA-PLA/CHNF/ZnO composite nanofibers (Keratin/polyvinyl alcohol-polylactic nanofibers/nanofibrillated chitosan/ZnONPs) composite	• Composite nanofibers were fabricated by using PLA and keratin/PVA as the matrix and nanofibrillated chitosan (CHNF)/ZnO as the nanofiller. • Incorporation of (CHNF)/ZnO resulted in reduction of the nanofiber size and enhancement in mechanical properties. • Wettability of the nanofiber composite varied according to the nanofiller type as it increase in case of using CHNF alone and decrease in the presence of ZnONPs. • Incorporation of CHNF/ZnONPs into the polymer matrix increased the cell viability significantly. • The fabricated nanocomposites can be proper candidate for tissue engineering applications.	Ranjbar-Mohammadi, Shakoori, and Arab-Bafrani (2021)
Antibacterial	Poly(lactic acid) /CsNP PLA/(Chitosan Nanoparticle) composite nanofibrous	• The developed PLA/chitosan composite was a highly porous structure in which CsNPs were homogenously distributed throughout the entire fiber. • PLA/chitosan fibrous membranes exhibited high antibacterial activity against *Escherichia coli* and *Staphylococcus aureus* that reach 99.4% and 99.5%, respectively.	Li, Wang, Zhang, and Pan (2018)

- The prepared lignin-g-PMMA was able to self-assemble into micelles (diameter approximately 100 nm) and successfully control the release of IBU (Ibuprofen). The results indicated that the drug release was faster in neutral pH (81% at pH 7.4) and slower in acidic media (16% release, at pH 1.5) (Cheng et al., 2020).
- Two types of chitin nanofillers namely nanocrystals (CHNC) and nanofibers (CHNF) incorporated into chitosan (CS) matrix were prepared by solvent evaporation-casting approach. The CS-based bionanocomposite films reinforced with CHNF exhibited remarkable mechanical properties compared to CHNC. The prepared CS/CHNF bionanocomposite showed significant antifungal activity against A. Niger (>80%). Conclusively, these bionanocomposite films exhibited great potential for medical devices (Salaberria, Diaz, Labidi, & Fernandes, 2015).

- Polycaprolactone/Nano-Hydroxyapatite/Chitin-Nano-Whisker nanocomposite scaffolds (Polycaprolactone/nHA/CNW) were fabricated, and their suitability for biomedical applications was investigated by using preosteoblast mouse bone cell line for cell proliferation and attachment assays. The results demonstrated that the existence of CNW in the filaments enhanced the mechanical properties of the 3D printed parts. CNW (3% content) showed remarkable improvement in the cell proliferation and attachment properties of the scaffolds. However, both nHA and CNW nanofillers improve the PCL biodegradation rate. Conclusively, the prepared PCL/nHA/CNW composite can notably enhance the mechanical biological and properties of the 3D printed products to be used in bone tissue scaffolds (Karimipour-Fard, Jeffrey, JonesTaggart, Pop-Iliev, & Rizvi, 2021).
- Chitin nanowhiskers (CNWs) were integrated into chitosan/β-glycerophosphate disodium salt for tissue engineering scaffold. The results revealed that tensile strength and elongation at break were both increased more than four compared with individual CS/GP hydrogel may be to that the CNWs acted as a cross-linker through hydrogen bond interaction in the gel formation process. The cytotoxicity of hydrogel in vitro studies confirmed a good biocompatibility of (CS/GP) injectable hydrogel containing CNWs (Wang, Chen, & Chen, 2017).

11.7   CONCLUSIONS

In this chapter, the role of organic nanofiller in polymer matrix for biomedical applications have been reviewed, which include drug delivery, tissue engineering, wound dressing, and medical. Organic nanofiller-based nanocomposite can be considered a green nanomaterial for biomedical applications. These organic nanofillers include but not exclude cellulose, chitin, chitosan, lignin and starch. Among the different types of organic nanofiller, nanocrystalline cellulose-based bionanocomposites have been thoroughly investigated due to their outstanding mechanical properties such as renewability, biodegradability and biocompatibility. However, the major challenge in utilizing organic nanofiller such as nanocrystalline cellulose in nanocomposite preparation is its hydrophilic nature, which results in poor dispersion in hydrophobic polymer matrices. That is why most of the conducted research work is mainly dependent on using inorganic nanofiller such as carbon nanostructure and metal or metal oxide nanoparticles. Therefore, much attention and R&D efforts regarding surface modifications should be paid to overcome these challenges and make them more attractive to be utilized in a wide range of biomedical fields.

REFERENCES

Abudula, T., Saeed, U., Memic, A., Gauthaman, K., Hussain, M. A., & Al-Turaif, H. (2019). Electrospun cellulose nano fibril reinforced PLA/PBS composite scaffold for vascular tissue engineering. *Journal of Polymer Research, 26*(5), 110. https://doi.org/10.1007/s10965-019-1772-y

Ahmed, R., Tariq, M., Ali, I., Asghar, R., Noorunnisa Khanam, P., Augustine, R., & Hasan, A. (2018). Novel electrospun chitosan/polyvinyl alcohol/zinc oxide nanofibrous mats with antibacterial and antioxidant properties for diabetic wound healing. *International Journal of Biological Macromolecules, 120*, 385–393. https://doi.org/10.1016/j.ijbiomac.2018.08.057

Akhlaghi, S. P., Berry, R. C., & Tam, K. C. (2013). Surface modification of cellulose nano-crystal with chitosan oligosaccharide for drug delivery applications. *Cellulose, 20*(4), 1747–1764. https://doi.org/10.1007/s10570-013-9954-y

Barbosa, A. I., Costa Lima, S. A., & Reis, S. (2019). Application of pH-responsive fucoidan/chitosan nanoparticles to improve oral quercetin delivery. *Molecules, 24*(2), 346.

Bharimalla, A. K., Deshmukh, S. P., Patil, P. G., & Vigneshwaran, N. (2015). Energy efficient manufacturing of nanocellulose by chemo-and bio-mechanical processes: a review. *World Journal of Nano Science and Engineering, 5*, 204–212.

Bhat, A. H., Dasan, Y., Khan, I., Soleimani, H., & Usmani, A. (2017). Application of nanocrystalline cellulose: Processing and biomedical applications *Cellulose-reinforced nanofibre composites* (pp. 215–240). Elsevier.

Bober, P., Liu, J., Mikkonen, K. S., Ihalainen, P., Pesonen, M., Plumed-Ferrer, C., … Latonen, R.-M. (2014). Biocomposites of nanofibrillated cellulose, polypyrrole, and silver nanoparticles with electroconductive and antimicrobial properties. *Biomacromolecules, 15*(10), 3655–3663.

Bondeson, D., & Oksman, K. (2007). Polylactic acid/cellulose whisker nanocomposites modified by polyvinyl alcohol. *Composites Part A: Applied Science and Manufacturing, 38*(12), 2486–2492.

Cai, Z.-X., Mo, X.-M., Zhang, K.-H., Fan, L.-P., Yin, A.-L., He, C.-L., & Wang, H.-S. (2010). Fabrication of chitosan/silk fibroin composite nanofibers for wound-dressing applications. *International Journal of Molecular Sciences, 11*(9), 3529–3539.

Chakraborty, A., Sain, M., & Kortschot, M. (2005). Cellulose microfibrils: A novel method of preparation using high shear refining and cryocrushing. *Holzforschung, 59*(1), 102–107.

Chauhan, V. S., & Chakrabarti, S. K. (2012). Use of nanotechnology for high performance cellulosic and papermaking products. *Cellulose Chemistry and Technology, 46*(5), 389.

Chen, X., Zhou, R., Chen, B., & Chen, J. (2016). Nanohydroxyapatite/cellulose nanocrystals/silk fibroin ternary scaffolds for rat calvarial defect regeneration. *RSC Advances, 6*(42), 35684–35691. https://doi.org/10.1039/c6ra02038k

Cheng, L., Deng, B., Luo, W., Nie, S., Liu, X., Yin, Y., … Chen, J. (2020). pH-responsive lignin-based nanomicelles for oral drug delivery. *Journal of Agricultural and Food Chemistry, 68*(18), 5249–5258. https://doi.org/10.1021/acs.jafc.9b08171

Chinga-Carrasco, G. (2011). Cellulose fibres, nanofibrils and microfibrils: The morphological sequence of MFC components from a plant physiology and fibre technology point of view. *Nanoscale Research Letters, 6*(1), 1–7.

Ciprari, D., Jacob, K., & Tannenbaum, R. (2006). Characterization of polymer nanocomposite interphase and its impact on mechanical properties. *Macromolecules, 39*(19), 6565–6573.

Coutinho, A. J., Costa Lima, S. A., Afonso, C. M. M., & Reis, S. (2020). Mucoadhesive and pH responsive fucoidan-chitosan nanoparticles for the oral delivery of methotrexate. *International Journal of Biological Macromolecules, 158*, 180–188. https://doi.org/10.1016/j.ijbiomac.2020.04.233

Crosby, A. J., & Lee, J. Y. (2007). Polymer nanocomposites: the "nano" effect on mechanical properties. *Polymer Reviews, 47*(2), 217–229.

Cui, H., Jiang, W., Wang, C., Ji, X., Liu, Y., Yang, G., … Ni, Y. (2021). Lignin nanofiller-reinforced composites hydrogels with long-lasting adhesiveness, toughness, excellent self-healing, conducting, ultraviolet-blocking and antibacterial properties. *Composites Part B: Engineering, 225*, 109316.

Czaja, W. K., Young, D. J., Kawecki, M., & Brown, R. M. (2007). The future prospects of microbial cellulose in biomedical applications. *Biomacromolecules, 8*(1), 1–12.

Divya, K., & Jisha, M. (2018). Chitosan nanoparticles preparation and applications. *Environmental Chemistry Letters, 16*(1), 101–112.

Duan, B., Yuan, X., Zhu, Y., Zhang, Y., Li, X., Zhang, Y., & Yao, K. (2006). A nano-fibrous composite membrane of PLGA–chitosan/PVA prepared by electrospinning. *European Polymer Journal, 42*(9), 2013–2022. https://doi.org/10.1016/j.eurpolymj.2006.04.021

Erbacher, P., Zou, S., Bettinger, T., Steffan, A.-M., & Remy, J.-S. (1998). Chitosan-based vector/DNA complexes for gene delivery: Biophysical characteristics and transfection ability. *Pharmaceutical Research, 15*(9), 1332–1339.

Fadaie, M., Mirzaei, E., Geramizadeh, B., & Asvar, Z. (2018). Incorporation of nanofibrillated chitosan into electrospun PCL nanofibers makes scaffolds with enhanced mechanical and biological properties. *Carbohydrate Polymers, 199*, 628–640. https://doi.org/10.1016/j.carbpol.2018.07.061

Ferrer, A., Filpponen, I., Rodríguez, A., Laine, J., & Rojas, O. J. (2012). Valorization of residual Empty Palm Fruit Bunch Fibers (EPFBF) by microfluidization: Production of nanofibrillated cellulose and EPFBF nanopaper. *Bioresource Technology, 125*, 249–255.

Figueiredo, P., Ferro, C., Kemell, M., Liu, Z., Kiriazis, A., Lintinen, K., … Kostiainen, M. A. (2017). Functionalization of carboxylated lignin nanoparticles for targeted and pH-responsive delivery of anticancer drugs. *Nanomedicine, 12*(21), 2581–2596.

Frone, A. N., Panaitescu, D. M., Donescu, D., Spataru, C. I., Radovici, C., Trusca, R., & Somoghi, R. (2011). Preparation and characterization of PVA composites with cellulose nanofibers obtained by ultrasonication. *BioResources, 6*(1), 487–512.

Gama, M., Dourado, F., & Bielecki, S. (2016). *Bacterial Nanocellulose: From Biotechnology to Bio-Economy.* Elsevier.

Gao, Y., & Wu, Y. (2022). Recent advances of chitosan-based nanoparticles for biomedical and biotechnological applications. *International Journal of Biological Macromolecules, 203*, 379–388. https://doi.org/10.1016/j.ijbiomac.2022.01.162

Gobi, R., Ravichandiran, P., Babu, R. S., & Yoo, D. J. (2021). Biopolymer and synthetic polymer-based nanocomposites in wound dressing applications: A review. *Polymers, 13*(12), 1962.

Heinze, T. (2016). Cellulose: Structure and properties. In O. J. Rojas (Ed.), *Cellulose Chemistry and Properties: Fibers, Nanocelluloses and Advanced Materials* (pp. 1–52). Springer International Publishing.

Henriksson, M., & Berglund, L. A. (2007). Structure and properties of cellulose nanocomposite films containing melamine formaldehyde. *Journal of Applied Polymer Science, 106*(4), 2817–2824.

Hule, R. A., & Pochan, D. J. (2007). Polymer nanocomposites for biomedical applications. *MRS Bulletin, 32*(4), 354–358.

Ibrahim, N. A., & Eid, B. M. (2017). Chitosan-based composite materials: Fabrication and characterization. In V. K. Thakur, M. K. Thakur, & M. R. Kessler (Eds.), *Handbook of Composites from Renewable Materials. Physico-Chemical and MechanicaL Characterization* (pp. 103–136). Scrivener Publishing LLC.

Ibrahim, N. A., Eid, B. M., & Sharaf, S. (2021). Nanocellulose: Extraction, surface functionalization and potential applications. In Vijay Kumar Thakur, Elisabete Frollini, & Janet Scott (Eds.), *Cellulose Nanoparticles: Volume 1: Chemistry and Fundamentals* (pp. 1–28). Royal Society of Chemistry.

Islam, M. T., Alam, M. M., Patrucco, A., Montarsolo, A., & Zoccola, M. (2014). Preparation of nanocellulose: A review. *AATCC Journal of Research, 1*(5), 17–23. https://doi.org/10.14504/ajr.1.5.3

Jayakumar, R., Prabaharan, M., Nair, S. V., & Tamura, H. (2010). Novel chitin and chitosan nanofibers in biomedical applications. *Biotechnology Advances, 28*(1), 142–150. https://doi.org/10.1016/j.biotechadv.2009.11.001

Jing, W., Chunxi, Y., Yizao, W., Honglin, L., Fang, H., Kerong, D., & Yuan, H. (2013). Laser patterning of bacterial cellulose hydrogel and its modification with gelatin and hydroxyapatite for bone tissue engineering. *Soft Materials, 11*(2), 173–180. https://doi.org/10.1080/1539445x.2011.611204

Johnson, R. K., Zink-Sharp, A., Renneckar, S. H., & Glasser, W. G. (2009). A new bio-based nanocomposite: Fibrillated TEMPO-oxidized celluloses in hydroxypropylcellulose matrix. *Cellulose, 16*(2), 227–238.

Karimipour-Fard, P., Jeffrey, M. P., JonesTaggart, H., Pop-Iliev, R., & Rizvi, G. (2021). Development, processing and characterization of polycaprolactone/nano-hydroxyapatite/chitin-nano-whisker nanocomposite filaments for additive manufacturing of bone tissue scaffolds. *Journal of the Mechanical Behavior of Biomedical Materials, 120*, 104583. https://doi.org/10.1016/j.jmbbm.2021.104583

Khalil, H. A., Davoudpour, Y., Islam, M. N., Mustapha, A., Sudesh, K., Dungani, R., & Jawaid, M. (2014). Production and modification of nanofibrillated cellulose using various mechanical processes: A review. *Carbohydrate Polymers, 99*, 649–665.

Koo, J. H. (2019). *Polymer Nanocomposites: Processing, Characterization, and Applications.* McGraw-Hill Education.

Li, H., Wang, Z., Zhang, H., & Pan, Z. (2018). Nanoporous PLA/(chitosan nanoparticle) composite fibrous membranes with excellent air filtration and antibacterial performance. *Polymers, 10*(10), 1085.

Li, J., Wei, X., Wang, Q., Chen, J., Chang, G., Kong, L., … Liu, Y. (2012). Homogeneous isolation of nanocellulose from sugarcane bagasse by high pressure homogenization. *Carbohydrate Polymers, 90*(4), 1609–1613.

Li, W., Guo, R., Lan, Y., Zhang, Y., Xue, W., & Zhang, Y. (2014). Preparation and properties of cellulose nanocrystals reinforced collagen composite films. *Journal of Biomedical Materials Research Part A, 102*(4), 1131–1139. https://doi.org/10.1002/jbm.a.34792

Li, X., Wang, C., Yang, S., Liu, P., & Zhang, B. (2018). Electrospun PCL/mupirocin and chitosan/lidocaine hydrochloride multifunctional double layer nanofibrous scaffolds for wound dressing applications. *International Journal of Nanomedicine, 13*, 5287–5299. https://doi.org/10.2147/ijn.s177256

Lin, N., & Dufresne, A. (2014). Nanocellulose in biomedicine: Current status and future prospect. *European Polymer Journal, 59*, 302–325.

Lin, S.-P., Loira Calvar, I., Catchmark, J. M., Liu, J.-R., Demirci, A., & Cheng, K.-C. (2013). Biosynthesis, production and applications of bacterial cellulose. *Cellulose, 20*(5), 2191–2219. https://doi.org/10.1007/s10570-013-9994-3

Liu, Y., Sui, Y., Liu, C., Liu, C., Wu, M., Li, B., & Li, Y. (2018). A physically crosslinked polydopamine/nanocellulose hydrogel as potential versatile vehicles for drug delivery and wound healing. *Carbohydrate Polymers, 188*, 27–36. https://doi.org/10.1016/j.carbpol.2018.01.093

Liu, Z., Jiao, Y., Wang, Y., Zhou, C., & Zhang, Z. (2008). Polysaccharides-based nanoparticles as drug delivery systems. *Advanced Drug Delivery Reviews, 60*(15), 1650–1662.

Malinen, M. M., Kanninen, L. K., Corlu, A., Isoniemi, H. M., Lou, Y.-R., Yliperttula, M. L., & Urtti, A. O. (2014). Differentiation of liver progenitor cell line to functional organotypic cultures in 3D nanofibrillar cellulose and hyaluronan-gelatin hydrogels. *Biomaterials, 35*(19), 5110–5121. https://doi.org/10.1016/j.biomaterials.2014.03.020

McCarthy, S. P. (2003). Biodegradable polymers. In A. Andrady (Ed.), *Plastics and the Environment* (pp. 359–377). John Wiley and Sons.

Mhd Haniffa, M. A. C., Ching, Y. C., Abdullah, L. C., Poh, S. C., & Chuah, C. H. (2016). Review of bionanocomposite coating films and their applications. *Polymers, 8*(7), 246. https://doi.org/10.3390/polym8070246

Mogoşanu, G. D., & Grumezescu, A. M. (2014). Natural and synthetic polymers for wounds and burns dressing. *International Journal of Pharmaceutics, 463*(2), 127–136.

Mondal, S. (2017). Preparation, properties and applications of nanocellulosic materials. *Carbohydrate Polymers, 163*, 301–316.

Mourya, V. K., & Inamdar, N. (2008). Chiosan-modifications and applications: Opportunities galore. *Reactive and Functional Polymers, 68*, 1031–1051.

Nasir, M., Gupta, A., Beg, M. D. H., Chua, G. K., & Asim, M. (2014). Laccase application in medium density fibreboard to prepare a bio-composite. *RSC Advances, 4*(22), 11520–11527.

Nasir, M., Hashim, R., Sulaiman, O., & Asim, M. (2017). 11-Nanocellulose: Preparation methods and applications. In M. Jawaid, S. Boufi, & A. K. H. P. S (Eds.), *Cellulose-Reinforced Nanofibre Composites* (pp. 261–276): Woodhead Publishing.

Nasir, M., Hashim, R., Sulaiman, O., Nordin, N. A., Lamaming, J., & Asim, M. (2015). Laccase, an emerging tool to fabricate green composites: A review. *BioResources, 10*(3), 6262–6284.

Nasrollahzadeh, M., Issaabadi, Z., Sajjadi, M., Sajadi, S. M., & Atarod, M. (2019). Types of nanostructures. *Interface Science and Technology, 28*, 29–80.

Pääkkö, M., Ankerfors, M., Kosonen, H., Nykänen, A., Ahola, S., Österberg, M., … Ikkala, O. (2007). Enzymatic hydrolysis combined with mechanical shearing and high-pressure homogenization for nanoscale cellulose fibrils and strong gels. *Biomacromolecules, 8*(6), 1934–1941.

Panthapulakkal, S., & Sain, M. (2012). Preparation and characterization of cellulose nanofibril films from wood fibre and their thermoplastic polycarbonate composites. *International Journal of Polymer Science, 2012*, 1–6. Article ID 381342. https://doi.org/10.1155/2012/381342.

Park, W.-I., Kim, H.-S., Kwon, S.-M., Hong, Y.-H., & Jin, H.-J. (2009). Synthesis of bacterial celluloses in multiwalled carbon nanotube-dispersed medium. *Carbohydrate Polymers, 77*(3), 457–463. https://doi.org/10.1016/j.carbpol.2009.01.021

Paul, D. R., & Robeson, L. M. (2008). Polymer nanotechnology: Nanocomposites. *Polymer, 49*(15), 3187–3204.

Peng, B. L., Dhar, N., Liu, H., & Tam, K. (2011). Chemistry and applications of nanocrystalline cellulose and its derivatives: A nanotechnology perspective. *The Canadian Journal of Chemical Engineering, 89*(5), 1191–1206.

Phanthong, P., Karnjanakom, S., Reubroycharoen, P., Hao, X., Abudula, A., & Guan, G. (2017). A facile one-step way for extraction of nanocellulose with high yield by ball milling with ionic liquid. *Cellulose, 24*(5), 2083–2093.

Phanthong, P., Reubroycharoen, P., Hao, X., Xu, G., Abudula, A., & Guan, G. (2018). Nanocellulose: Extraction and application. *Carbon Resources Conversion, 1*(1), 32–43.

Pinkert, A., Marsh, K. N., Pang, S., & Staiger, M. P. (2009). Ionic liquids and their interaction with cellulose. *Chemical Reviews, 109*(12), 6712–6728.

Podsiadlo, P., Choi, S.-Y., Shim, B., Lee, J., Cuddihy, M., & Kotov, N. A. (2005). Molecularly engineered nanocomposites: Layer-by-layer assembly of cellulose nanocrystals. *Biomacromolecules, 6*(6), 2914–2918.

Poonguzhali, R., Basha, S. K., & Kumari, V. S. (2017). Synthesis and characterization of chitosan-PVP-nanocellulose composites for in-vitro wound dressing application. *International Journal of Biological Macromolecules, 105*, 111–120. https://doi.org/10.1016/j.ijbiomac.2017.07.006

Rahmani, S., Maroufkhani, M., Mohammadzadeh-Komuleh, S., & Khoubi-Arani, Z. (2022). Chapter 7-Polymer nanocomposites for biomedical applications. In A. Barhoum, J. Jeevanandam, & M. K. Danquah (Eds.), *Fundamentals of Bionanomaterials* (pp. 175–215). Elsevier.

Rahmat, M., Karrabi, M., Ghasemi, I., Zandi, M., & Azizi, H. (2016). Silane crosslinking of electrospun poly (lactic acid)/nanocrystalline cellulose bionanocomposite. *Materials Science and Engineering: C, 68*, 397–405.

Rallini, M., & Kenny, J. M. (2017). 3-Nanofillers in polymers. In C. F. Jasso-Gastinel & J. M. Kenny (Eds.), *Modification of Polymer Properties* (pp. 47–86). William Andrew Publishing.

Ranjbar-Mohammadi, M., Shakoori, P., & Arab-Bafrani, Z. (2021). Design and characterization of keratin/PVA-PLA nanofibers containing hybrids of nanofibrillated chitosan/ZnO nanoparticles. *International Journal of Biological Macromolecules, 187*, 554–565. https://doi.org/10.1016/j.ijbiomac.2021.07.160

Rebouillat, S., & Pla, F. (2013a). State of the art manufacturing and engineering of nanocellulose: A review of available data and industrial applications. *Journal of Biomaterials and Nanobiotechnology, 4*(02), 165.

Rebouillat, S., & Pla, F. (2013b). State of the art manufacturing and engineering of nanocellulose: A review of available data and industrial applications. *Journal of Biomaterials and Nanobiotechnology, 4*, 165–188.

Rescignano, N., Fortunati, E., Montesano, S., Emiliani, C., Kenny, J. M., Martino, S., & Armentano, I. (2014). PVA bio-nanocomposites: A new take-off using cellulose nanocrystals and PLGA nanoparticles. *Carbohydrate Polymers, 99*, 47–58. https://doi.org/10.1016/j.carbpol.2013.08.061

Ribeiro, D. M. L., Carvalho Junior, A. R., Vale de Macedo, G. H. R., Chagas, V. L., Silva, L. D. S., Cutrim, B. D. S., … de Miranda, R. D. C. M. (2019). Polysaccharide-based formulations for healing of skin-related wound infections: Lessons from animal models and clinical trials. *Biomolecules, 10*(1), 63.

Saito, T., Kimura, S., Nishiyama, Y., & Isogai, A. (2007). Cellulose nanofibers prepared by TEMPO-mediated oxidation of native cellulose. *Biomacromolecules, 8*(8), 2485–2491.

Saito, T., Nishiyama, Y., Putaux, J.-L., Vignon, M., & Isogai, A. (2006). Homogeneous suspensions of individualized microfibrils from TEMPO-catalyzed oxidation of native cellulose. *Biomacromolecules, 7*(6), 1687–1691.

Salaberria, A. M., Diaz, R. H., Labidi, J., & Fernandes, S. C. (2015). Preparing valuable renewable nanocomposite films based exclusively on oceanic biomass–Chitin nanofillers and chitosan. *Reactive and Functional Polymers, 89*, 31–39.

Schadler, L., Brinson, L., & Sawyer, W. (2007). Polymer nanocomposites: A small part of the story. *Jom, 59*(3), 53–60.

Schenzel, K., Fischer, S., & Brendler, E. (2005). New method for determining the degree of cellulose I crystallinity by means of FT Raman spectroscopy. *Cellulose, 12*(3), 223–231.

Schmidt, G., & Malwitz, M. M. (2003). Properties of polymer–nanoparticle composites. *Current Opinion in Colloid & Interface Science, 8*(1), 103–108.

Shiraishi, S., Imai, T., & Otagiri, M. (1993). Controlled release of indomethacin by chitosan-polyelectrolyte complex: Optimization and in vivo/in vitro evaluation. *Journal of Controlled Release, 25*(3), 217–225.

Siddiqui, N., Mills, R. H., Gardner, D. J., & Bousfield, D. (2011). Production and characterization of cellulose nanofibers from wood pulp. *Journal of Adhesion Science and Technology, 25*(6–7), 709–721.

Singh, M. R., Patel, S., & Singh, D. (2016). Natural polymer-based hydrogels as scaffolds for tissue engineering. In Alexandru Mihai Grumezescu (Ed.), *Nanobiomaterials in Soft Tissue Engineering* (pp. 231–260). Elsevier.

Siró, I., & Plackett, D. (2010). Microfibrillated cellulose and new nanocomposite materials: A review. *Cellulose, 17*(3), 459–494.

Sothornvit, R. (2019). Nanostructured materials for food packaging systems: New functional properties. *Current Opinion in Food Science, 25*, 82–87.

Sukul, M., Min, Y.-K., Lee, S.-Y., & Lee, B.-T. (2015). Osteogenic potential of simvastatin loaded gelatin-nanofibrillar cellulose-β tricalcium phosphate hydrogel scaffold in critical-sized rat calvarial defect. *European Polymer Journal, 73*, 308–323. https://doi.org/10.1016/j.eurpolymj.2015.10.022

Tang, L., Li, X., Du, D., & He, C. (2012). Fabrication of multilayer films from regenerated cellulose and graphene oxide through layer-by-layer assembly. *Progress in Natural Science: Materials International, 22*(4), 341–346.

Tayeb, A. H., Amini, E., Ghasemi, S., & Tajvidi, M. (2018). Cellulose nanomaterials—Binding properties and applications: A review. *Molecules, 23*(10), 2684.

Tripathy, J. (2017). Polymer nanocomposites for biomedical and biotechnology applications. In Deba Kumar Tripathy & Bibhu Prasad Sahoo (Eds.), *Properties and Applications of Polymer Nanocomposites* (pp. 57–76). Springer.

Wang, Q., Chen, S., & Chen, D. (2017). Preparation and characterization of chitosan based injectable hydrogels enhanced by chitin nano-whiskers. *Journal of the Mechanical Behavior of Biomedical Materials, 65*, 466–477. https://doi.org/10.1016/j.jmbbm.2016.09.009

Xu, L., Huang, Y.-A., Zhu, Q.-J., & Ye, C. (2015). Chitosan in molecularly-imprinted polymers: Current and future prospects. *International Journal of Molecular Sciences, 16*(8), 18328–18347.

Yang, S., Zhang, X., & Zhang, D. (2019). Electrospun chitosan/poly (vinyl alcohol)/graphene oxide nanofibrous membrane with ciprofloxacin antibiotic drug for potential wound-dressing application. *International Journal of Molecular Sciences, 20*(18), 4395. https://doi.org/10.3390/ijms20184395

Yuan, N., Xu, L., Zhang, L., Ye, H., Zhao, J., Liu, Z., & Rong, J. (2016). Superior hybrid hydrogels of polyacrylamide enhanced by bacterial cellulose nanofiber clusters. *Materials Science and Engineering: C, 67*, 221–230. https://doi.org/10.1016/j.msec.2016.04.074

Zhang, X., Huang, J., Chang, P. R., Li, J., Chen, Y., Wang, D., … Chen, J. (2010). Structure and properties of polysaccharide nanocrystal-doped supramolecular hydrogels based on Cyclodextrin inclusion. *Polymer, 51*(19), 4398–4407. https://doi.org/10.1016/j.polymer.2010.07.025

Index

γ rays 127
ζ-potential 180

absorption bands 159
active agents 189
Ag nanoparticles 217
agglomeration 11
alginate-based nanomaterials 247
anisotropic shape 302
antimicrobial activity 297
arc discharge method 26
asbestos 282

bacterial cellulose 239
bacterial nanocellulose (BNC) 83
bactericidal and bacteriostatic 183
bio-based composites 146
biodegradability 3
biological molecules 22
biological tissues 186
biomedical applications 3
biopolymer 241

carbon nanotubes (CNTs) 9, 121, 200
catalytic efficiency 23
cationic polymer 310
cell–hydrogel matrix 187
cell membrane 216
cellulose 3
cellulose nanofibers 293
cellulose nanowhiskers 95
cellulosic materials 148
chemical treatments 235
chemical vapor deposition (CVD) 37
chitin 90
chitin nanocrystal (ChNC) 86
chromatography 210
colloidal stability 212
colloidal stabilizers 181
contact angle 165
coupling agents 128
cross-linking 176
cryocrushing 150
crystallinity 158
CVD (chemical vapor deposition) 121
cytotoxicity 213

drug delivery 182

ecotoxicity 201
elastomers 147

electrochemical deposition 34
electrolytes 10
electron beam lithography 25
electrospinning 30
encapsulation 126
energy conversion and storage 187
environmentally friendly 166
enzymatic pretreatment 303
esterification 152

flammability 234
flax fiber (Linum usitatissimum) 288
Fourier-transform infrared (FTIR) 263
fracture mechanism 163
functionally graded materials (FGMs) 65

graft monomer particles 128
grafting 155
green nanotechnology 218
green solvents 245

hemp fibers 285
high water holding capacity 310
hydrogels 175
hydrophobicity 153
hydrothermal 31
hydroxyl groups 149

in vitro nanotoxicity 202, 203
inorganic nanomaterials 8
ionic gelation 184
ionotropic gelation 311

kenaf fibers 287

laser ablation 26
lignocellulosic biomass 236
linear glucans 242
lithographic patterning 24
low-carbon footprint 237
low-pressure plasma 127

mechanical and thermal properties 72
mechanical fibrillation 304
mechanical grinding 244
mechanical properties 2
mechanochemistry 27

Nano Chitosan 7
Nano Lignins 7
nanocarbon 190

nanocellulose 74
nanochitin 243
nanoclays 122
nanocomposite 72
nanoelectronics 174
nanofabrication 23
nanolignin 120
nanomaterial emissions 220
nanoparticle dispersion 117
nanorevolution 21
nanoscience and nanotechnology 21
natural fibers 282
network morphology 157

organic matrices 298
organic nanofillers 1, 73

packaging materials 188
pH-responsive delivery 248
phosphorylation 154
photocatalytic activity 214
photodegradation 215
physicochemical properties 246
piezoelectric capabilities 92
plasma treatment 127
poly(dl-lactide-co-glycolide) nanoparticles 7
poly(vinyl alcohol) 5
polycarbonates 5
polyethylene terephthalate (PET) 9
Polymer composites 116
polymer hydrogels (PHGs) 259
polymer matrix 6
polymer nanocomposites 1
polymerization 4
polysaccharide 8, 179
polytetrafluoroethylene nanoparticles 9
potentially mineralizable 262

radical polymerization 265
Raman spectroscopy 137
renewable 233
respiratory tract 204
rheological properties 160

self-adhesion 188
self-healing 11
silk fibers 283, 284
single-walled carbon nanotubes (SWCNTs) 26
SiO_2 nanoparticles 207
sisal fiber 289
soil fertility 258
sol-gel process/ing 33, 122
starch 93, 240
stress transfer 161, 162
structure–property relationship 156
surface modification 125
surfactant 260
synthetic polymers 299

TEMPO-oxidised cellulose nanofibers 238
thermal stability 164
TiO_2 nanoparticles 206
top–down approach 24
transmission electron microscopy (TEM) 266

ultrahigh-vacuum 28
UV radiation 208

viscoelastic reaction 90

wastewater treatment 209
water fluxes 10
well-dispersed 178
wettability 165
wood fibers 292
wound healing 185